Jan Harmsen, André B. de Haan, Pieter L. J. Swinkels
Product and Process Design

Also of interest

Multiphase Reactors
Harmsen, Bos, 2023
ISBN 978-3-11-071376-3, e-ISBN 978-3-11-071377-0

Process Intensification.
Breakthrough in Design, Industrial Innovation Practices, and Education
Harmsen, Verkerk, 2020
ISBN 978-3-11-065734-0, e-ISBN 978-3-11-065735-7

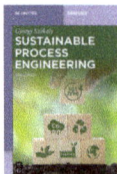

Sustainable Process Engineering
Szekely, 2024
ISBN 978-3-11-102815-6, e-ISBN 978-3-11-102816-3

Product-Driven Process Design.
From Molecule to Enterprise
Zondervan, Almeida-Rivera, Camarda, 2023
ISBN 978-3-11-1,01490-6, e-ISBN 978-3-11-101495-1

Sustainable Process Integration and Intensification.
Saving Energy, Water and Resources
Klemeš, Varbanov, Wan Alwi, Manan, Fan, Chin, 2023
ISBN 978-3-11-078283-7, e-ISBN 978-3-11-078298-1

Jan Harmsen, André B. de Haan,
Pieter L. J. Swinkels

Product and Process Design

—

Driving Sustainable Innovation

2nd, Completely Revised and Extended Edition

DE GRUYTER

Authors
Ir. Jan Harmsen
Harmsen Consultancy B.V
Hoofdweg Zuid 18
2912 ED Nieuwerkerk aan den IJssel
The Netherlands

Prof. Dr. Ir. André B. de Haan
Section Process Systems Engineering
Department of Chemical Engineering
Delft University of Technology
Van der Maasweg 9
2629 HZ Delft
The Netherlands

Ir. Pieter L. J. Swinkels
Department of Chemical Engineering
Faculty of Applied Sciences
Delft University of Technology
Van der Maasweg 9
2629 HZ Delft
The Netherlands

ISBN 978-3-11-078206-6
e-ISBN (PDF) 978-3-11-078212-7
e-ISBN (EPUB) 978-3-11-078226-4

Library of Congress Control Number: 2024930105

Bibliographic information published by the Deutsche Nationalbibliothek
The Deutsche Nationalbibliothek lists this publication in the Deutsche Nationalbibliografie;
detailed bibliographic data are available on the Internet at http://dnb.dnb.de.

© 2024 Walter de Gruyter GmbH, Berlin/Boston
Cover image: MR.Cole_Photographer/Moment/Getty Images
Typesetting: Integra Software Services Pvt. Ltd.
Printing and binding: CPI books GmbH, Leck

www.degruyter.com

Preface to the second edition

This second edition is a major improvement over the first edition. It now contains examples and exercises in most chapters. With these the book can now more easily be used as a textbook in education. In addition to updating all chapters, the content has been extended for making sustainable designs; hence the addition of "Sustainable" to the subtitle. The authors therefore proudly present this second edition to the readers.

We thank Dr. Ir. Farzad Mousazadeh, Senior Design Supervisor/Lecturer, Faculty of Applied Sciences, TU Delft, for his comments and advise on Chapter 9 about modeling.

https://doi.org/10.1515/9783110782127-202

Preface to the first edition

We, the authors, have extensive experience in designing and developing innovative products and processes, for a broad variety of companies. Some of these companies are strongly product-driven, while others are strongly process-driven, and some by both products and processes. The companies belong to various industries, such as oil and gas, consumer products, chemicals, life sciences, food, and engineering, procurement, and construction.

We also have extensive experience in teaching design methodology for products and processes both in universities and at companies. Coming from these combined industrial and academic experiences we felt a strong need to write a book covering all relevant design methods and heuristics focused on innovative solutions to establish breakthrough as well as adapted innovations. This book can be used as a textbook not only for teaching in academia and industries but also for inspiring industrial product and process engineers to create novel designs.

From our industrial experiences, we have gleaned that industrial innovation can be tremendously accelerated by incorporating design activities throughout the innovation process to drive projects from early stage idea generation to commercial implementation.

Our academic experience has taught us how to unlock and develop the design skills of students and enable them to generate novel concepts for products and processes.

These experiences, complemented by recent insights into design and innovation in the literature, have been used to create this book. During this process, a lot of discussion took place regarding the content, the connections between the chapters, and the format and structure of the book. We felt privileged that such a large group, with different knowledge and experiences but with the same conviction and motivation, were in a position to cooperate with us to complete this book.

We would like to acknowledge the people who contributed to this book: Johan Grievink for writing Chapter 9 and contributing to Chapters 4 and 8; Ido Pat-El for writing Chapter 11 on health and safety and Section 5.7 on designing for human factors; Jos van Reisen for writing Section 5.8 on designing for process energy efficiency; Christhian Almeida Rivera for contributing to Chapters 4 and 8 and proofreading Chapter 9; Rob van der Lans for contributing to the characteristic times of Chapter 8; Gabrie Meesters for contributing to Chapters 3 and 8; Mark Toonen, Innovation Manager at Cosun, for providing information on innovation management in practice; Tom Brooijmans of Cosun Biobased Products for reviewing Chapter 10; Gijsbert Korevaar, Asst. Prof. of energy, services, and systems, TU Delft, for reviewing Chapter 12; Maarten Verkerk, Professor at TU Eindhoven, for reviewing the use of modal aspects in Chapter 3; and Yvette van Herwijnen, of AMI consultancy, for providing input on Chapter 3 on project management.

We thank Giljam Bierman for many years of contribution in design projects and design methods at the (post-MSc) PDEng (now EngD) level at TU Delft, and for contributing this expertise in Chapter 4. We thank the following people for their inspiration

https://doi.org/10.1515/9783110782127-203

and cooperation in product and process design at TU Delft over the last 25 years: Hans Wesselingh, Peter Appel, Gert Frens, Johan Grievink, Cees Luteijn, Giljam Bierman, Peter Jansens, Jacob Moulijn, Freek Kapteijn, Ernst Sudholter, Michiel Kreutzer, Ruud van Ommen, Ger Koper, Margot Weijnen, Paulien Herder, Peter Bongers, Cristhian Almeida Rivera, Gabrie Meesters, Rafiqul Gani, Kees van Overveld, Adrie Huesman, Henk van den Berg, Jan Dijk, Jeu Lambrichts, Peter Daudey, Urjan Jacobs, Henk Nugteren, Peter Hamersma, Jos van Reisen, Sorin Bildea, Tony Kiss, Karel Asselbergs, Luuk van der Wielen, Adrie Straathof, Mark van Loosdrecht, Sef Heijnen, Marcel Ottens, Maria Cuellar Soares, Geert Jan Witkamp, Henk Noorman, Peter Verheijen, as well as many MSc students in chemical engineering, PDEng (now EngD) trainees in process and equipment design, bioprocess engineering, chemical product design, and PhD students in process systems engineering over the last 25 years.

Contents

Part A: Innovation and industry

Part B: **Design generation**

Part C: **Design optimization**

Part D: **Education**

Author's biographies

Ir. Jan Harmsen
Harmsen Consultancy BV
Hoofdweg Zuid 18, 2912ED, Nieuwerkerk ad IJ., The Netherlands
E-mail: jan@harmsenconsultancy.nl

Jan Harmsen is a consultant on sustainable process and product innovation. He provides advice and courses to industry and academia. After his graduation in chemical technology at Twente University in 1977 he joined Shell. There he held professional positions in process research, notably on three-phase reactors, biotechnology, process development, reaction engineering, process concept design, process implementation, and finally on process intensification till 2010. He became part-time Hoogewerff-Professor of Sustainable Chemical Technology in 1997, first at Delft University of Technology and later at Groningen University till 2013.

He authored books: Jan Harmsen and René Bos, *Multiphase Reactors* (De Gruyter, 2023); Jan Harmsen and Maarten Verkerk, *Process Intensification* (De Gruyter, 2020); Jan Harmsen, *Industrial Process Scale-Up: A Practical Innovation Guide from Idea to Commercial Implementation*, 2nd ed. (Elsevier, 2019); Gerald Jonker and Jan Harmsen, *Engineering for Sustainability: A Practical Guide for Gustainable Design* (Elsevier, 2012); Jan Harmsen and Joseph Powell (eds.), *Sustainable Development in the Process Industries: Cases and Impact* (John Wiley & Sons, 2010).

Prof. dr. ir. André B. de Haan
Department of Chemical Engineering, Faculty of Applied Sciences
Delft University of Technology
Van der Maasweg 9, 2629 HZ Delft, The Netherlands
E-mail: a.b.dehaan@tudelft.nl

Prof. André de Haan currently combines a position as Principal Technologist at the Cosun Innovation Centre with a part-time professor position at Delft University of Technology (DSM). After finishing his PhD, he was employed in various positions at DSM (1991–1999), held the chair in separation technology at the University of Twente (1999–2006), and the chair in Process Systems Engineering at the Technical University of Eindhoven (2006–2010). He worked for Corbion-Purac as corporate scientist process technology (2010–2016), first in combination with a part-time position at the Technical University of Eindhoven (2010–2013), which he exchanged for Delft University of Technology in 2014. He authored the books *Industrial Separation Processes* and *Process Technology*, both published with De Gruyter.

https://doi.org/10.1515/9783110782127-205

Ir. Pieter L. J. Swinkels
Department of Chemical Engineering, Faculty of Applied Sciences
Delft University of Technology (TU Delft)
Van der Maasweg 9, 2629 HZ Delft, The Netherlands
E-mail: p.l.j.swinkels@tudelft.nl

Ir. Pieter Swinkels is Director of the post-MSc Engineering Doctorate (EngD) designer programs in chemical product and process design at Delft University of Technology (TU Delft). After obtaining his Ir. (MSc) degree at Technical University Eindhoven, with specialization in bioprocess engineering at TU Delft, he was employed by Unilever. He was a member of the R&D team that invented novel detergent powder formulations and manufacturing processes (1986–1991). Within Unilever's specialty chemicals businesses (later part of National Starch and Chemical Company and ICI) he worked in various positions in manufacturing, product/process design and development, and as divisional process development manager (1992–2001). In 2001, he moved to TU Delft as Assistant Professor of Product and Process Design and Engineering and became PDEng (now EngD) program director. He develops (bio)chemical product/process design methodologies, teaches them, and applies these in MSc and EngD design projects in cooperation with companies from his extensive industrial network. For his exceptional contribution to the post-MSc Engineering Doctorate programs at the four Dutch technical universities (4TU) he was awarded the Stan Ackermans Medal in 2023.

Prof. em. ir. Johan Grievink
Department of Chemical Engineering, Faculty of Applied Sciences
Delft University of Technology
Van der Maasweg 9, 2629 HZ Delft, The Netherlands
E-mail: j.grievink@tudelft.nl

Johan Grievink is Professor Emeritus of Chemical Engineering at Delft University of Technology. He was educated as a chemical engineer and applied mathematician at Twente University, after which he joined Shell Research, Amsterdam, in 1971. There, he contributed in various positions to the rapidly evolving field of mathematical modeling and computing for development, design, and operations of chemical processes. In 1992 he was transferred to TU Delft, to teach process modeling, design, and optimization of operations as well as to develop educational policies for MSc and PDEng programs. The focus of his research was on processes for conversion of energy carriers (e.g., natural gas-to-liquid fuels) and for industrial crystallization. His research gradually broadened from process engineering to the interactions between product and process engineering. Having retired in 2008 he continues to perform some academic duties and to offer industrial consultancy in process modeling and optimization.

Ir. Ido E. Pat-El
Royal HaskoningDHV
Industry & Buildings
Contactweg 47,1014 AN Amsterdam, The Netherlands
E-mail: ido.pat-el@rhdhv.com

Ido Pat-El graduated from the Delft University of Technology in 2007 with an MSc (ir) degree in Chemical Engineering. During his college years, he was involved in safety-related projects for various organizations. These include TNO, the Dutch State Supervision of Mines, and the Centre of Industrial Fire Safety of the Rotterdam-Rijnmond Port District Fire Services. He is currently a process safety specialist at Royal HaskoningDHV, a Dutch engineering and consultancy firm. In this role, he has worked on engineering projects across various sectors, including nuclear, hydrogen, chemical, and offshore. He has also served as the Lead Technical Safety Engineer in the design of offshore projects. Ido has authored papers in his field and presented on topics in process safety at national and international forums. He also gives guest lectures on safety in design at the Delft University of Technology for master's students.

dr. ir. Jos L. B. van Reisen, EngD
McDermott Engineering and Construction
Prinses Beatrixlaan 35,
2595 AK The Hague, The Netherlands
Email: jvanreisen@mcdermott.com

Jos van Reisen holds the position of Senior Principal Process Engineer at McDermott, an engineering and construction group, located in The Hague. As Guest Lecturer, he teaches energy efficient process design at Delft University of Technology since 1993. He has an MSc (ir) degree and an Engineering Doctorate (EngD) in chemical engineering from Delft University of Technology. He has also received a PhD in chemical engineering from Delft University of Technology for his research on a design method for the application of multistream heat exchangers in the energy-saving retrofit of chemical processes.

He worked for ABB Lummus Global from 1998 until its merger with CB&I and later McDermott, contributing to a wide range of projects from basic to detailed engineering and construction supervision for clients throughout the energy chain, both up- and downstream with focus on energy analysis, energy-efficient design, and energy transition technologies.

Part A: **Innovation and industry**

1 Introduction goal, scope, and structure

1.1 Purpose of this book

The purpose of this book is to help designers make designs that contribute to the Sustainable Development Goals (SDGs) of the United Nations [1]. To achieve these goals the world requires novel products and novel processes to increase human health, to reduce the burden on the environment, and to increase prosperity, particularly in deprived areas.

This need to innovate for a better world has also been identified by industrial companies, notably organized in the World Business Counsel for Sustainable Development, who subscribed to the SDGs. For each stage of industrial innovation projects, this book provides an innovation stage-gate model. For each stage design methods are provided.

In this second edition all chapters on design now have examples and exercises. This should make the book suitable as a textbook for educational purposes, both in academia and in industry.

The design methods provided are based on heuristics as well as on systems engineering approaches. The methods are mostly applicable to processes and homogeneous products. The latter means that the product composition is homogeneous down to the micron scale and often down to the molecular scale. The design methods, however, appeared to be also useful for mechanical engineering products. To that end, an introduction to mechanical and mechatronic design procedures is provided in Chapter 5.

The book also provides descriptions on how to manage innovation projects so that they are effective and efficient. Effective means that the goal – a commercial implementation – is achieved; efficient means that the novel product and/or technology require minimal resources and that the innovation project is carried out with minimal effort, time, and cost.

Innovation projects can only achieve all these goals by having trained and experienced people skilled in executing innovation projects. The focus of this book is, therefore, twofold. It is written as a textbook for design courses and as a guideline for industrial innovators to develop better products and processes. This book can help:

- educate student engineers for their future careers in designing new products, processes, and systems that address the needs of people, save the environment, and create prosperity locally and globally;
- inspire educators with novel methods in educating students in the art of design;
- inspire industrial researchers to invent, design, and develop novel products and processes;
- support start-up entrepreneurs in organizing their endeavors with the right mix of people in terms of education, experience, and behavior, through cooperation with companies and institutes in open innovation projects;

https://doi.org/10.1515/9783110782127-001

–   assist managers of research and development departments in getting a good mix
    of people with specific skills and providing them with innovation methods that
    have been proven to work;
–   organize and communicate an integrated approach to R&D, design, manufactur-
    ing, and marketing of novel products and processes for successful launches; and
–   inform organizations that want to enter the market of process-product industries
    about how these industries are structured and how their innovation is organized.

This last point may need some clarification. The authors have noticed that what is
going on in this set of industries is not known or understood by people outside this set
of industries. Even for people working in these industries but not directly involved in
the actual manufacturing process, it is still difficult to understand how they function.

There are several reasons for this lack of understanding:

(1) The first reason is that most people have never been in contact with these indus-
    tries because the manufacturing sites are at locations remote from cities and of-
    fice buildings.
(2) A second reason is that these industries relate primary resources via long supply
    chains of intermediate products to the final consumer. People only see that final
    product, but not the intermediate products and their processes.
(3) A third reason is that these industries are complex. They have many system levels.
(4) A fourth reason is that there are many industry branches related to processes
    and products. Often people within a certain branch are not aware of the existence
    of most other branches.

For companies wanting to enter this set of industries, or for companies wanting to
change their supply chain, or for people wanting to work in these industries, this
book should be useful. The focus of the book is, therefore, on methods, guidelines,
and heuristics for novel products and novel process design, and for their development
to the market. To that end the business case development and the experimental data
generation are systematically integrated.

Since major and decisive effects on final products and processes occur in the concept stages, most emphasis is placed on those stages. However, the other stages in
commercial implementation are also discussed.

The content of this book is partially derived from the tacit knowledge of the authors, shaped by experiences in design and innovation. This tacit knowledge has been
transformed into coded knowledge using published structures on design steps and innovation stages and based on findings and categories in design knowledge as explained in Chapter 4.

The content of this book is validated as much as possible with available scientific
literature. In many parts, the information is made plausible by engineering reasoning.
In several parts, the content is based on one or a few cases. As for the education part,

the validation is that students using the theory were able to generate novel and practical designs.

1.2 Book scope

The scope of this book is to provide general knowledge on developing novel products and their processes through a combination of designing and experimentation in combination with an innovation stage-gate management method, from idea to commercial-scale implementation. The book does not provide detailed descriptions of specific process technologies or specific process steps such as unit operations. For those, the reader is referred to other books, such as *Process Technology* by de Haan [2], and for descriptions of specific processes, to books such as by Moulijn et al. [3].

1.2.1 Product classification and description

The class of products treated in this book is homogeneous down to the micron scale and to the molecular scale. Typical products belonging to this class are transport fuels (petrol, diesel, and kerosene) chemicals, plastics, resins, medicines, healthcare products, detergents, food products, feed products, and ceramics.

A product is called a "chemical product" when it is formed by molecular-scale transformations by means of reactions and by creating multiscale physical structures, such as dispersed or stacked thermodynamic phases. All such products are manufactured by the chemical process industry and related manufacturing sectors, such as food, energy, and electronic materials.

A special subclass is a set of products that are homogeneous down to the micron scale but become heterogeneous below this scale and have two or more phases. These are called "structured products." An example is mayonnaise, which is a stable dispersion of water in an oily phase. Another example is a heterogeneous catalyst. Designing these types of products requires special knowledge as described in the section on the design of structured products.

A special class of products is in fact microprocesses that operate on chemical engineering principles (reactions, separations, and fluid flow) and have information as the primary output and not a chemical product or energy as such. The information concerns features of the feed material(s) being processed. Examples of such micro-process products are integrated labs-on-a-chip, where the purpose is rapid screening of many samples and finding the ones with best performance or generating medical diagnostics from samples of body fluids.

The design of discrete products and devices such as cars do not belong to the core of this book, but many design and innovation methods provided in this book are also useful for designing such products and devices.

1.2.2 Description of process industries

Industries that produce homogeneous products are often called process industries. There are many process industry branches. Here is a list to give an idea: oil and gas (petroleum), chemicals, plastics, rubber, pharmaceuticals, food, beverages, ceramics, base metals, coal, textiles, coating and paint, paper and pulp, electricity (power), and (drinking) water.

Most process companies are large and have their own research and development, design department, manufacturing processes, marketing and sales, and a head office with supporting departments such as legal and public affairs. Nowadays, research and development (R&D) are combined with the design department into a technology center so that design works more closely with R&D and can be better integrated into the whole innovation trajectory. Apart from these manufacturing companies there are many other companies and institutes relevant to innovation in the field of process industries. Those are described in detail in Chapter 3.

1.3 Book structure

The structure of this book follows a systematic top-down systems approach.

– Part A is about the what, why, and how of industrial innovation. It provides context for any innovation or design by a description of all system levels in nature, down to the smallest molecular scale. It also describes innovation structures, methods, and collaboration options with other organizations and elaborates on the innovation stage-gate approach with all major steps included. These stages are referred to throughout other parts of the book. This section also contains management guidelines for governing innovation portfolios and innovation projects.
– Part B is about the why and how of concept design for innovation. This section focuses on generating novel ideas and concepts. It provides, in a central chapter, a comprehensive method for concurrent product and process design for each innovation stage. This chapter will be particularly useful to experienced designers and innovators to generate, plan, and manage complex radical innovations. This section also contains other chapters that systematically guide students and less experienced designers from design scoping through design execution. These chapters focus on discovery and concept stages.
– Part C is about how to optimize designs through modeling, evaluations, and reporting.
– Part D is about how to teach design for innovation. It describes established education engineering programs and courses at Delft University of Technology.

References and further reading

[1] Negro C, Garcia-Ochoa F, Tanguy P, Ferreira G, Thibault J, Yamamoto S, Gani R. Barcelona declaration 10th World Congress of Chemical Engineering 1–5 October 2017, Chemical Engineering Research and Design, 129, 2018, A1–A2.

[2] De Haan AB. Process Technology An Introduction, Berlin: De Gruyter, 2015.

[3] Moulijn JA, Makkee M, van Diepen AE. Chemical Process Technology, 2nd edn, Hoboken: J. Wiley, 2013.

2 All system levels relevant to design for innovation

2.1 Introduction

For process and product designers, it is important to know systems that are relevant to their design. The largest system level, Earth, is relevant for life cycle assessments of designs. The smallest system level, molecules, is relevant to chemical composition changes to make products from feedstocks. An overview of all system levels facilitates choices to be made on what system levels design and modeling will take place and how the design will fit in higher system levels.

To this end, Tab. 2.1 provides an overview of system levels relevant to product process innovations. This overview has been derived from Allenby [1] and Hartmann [2]. The wording of each level however has been modified in such a way that it should be easy to understand for all involved in design and innovation.

Each system level is connected to the neighboring levels via mass and energy streams. Changes inside a system level thereby affect all other system levels. Products and their manufacturing processes are therefore not islands, but are connected to larger and smaller systems.

Tab. 2.1: All system levels from Earth to elementary.

System Level	Size M	Main elements
Earth	40×10^6	Sun, nature, Society
Society	10×10^6	Politics, industry techology
Value chain	10×10^6	Value creation steps
Industrial symbiosis network	40×10^3	Connections between industry and society
Industrial complex	10×10^3	Factories connected at industrial site
Factory	1×10^3	Processes and its owner at a location
Process	400	Manufactoring set transforming feed to product
Process step	100	Functional subprocess
Unit operation	10	Generalized transformation operation
Main equipment	1	Mechanically defined artifact
Characteristic subprocess	0.1	Defined subsection of unit operation
Microelement	10^{-3}	Elementary element
Elementary	10^{-9}	Molecular scale

Partly derived from Allenby [1] and Hartmann [2].

These system-level descriptions can help various actors in design and innovation as follows:
- Innovators can get a big picture of how innovation objects are connected to other systems.
- Designers see how higher and lower system levels play a role in their design scope.

https://doi.org/10.1515/9783110782127-002

- Innovation managers get a framework to communicate with other parties and stakeholders involved in the innovation.
- All involved in innovation see how communication between stakeholders can be facilitated by having a discussion on all the system levels and which levels are particularly relevant.

2.2 Earth system

Major system elements and streams of the Earth system are represented in Fig. 2.1.

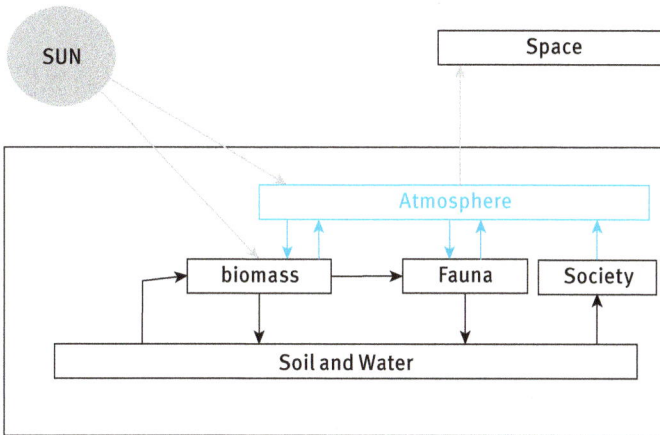

Fig. 2.1: Earth system for energy and mass flows.

The Earth system is approximately closed for mass.

For energy, the Earth is in dynamic equilibrium. Sunlight energy, thermodynamically a high-quality energy input equivalent to a temperature of 6,000 K, enters the system and is radiated back from the Earth's surface with a black mass radiation equilibrium temperature of 256 K. The lower frequency radiation is absorbed by gases in the atmosphere, notably water vapor, carbon dioxide, methane, and other components. These in turn radiate heat back into space at an average equilibrium temperature of 288 K [3]. The temperature difference between the black body radiation from the Earth's surface and the radiation from the Earth's atmosphere of 32 K is called global warming.

For the last 10,000 years, till the industrial revolution in the nineteenth century, this system level has been dynamically stable with no significant accumulation of materials in any of the system elements [3].

The system element, biomass stores a fraction of the solar radiation by photosynthesis of water and carbon dioxide into carbohydrates (and some other components).

The system element, fauna, converts carbohydrates back to carbon dioxide and water, while using the released high quality energy for muscle activities.

The system element, society, contains all human activities that cause input from and output streams to the Earth's system. This Earth system is an inspiration for engineers, as it shows that a nearly steady state on a global scale is obtainable for mass transformations driven by solar energy. This inspiration is an important part of design for industrial ecology and is described in Chapter 5.

2.3 Society system

2.3.1 Present society system and material flows

Figure 2.2 shows the present society system level and its material flows with the Earth.

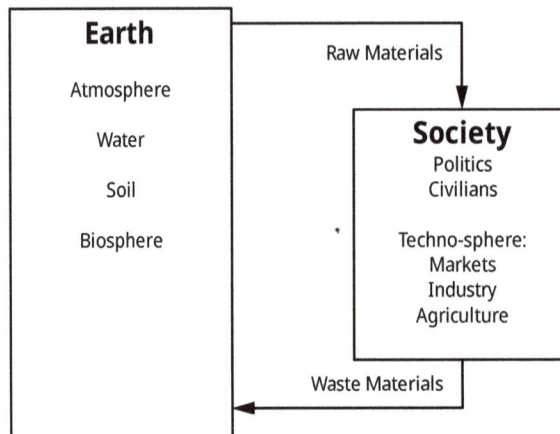

Fig. 2.2: Present society system and material flows.

The society model contains politics and civilians. It also contains the technosphere consisting of markets, industry, and agriculture. The technosphere is a useful collective term for all activities that cause transformations of raw inputs to consumer products and beyond.

Civilians and political systems influence the technosphere via purchasing products and by laws. Major global actors are the United Nations (UN), Non-governmental Organizations (NGOs), the World Business Counsel for Sustainable Development (WBCSD), and the collective worldwide universities.

Example: society and technosphere interaction

An example of influences on the technosphere are the Paris agreement in on climate change, set up by the United Nations, by which governments of 195 countries agreed to take measures to reach net zero global warming gas emissions by 2050 [4] and the formulation of Sustainable Development Goals (SDG) by the United Nations and the WBCSD [5].

These global agreements are turned into policies by governments and by industries, which stimulate innovation to create technologies that are embedded in the Earth system and society in such a way that they keep these surrounding systems intact, while creating prosperity.

On lower society levels too, stakeholders' interactions influence the technosphere. Governments provide laws and budgets for research institutes and universities to generate knowledge novel technologies. Governments in general also provide budgets for education, which in turn generate the next generation of skillful people to develop society. Chapter 3 provides more information on important parties and stakeholders in society relevant to innovation.

The material flows between society and the Earth are clustered in Fig. 2.2 into single streams. To illustrate these streams the carbon flows are described. The present society takes in carbon from fossil sources and burns into carbon dioxide, which is released to the atmosphere.

This fossil carbon conversion happens at such a scale that the carbon dioxide in the atmosphere increases and thereby the back radiation from the Earth's surface is absorbed more. This causes temperature increase of the atmosphere by one or more degrees centigrade. This is called enhanced global warming. This in turn means that more water evaporates from the ocean, which condenses elsewhere, causing more wind and more rain and thereby changing climate behavior in various parts of the Earth [3]. The next section describes futuristic alternative flows of carbon in closed cycles.

2.3.2 Society system: futuristic closed material flows

Figure 2.3 describes a futuristic system of closed carbon mass flows via nature and via technical transformation steps.

Figure 2.3 is derived from the Ellen MacArthur Foundation [6] with a few changes. Farming is placed inside society as farming is a human activity. Furthermore the term biosphere is replaced by nature so that the Earth's atmosphere and water streams can be also seen as part of that system.

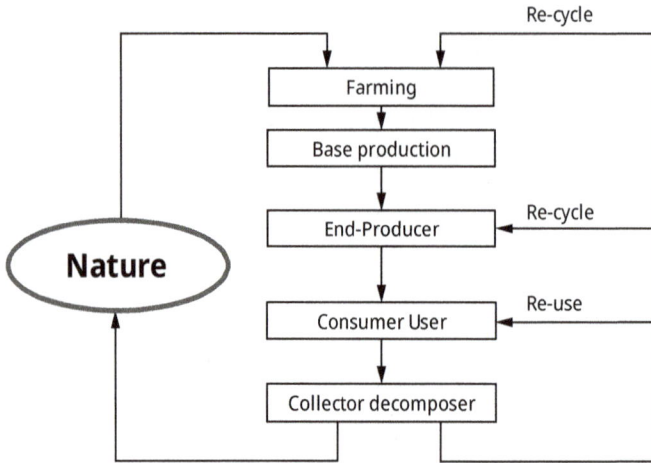

Fig. 2.3: Furturistic closed mass flows society and nature.
Derived from Ellen MacArthur Foundation.

Example: carbon cycle

By policy changes and by innovation, increasingly renewable sources of energy are being developed, notably water, wind, solar and biomass, in such a way that in the end the technosphere will be connected to nature via mass and energy in a steady state flow, hence, with no accumulation of carbon in one subsection and depletion in a different subsection of nature.

2.4 Value chain system

The value chain system is about the economic value increase of material flows from primary natural resources transformed to consumer products via industrial upgrading steps. Value chains are a subset of society. Value chains are also called supply chains by companies further down the value chain.

A major current value chain for the chemical industry is shown in Fig. 2.4. It starts from fossil source of crude oil and ends in consumer products.

Another supply chain starts from biomass such as sugar beet, corn, or sugar cane, which is converted to glucose and then to food, plastics (polylactic acid), and also to pharmaceutical products, as indicated in Fig. 2.4.

This value chain from biomass shows that by innovations in product and process design, not only fossil resource-based value chains have to be considered but also these renewable value chains. By breakthrough innovation more products are likely to be produced from renewable resources at prices competing with products from fossil sources. Moreover, new value chains need to be created that close the materials

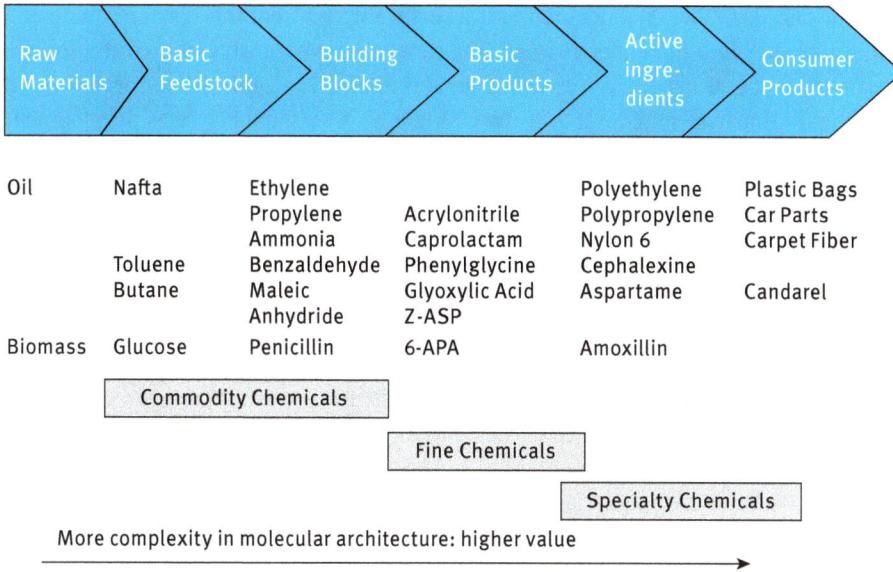

Raw Materials	Basic Feedstock	Building Blocks	Basic Products	Active ingredients	Consumer Products
Oil	Nafta	Ethylene Propylene Ammonia		Polyethylene Polypropylene Nylon 6	Plastic Bags Car Parts Carpet Fiber
	Toluene Butane	Benzaldehyde Maleic Anhydride	Acrylonitrile Caprolactam Phenylglycine Glyoxylic Acid Z-ASP	Cephalexine Aspartame	Candarel
Biomass	Glucose	Penicillin	6-APA	Amoxillin	

Commodity Chemicals

Fine Chemicals

Specialty Chemicals

More complexity in molecular architecture: higher value

Fig. 2.4: Chemical industry value chain.

stream. These are called circular economy value chains. This type of innovation is discussed in Section 3.2.2.3 circular economy and also in Chapter 4 where for product and process design, novel value chains are taken into account.

2.5 Industrial symbiosis system level for mass and energy

Industrial symbiosis systems are about connections between process factories and other society elements such as cities. District heating of houses by industrial waste heat is an example of industrial symbiosis.

In this system level, different parts of society such as cities and industrial complexes meet and also different disciplines meet. Sustainable housing design for example, means that architects and process engineers interact to get an integrally optimized housing system where washing, waste reuse, heating and lighting, and the use of sunlight as energy source are combined. Industrial symbiosis is therefore a level in which innovation can flourish.

2.5.1 Example: domestic wastewater to industrial boiler feed water

An example of industrial symbiosis is the City Terneuzen municipal wastewater conversion to boiler steam water for Dow Chemical, Terneuzen, Netherlands.

For decades, Dow Chemical took salty surface water from the Westerschelde and converted that into boiler feed water. The most energy-intensive and costly process step was the desalting using ion-exchange resin beds. The central Dow organization embraced industrial symbiosis as a useful method. A methodology was development and local factories were invited to come up with industrial symbiosis solutions.

Dow Chemical, Terneuzen, picked up the idea, looked around for options, and identified the wastewater stream of city Terneuzen as this water contains far less salt then the surface water. By cooperation between the city of Terneuzen, the Zeeuws-Vlaanderen Water Board, Elides, and Dow Chemical, the regional infrastructure was adapted by making a 6-km long pipeline from the municipal wastewater treatment plant to the Elides pumping station. The Reverse Osmosis (RO) plant for treating brackish estuary water was partly changed to biologically treated wastewater. The implemented solution meant the reuse of 2.5 million km^3 municipal wastewater, rather than this as an effluent was charged to the surface water, 65% energy savings, and reduction of chemicals intake for cleaning the RO membranes. In 2007, the novel solution obtained the EU Responsible Care Award [7].

This new industrial symbiosis system is illustrated in Fig. 2.5.

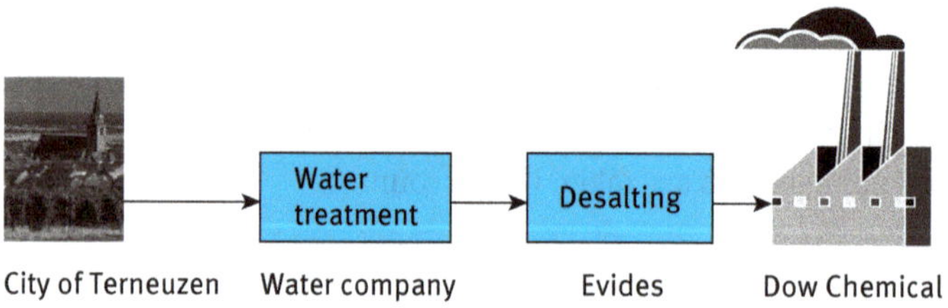

City of Terneuzen   Water company   Evides   Dow Chemical

Fig. 2.5: Industrial symbiosis city Terneuzen & Dow Chemical municipal waste water to industrial boiler feed water wu [6].

This industrial symbiosis case also demonstrates the power of thinking in more than one system level. Only by going to the society system level, the industrial symbiosis solution was found. It is also an example of social innovation. Chapter 3 describes these types of innovation. Chapter 5 shows a design approach based on industrial symbiosis.

2.6 Industrial complex system level

An industrial complex contains many factories connected by networks of many different mass and energy flows. Often, certain shared utilities such as steam generation and wastewater treatment are common. In Germany, these complexes are often called Verbund.

Designing a process to be incorporated in an existing industrial complex is called designing for a brown field, while designing a process as a stand-alone with no common utilities is called designing for a green field. Designing for a brown field situation has many advantages as the major facilities such as utilities and infrastructures for all kinds of energy, such as electricity and steam, base chemicals, and wastewater treatment are available.

Designing a process for a so-called greenfield situation, for a location with nothing there, means that the infrastructure must also be designed. This means that a concept for the whole complex has to be generated. This, on the other hand also gives many opportunities to design a very efficient and environmentally friendly complex, where industrial symbiosis level design thinking is applied.

2.6.1 Example: Rotterdam harbor industrial complex

The Rotterdam harbor industrial complex is about 40 km long and about 2 km wide. It contains oil refineries, chemical factories, and solid waste conversion plants, wastewater treatment plants, connected by steam, electricity, waste heat, hydrogen, HCl, compressed air, and many other network streams. The Huntsman chemicals plants, for instance, are connected by 20 streams to other plants [8].

2.7 Factory system level

A factory consists of one or more manufacturing processes of the same owner at the same location. It also includes storage facilities for feeds and products. In most cases it is a subsystem of an industrial complex. It converts input materials into higher value output products. Other outputs such as by-products and energy may be sent to other parts of the industrial complex.

In designing a factory, it is always beneficial to take the higher system levels into account. Often, facilities can be shared and by-products used in other factories of the industrial complex.

2.8 Process system level

2.8.1 Process system description

A process is part of a factory. It is a set of steps to transform input materials to a market product. Some companies call this a (chemical) plant. Often, the process (plant) is named after the product it manufactures.

Process design is one of the core topics dealt with in this book, notably in Chapters 4–13.

2.8.2 Products as part of processes

The term product in general means something that is manufactured and sold to customers. If it is sold to industrial customers it is often called an intermediate product. If it is sold to consumers, it is called a consumer product. This book, see Chapter 1, focusses on homogeneous products. These are for instance intermediate chemicals, such as solvents and olefins and consumer products such as food products, detergents, and plastics. Chapter 1 provides examples of product types from various industry branches.

The majority of these products have a special function in their use by the consumers. Most of these products react chemically or physically when used. Examples are: food products that react inside humans and animals to other components that are used to grow and maintain a healthy body; also, detergents that interact with dirt and clothing during washing, such that the dirt is separated from the clothing.

Due to these special functional purposes novel product design is a complex act, as the behavior of the client with the novel product and the product behavior have to be considered. The product design is even more complex as design of the manufacturing process belonging to the product has to be taken into account and also the higher system levels, such as the supply chain have to be considered. Chapter 4 provides an elaborate method for doing so.

2.9 Process step system level

The process step is a subsystem of a process. It is cluster of process unit operations that perform a desired action. Examples are feedstock pre-treatment, chemical conversion, distillation train, product purification, and product packaging.

This process step level is very useful when designing a large novel process or an industrial complex with several large processes. It keeps the representation of the whole design simple with a limited number of block flow diagrams in which each block represents a process step. Changing connections between process steps, for instance, is then easily done. Communicating about the whole design is then easy too.

2.10 Unit operation system level

A unit operation is characterized by the same physical and chemical design rules for any application, so that the same unit operation can be applied to many different applications using the same rules. It was defined by Arthur D. Little in 1916 [9]. Its definition resulted in enormously more efficient process designs because it allowed for: a) educating students on unit operations applicable to any process design, b) improving the generic rules by research, and c) improving actual process designs by applying the rules. Prior to this unit operation invention, only particular technologies existed such as sugar technology, coal technology, and crude oil refinery technology. Knowledge generated was only applicable for the particular application field.

An enormous amount of information about designing with unit operations can be found in Perry's *Chemical Engineers' Handbook* [10]. An introduction to unit operations is provided by de Haan [11].

Process design based on unit operations still dominates the design method of chemical processes. In fact a whole new set of unit operations has now been researched and developed based on process intensification principles [12 Harmsen Verkerk article]. Both the principles and the established new unit operations are provided by Stankiewicz [13].

The advantages of process designing with unit operations have been proven in practice over the last 100 years. However, this design approach does not always deliver the best results. Process design based on functions can deliver results far better than any design based on unit operations. The design and development method are described in detail for the process design of methyl acetate by Eastman chemical and for the process design of biomass to biofuel [14]. This new way of designing however has been hardly applied so far in industry, probably because unit operations design is the method taught at universities and also this way of designing crosses several system boundaries.

This new way of designing may indicate that system-level descriptions are useful but should not limit the process designer.

2.11 Main equipment system level

Main equipment means all individual major apparatus required for the processes. A unit operation distillation for instance will consist of the following main pieces of equipment: a column with internals, a top condenser with a heat exchanger, a top vessel to separate vapor from liquid, a bottom boiler heat exchanger, and a pump to feed liquid reflux back to the column.

2.12 Characteristic subprocess system level

A characteristic subprocess is a subsection of a unit operation. It is characterized by well-described physical and or chemical phenomena and clear boundary conditions. An example of a subprocess is the gas and liquid flow and mass transfer on a tray in a distillation column.

A subprocess allows for dedicated modeling and allows for connecting the models of each subprocess to obtain an overall model for the whole process unit operation. The subprocess model is often standardized as modules. The modules of the subprocesses can then be connected to each other by energy and mass flows. Computer modeling packages for these subprocesses and their connections to form integral models can be purchased from companies.

2.13 Microelement system level

A microelement, such as catalyst particle or a droplet in a liquid extraction unit, is governed by boundary conditions (such as interphase equilibrium), basic heat transport, and mass transport rules. The innovation at this level can involve catalysis, microbiology, interface chemistry, particle technology, and chemical engineering reaction engineering.

The conditions at the interface are part of the microelement modeling. In this way the microelement can be studied separately from the hydrodynamics of the fluids surrounding the microelement. It can therefore be studied prior to the selection of the process equipment with dedicated laboratory equipment under controlled uniform conditions outside the microelement.

2.14 Elementary system level

The elementary system level is uniform over its scale and governed by basic physical and chemical properties, and chemical reactions. All these properties and reactions are scale-independent; hence, they can be studied separately by experts in chemistry and physical properties.

The elementary system level of a reactor is the molecular reaction. The elementary system level of a distillation is the gas-liquid equilibrium.

Example: Ethylene glycol for all system levels
An example of a description for all system levels for a given case is shown in Tab. 2.2 for ethylene glycol production. A short explanation for each level is given below.

The system level, Earth, for this case is about crude oil extraction from the soil to society and carbon dioxide emissions from society to the atmosphere.

The system level, society, is first of all about making and enforcing health, safety, and environmental laws by politics for processes in general and for ethylene glycol, specifically. The Paris agreement about reducing climate change by reducing global warming gas emissions is also relevant to ethylene glycol, as it emits carbon dioxide over its life cycle.

Tab. 2.2: All system levels of ethylene glycol production.

System Level	Size M	Main elements
Earth	40×10^6	Crude oil extraction, carbon dioxide emission
Society	10×10^6	Laws, markets, technosphere
Value chain	10×10^6	Ethylene glycol, plastics, consumer products
Industrial symbiosis network	40×10^3	PCS – Jurong Island, Singapore
Industrial complex	10×10^3	PCS – Jurong Island, Singapore
Factory	1×10^3	Shell chemicals factory
Process	400	Shell ethylene glycol
Process step	100	Ethylene oxide reactor
Unit operation	10	Fixed bed gas/solid catalyst reactor
Main equipment	1	Reactor vessel
Characteristic subprocess	0.1	Single pipe on EO reactor
Microelement	10^{-3}	Catalyst particle
Elementary	10^{-9}	Reaction: Ethylene + O_2 = ethylene oxide

The value chain contains naphtha, ethylene, ethylene glycol, (conversion to) plastics, and (conversion to) consumer products.

The industrial complex, Jurong Island, Singapore houses more than 100 companies. Here many common services are shared. It has for instance a common security organization, common logistics, and several common networks, so it can be seen as and industrial symbiosis system [15]. The Shell ethylene glycol factory is part of this industrial symbiosis [16].

This ethylene glycol factory contains several process steps. The ethylene oxide process step is now taken as the case for the remainder of the system levels.

The process step inside this process is the ethylene oxide reactor with its auxiliary equipment.

The unit operation is the ethylene oxide reactor. It is a multi-tubular cooled gas/solids fixed bed reactor [17].

A main piece of equipment is the reactor vessel.

A characteristic subprocess is a single tube containing the catalyst.

A micro element is a catalyst particle surrounded by flowing gas.

The elementary level is the chemical reaction of ethylene with oxygen to ethylene oxide.

2.15 Exercises

Exercise 2.1 Transform Guitang case information into closed material cycles
Given: The industrial symbiosis complex of Guitang China [18] as shown in Fig. 2.6.

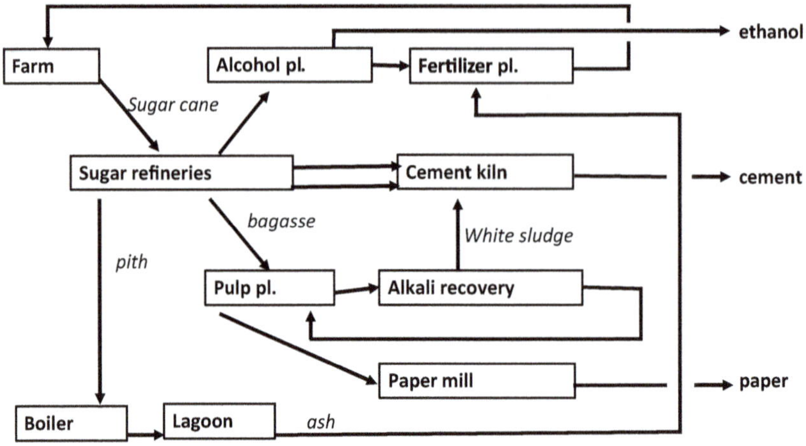

Fig. 2.6: Industrial symbiosis of Guitang China [18].

Task 1: Place blocks and streams of Fig. 2.6 into closed material diagram as in Fig. 2.3.

Exercise 2.2 Choose system levels for flow sheet model of fossil based ethylene glycol
Given: A company has an ethylene glycol process with ethylene and oxygen as feedstocks. The company wants to make a flow sheet model to optimize the process conditions. This task is given a young chemical engineer, who just joined the company.
Task: What system levels should the engineer incorporate in his modeling?

Exercise 2.3 Choose system levels for ethylene glycol from renewable resources
Given: A chemical route from biomass via sugars and lactic acid to ethylene glycol is reported in literature. The young chemical engineer is asked to make a flow sheet model for this route.
Task: What system levels should he incorporate in his modeling?

References and further reading

[1]   Allenby B. Industrial Ecology: Policy Framework and Implementation, Saddle River New Jersey: Prentice Hall Inc., 1999.
[2]   Hartmann K. Analysis and Synthesis of Chemical Process Systems, Hoboken: Elsevier, 2013.
[3]   Boeker E and Grondelle R. Environmental Physics, 2nd edn, Chicester: J. Wiley, 1995.
[4]   UN, Paris Agreement, United Nations Treaty Collection. 8 July 2016. Resourced from: https://treaties. un.org/pages/ViewDetails.aspx?src=TREATY&mtdsg_no=XXVII-7-d&chapter=27&clang=_en.
[5]   UN, Transforming our world: the 2030 Agenda for Sustainable Development, Resolution adopted by the General Assembly on 25 September 2015, resourced on www 5-5-2017. Resourced from: http://www.un.org/sustainabledevelopment/sustainable-development-goals/.

[6] Ellen MacArthur Foundation. sourced 14-8-2023 https://ellenmacarthurfoundation.org/.
[7] Wu Q. Industrial ecosystem principles in industrial symbiosis: By-product synergy. In Harmsen J, Powel J (eds). Sustainable Development in the Process Industries, Cases and Impact, Hoboken: J. Wiley, 2010, p. 249–263.
[8] Baas LW, Korevaar G. Eco-industrial parks in the Netherlands: The Rotterdam harbor and industrial complex. In Harmsen J, Powell JB (eds). Sustainable Development in the Process Industries, Cases and Impact, Hoboken: Wiley, 2010, p. 59–80.
[9] Little AD, et al., Chemistry in history, Chemical Heritage Foundation, Retrieved 13 Nov. 2013. Resourced from https://wikipedia.org/w/index.php?title=Unit_operation&oldid=709344969.
[10] Maloney J. Perry Chemical Engineers Handbook, New York: The McGraw-Hill Companies, Inc., 2008.
[11] De Haan AB. Process Technology: An Introduction, Berlin: Walter de Gruyter GmbH, 2015.
[12] Harmsen J, Verkerk M. A new approach to industrial innovation, Chemical Engineering Progress, 117(3), 1 Mar 2021, 50–53.
[13] Stankiewicz A, van Gerven T, Stefanidis G. The Fundamentals of Process Intensification, Weinheim: John Wiley & Sons, 2019.
[14] Harmsen J, Verkerk M. Process Intensification: Breakthrough in Design, Industrial Innovation Practices, and Education, Berlin: de Gruyter, 2020.
[15] Jurong Island Singapore. Sourced: 14-8-2023 https://www.jtc.gov.sg/find-land/land-for-long-term-development/jurong-island.
[16] Shell Ethylene Glycol process locations. Sourced 14-8-2023, https://www.shell.com/business-customers/chemicals/manufacturing-locations.html.
[17] EOEG Manufacturing Technology, Shell white paper. sourced 14-8-2023 https://catalysts.shell.com/hubfs/Content%20Library/SCT%20-%20Enhancements%20in%20EOEG%20Manufacturing%20Technology.pdf.
[18] Zhu Q, et al. Industrial symbiosis in China: A case study of the Guitang Group, Journal of Industrial Ecology, 11, 2007, 31–42.

3 Managing innovation

3.1 Overview

3.1.1 Innovation terms

The word innovation in this book is used in three ways. First, it means the successful introduction of a new product or process into the market. Second, it means management methods and systems in place to always have a consistent portfolio of innovation projects in progress. This is called innovation portfolio management. Third, it means the management of individual innovation projects into successful market introductions. This is called innovation project management. The overall meaning of this chapter is to show that innovation pays off and is even more beneficial to companies when managed by proven portfolio and project methods.

Historic business trends in product and process innovations, leading to concurrent product-process innovation, including supply chain and product market, are shown in Section 3.2. This section also provides business motives for innovation.

Various types of innovation are described in Section 3.3. Innovation partners of industrial innovation are described in Section 3.4. Portfolio management of innovation inside enterprises is described in Section 3.5. The remaining sections deal with project management in the innovation stages. An overview of the stages and key activities is provided in the next section.

This chapter treats management of innovation in a limited way. Far more information on innovation management is found in dedicated textbooks such as those written by Goffin and Mitchell [1] and Cooper [2].

3.1.2 Stage-gate approach

Innovation project management, as described in this book, is based on a stage-gate approach. An overview of the stages applied in this book is presented in Fig. 3.1. For each of the stages, Tab. 3.1 provides some key characteristics. The same stages may have different names in companies, but the activities during each stage will nearly be the same and the sequence of stages will be similar. Gates are placed in between the stages. At these gates, a decision is taken to stop or to continue the project.

This stage-gate approach ensures that in the early stages, with minimum effort, only the most critical aspects are investigated so that a project can be stopped timely. A full rationale of the stage-gate method is provided in Section 3.6. More detailed descriptions of the stages and their key activities/deliverables are provided in Sections 3.7–3.12.

https://doi.org/10.1515/9783110782127-003

Discovery (Ideation) Stage
- Novel product/process sketch
- Opportunity assessment/business justification
- Proof of principle
- Project charter

Find/develop:
- **materials** that have desired properties and performance
- required **process/ manufacturing** technology

Yes ← Materials and/or technology invention required?

No ↓

Initiate Project — No → Discard Project Charter

Yes ↓

Concept Stage
- Economic potential analysis
- Customer perception, technical requirements & critical success factors
- Customer value proposition
- Superior product concepts
- Database creation
- Process concept generation
- Process concept design (base case)
- Proof of concept (lab/bench scale)
- First SHE/sustainability assessment

Concept Stage Gate Review → Fail

Feasibility Stage
- Product prototypes
- Performance testing methods
- Preliminary product evaluation
- Scale up strategy
- Mini plant/pilot-scale validation
- SHE/sustainability assessment
- Business case
- Improve base case design(s)
- Process synthesis/simulation
  Reactor, recycles, separations
  Energy & mass integration
  Raw material handling
  Product packaging & Batch sequencing
- Plantwide controllability assessment

Feasibility Stage Gate Review → Fail

Development Stage
- Selected customer evaluations
- Update SHE/sustainability assessment
- Update business case
- Detailed design & Equipment sizing
- Mini/Pilot plant testing
- Development startup strategies

Development Stage Gate Review → Fail

Detailed Design Stage
- Detailed plant design
- Procurement process equipment
- Construction
- Demonstration plant decision

Detailed Design Stage Gate Review → Fail

Process Startup & Product Launch Stage
- Start up plan & manual
- Plant start up & operation
- Advertising, customer introduction
- Pricing, product literature

Process Startup & Product Launch Stage Gate Review → Fail

↓ Succes

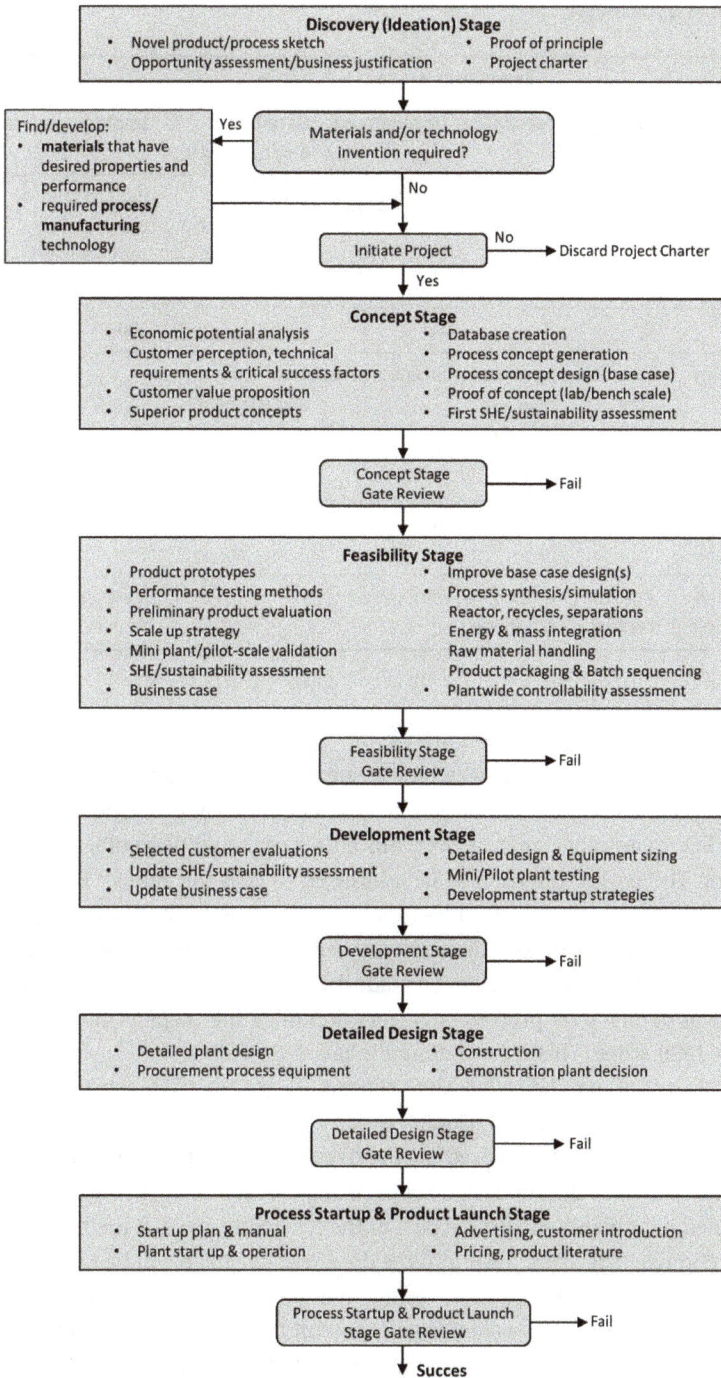

Fig. 3.1: Stage-gate approach used in this book.

Tab. 3.1: Overview of innovation stages.

	Innovation stages					
	Discovery	Concept	Feasibility	Development	Detailed engineering	Implementation
Purpose	Idea generation	Screening	Feasibility concept	Validation	Engineering procurement construction	Product launch process, commissioning, and start-up
Invested (%)	0.01	0.1	2	10	94	100
Stage-gate decision	Strategy fit	Business case	Business case	Business case	Top management	
Duration (years)	0.3–1	1–2	1	1–5	1–2	0.1–3
Design scope (%)	1	10	50	90	100	
Experimental	Proof of principle	Proof of concept	Product prototype	Pilot plant	–	Start-up

3.1.2.1 Short description of stage characteristics

The discovery stage comprises ideas generation. In some companies, this is also called the ideation stage. Very little design is carried out. It will be sketches of the novel product and process. The investment on a project will be a very small percentage of the overall investment. The critical question to be addressed to pass the gate is: Does the idea fit the business strategy? Successful ideas that pass the gate to the concept stage will be followed up by design and experimentation to such a level that a crude estimate of the commercial-scale investment cost can be made. In the feasibility stage, a commercial-scale design will be made by process engineers, in which the scope of the design will be 50% of the total scope. This type of design is called Front-End Loading 1 (FEL-1) in America and Asia and it may be called "50% concept design" or "50% concept engineering," or "50% base engineering" in Europe.

The investment estimate will be based on the design inside the scope and on an estimate of the elements outside the design scope, using correlations. The design and cost of a downscaled pilot plant will also be part of the feasibility stage. If the business case is attractive enough, a decision to go into the development stage will be taken.

In the development stage, new product testing and pilot plant testing will be carried out, followed by a commercial-scale design based on the product testing and the pilot plant testing. The design scope will now cover 90% of all items of the commercial-scale plant. The investment cost estimate thus has an uncertainty of a maximum of 10%, but is likely to be only a few percent. This is called Front-End Loading 3 (FEL-3).

For very large investments, a design with a scope of 80% definition will be made, as this requires less design cost to be made and is still accurate enough to take a decision to continue or not. This is called Front-End Loading 2 (FEL-2).

If the decision is taken for the commercial scale, an engineering, procurement, and construction (EPC) contractor will be asked to make a detailed engineering design, and execute procurement of equipment and construction of the process. When the construction is finished, the new plant will be commissioned and will be started up. The new product will then be launched into the market.

The duration of each stage strongly depends on the project innovation class (see Section 3.3). Break-through innovation class 6 can take up to 19 years. Class 0 may take 1 year. The figures provided are indications only.

3.2 Business focus and motives for innovation

3.2.1 Business focus trends on product and process innovation

The trends in business focus in chemical product and process innovation of the last 100 years are summarized in Tab. 3.2. The information is from Arora [3].

Tab. 3.2: History of product and process innovations.

Period	1920–1950	1950–1980	1980–2015
Business driver	Novel products	Global scale-up	Global competition Global environment
Design	Unit operations	Unit operations	Function integrations
Methods	Monographs	Phenomenae models	Integrated models simulations
Scale-up	Empirical	Pilot plants	Pilot plant validation

Source Business Driver: Ashish Arora, Implications for energy innovation from the chemical industry, NBER, 2010.

In the period 1920–1950, the innovation focus was on novel products such as plastics. Process design became more systematic by the introduction of the concept of unit operations. In the period 1950–1980, the focus shifted to far larger processes to improve on efficiency and reduce investment cost per ton of product.

In the next two periods, 1950–1980 and 1980–2015, global competition and concerns about the environment (acid rain, ozone layer depletion, and global climate change) induced process innovation, with process intensification as the major leading principle. Computer modelling greatly facilitated this innovation [4]. This is also the reason that Chapters 8 and 9 on product and process modeling are included in this book.

In all the design methods of the past, product design and process design were carried out separately. In some companies, their research and development were also carried out separately. The product was developed under the assumption that the process would consist of the available standard unit operations and that no interaction between product development and process development was needed. The textbook *Product and Process Design Principles* of Seider shows how these designs can be made [5].

Recently, a simultaneous product and process design methods based on simulation and optimization has been proposed. However, the method requires a complete mathematical description of all physical and chemical phenomena of the product (and of potential processes), so it can only be used in limited cases [6].

Almeida and others developed a so-called anticipating reciprocal approach to product and process design, which facilitates an optimum combined design of product and process. His method is the base for our book, as found in Chapters 4, 6, 7, 8, and 9. The advantages are briefly [6]:

1. It incorporates a larger solution space than sequential approaches, by which more optimum combined solutions of product and process design can be determined.
2. It allows for qualitative and quantitative design approaches.
3. It has been successfully applied in industry.

3.2.2 Business motives for innovation

Companies, in general, differ in their motivation to innovate. The sections hereafter describe the most encountered innovation drivers.

3.2.2.1 Competition as innovation driver

Arthur D. Little carried out a global innovation benchmark study of many companies, covering both processes and consumer products. The bench study revealed that innovation is regarded by all companies as the major pillar to stay in business. For most companies, innovation is regarded as the major pillar of sustaining competitive advantage and long-term increase of company value [7].

Top innovators actively try to open up new business areas and product development, which are significantly important to them. The following statistics of these top innovators indicate the importance [7].

Top innovators:
–   Realize up to 2× higher sales from new products/services
–   Have up to 2× times higher earnings from new products/services
–   Have half the time for breakeven of earning to innovation cost

What happens if companies lag behind in innovation? The answer is indicated by statistical results of a large worldwide bench mark study on energy efficiency of chemical processes by PDC. They gathered data of different process in commercial operation for the same product, and did that for 100 different chemical products. The results are shown in Tab. 3.3 [8]. For each product, the most efficient process is set at 100% energy efficiency. Energy consumption per ton of product (for the same product) is calculated and divided by the figure for the most efficient process and expressed as a percentage called the energy efficiency ratio (EER). The average EER per world sector has also been calculated and is shown in Tab. 3.3. It shows that the averaged EER is a factor 2 higher than the best processes. It also means that a lot of processes are even less energy efficient. These far less efficient processes thus lag in innovation and will struggle to survive in a competitive world.

Tab. 3.3: Energy efficiency ratio of processes in various world sectors [9].

World sector	Average energy efficiency ratio (%)
Western Europe	176
North America	207
Rest of the world	224
Asia	230

3.2.2.2 Learning curve as innovation driver

Learning curves (also called experience curves or Boston learning curves) are plots of the market prices of a product (vertical axis) plotted against all cumulative sales of that product (horizontal axis), taking the logarithmic values on both axes, so drawing up a log-log plot. In most cases, a downward straight line is obtained and its slope is determined. These learning curves thus quantify the rate at which the market price comes down with increasing cumulative sales [9].

These experience curves often hold for many decades and become a self-fulfilling prophecy as companies keep innovating to reduce the cost, knowing that the competition also does so. A good example of such a learning curve is the case of photovoltaic solar cell, as provided by Swanson. It has been determined from 1979 till 2005 and it shows a cost reduction of 20% per doubling of installed capacity over that period [10].

The beginning of a learning curve is when commercial-scale production starts and the product enters the market. The time prior to that beginning can be very long. The basic principle of a solar cell is known since the eighteenth century. The breakthrough technology for commercial production arrived in the seventies with the silicon photovoltaic technology of both, the product and its required manufacturing process. That technology breakthrough resulted in a breakthrough innovation. Commercial-scale implementation was then enormously successful, with a rapid increase of installed solar cells. This success in turn caused a strong drive to reduce the manufacturing cost. This cost reduction follows a learning experience curve of 20% cost reduction for every doubling of installed capacity, price expressed in ($/w) from 1979 till present, called Swanson's law [10].

The main implication of innovation for products for which learning curves are available is that the innovation must aim at the moving future cost target. A book about learning curves is provided by Jaber [9].

3.2.2.3 Circular economy as innovation driver

Contributing towards a circular economy is, for many companies, an innovation driver. It means that products and processes that fit to the circular economy concept are invented. Chapter 5 contains a section on the definition of circular economy.

The World Business Council for Sustainable Development (WBCSD) provides the following 5 innovative business models for this purpose [11]:

- Circular supplies: meaning that all inputs are bio based or fully recyclable and that only renewable energy is used for driving the processes
- Resource recovery: meaning that useful resources are recovered from used materials, by-products, and waste
- Product life extension: meaning that product lifecycles are extended by repairing, upgrading, and reselling, as well as through innovative product redesign
- Sharing platform: meaning that product users are connected to one another and are encouraged to share its use and access or have ownership to increase product use
- Product as a service: meaning, moving away product ownership and offering customers paid access to products, allowing companies to retain the benefits of circular resource productivity or ownership to increase product use

The WBCSD also provides three disruptive technologies facilitating circular economy [11]:

- Digital technologies such as Internet of things (IoT), big data, blockchain, and RFID help companies track resources, and monitor utilization and waste capacity
- Physical technologies such as 3D printing, robotics, energy storage and harvesting, modular design technology and nanotechnology help companies reduce production and material costs and reduce environmental impact
- Biological technologies such as bioenergy, bio-based materials, biocatalysis, hydroponics and aeroponics help companies move away from fossil-based energy sources

3.2.2.4 Sustainable development as innovation driver

The UN General assembly of September 25, 2015, defined 17 specific measurable targets, named the Sustainable Development Goals (SDGs), to be achieved by 2030. These are listed in Chapter 13.

Chakravarty mentions three reasons for adopting the SDGs of the UN for business [12]. "First, the global goals campaign represents a significant new opportunity for companies that view emerging and frontier markets as their source of long-term growth. According to estimates from McKinsey, consumers in these markets could be worth $30 trillion by 2025; a significant step up from the 2010 value of $12 trillion. Since 2011, as emerging markets have suffered from slower growth and fresh social unrest, that $30 trillion prize seems more distant. Acting on the global goals could help address several of these obstacles that are giving rise to "trapped value" in the emerging markets.

Second, with the public declarations by many companies to help with the goals, there is likely to be competitive pressure. Some companies could get a jumpstart in their industry in organizing partnerships and even positioning themselves as leaders in sustainable development, using the goals as a branding anchor. Being slow to act could lead to the risk of being left out of these relationships. It then could form a competitive disadvantage from a brand equity perspective.

Third, the goals cannot be realized without business participation. The price tag for accomplishing these global goals is estimated to be up to $3 trillion a year, for 15 years. For most governments, financing the global goals campaigns will be a stretch; governments have already committed in the past for similar targets."

There are special links between the business drivers for sustainable development and value chains. For the environmental impact of products, life cycle assessment is now common practice. The value chain is a subpart of the life cycle chain; so, inevitably, other companies of the value chain must be involved to reduce the environmental impact. For renewables as feedstock, even completely new value chains must be established.

For the social dimension of sustainable development also, partners along the value chain have to be involved. This is the case when the early part of the value chain is in developing countries. A major part of the value should be created in those countries and negative impacts, such as child labor, must be avoided.

3.2.2.5 Corporate social responsibility as innovation driver

In the business realm, sustainable development is often called corporate social responsibility (CSR). It is founded on the United Nations description of sustainable development with environmental, social, and governance (ESG) dimensions. One motive for firms to endorse a CSR-driven strategy is to obtain capital at lower cost.

In a large and recent empirical research on firms' abilities to access finance in financial markets, Cheng proves that companies with some better CSR performance indeed face lower capital constraints.

Cheng investigated data from 2,439 public listed firms during the period 2002 to 2008. As a measure for CSR performance, he used a panel dataset with ESG performance scores obtained from Thomson Reuters ASSET4, a Swiss-based company that specializes in providing objective, relevant, auditable, and systematic ESG information and investment analysis tools to professional investors, who built their portfolios by integrating ESG (nonfinancial) data into their traditional investment analysis. It is estimated that investors representing more than €2.5 trillion assets under management use the ASSET4 data, including prominent investment houses such as BlackRock [13]

The CSR measurements of Thomson Reuters ASSET4 involve 220 criteria grouped under 4 pillars, each with main categories as shown in the Tab. 3.4 [14].

Tab. 3.4: Corporate social responsibility ASSET4 performance information structure [14].

Economic	Environmental	Social	Corporate governance
Client loyalty	Resource reduction	Employment quality	Board structure
Performance	Emission reduction	Health and Safety	Compensation policy
Shareholder loyalty	Product innovation	Training and development	Board functions

CSR is a process for companies to integrate social, governance, environmental, and supply chain sustainability into operations and corporate strategy. This subject is beyond the scope of the book. However, assessing a novel technology on contributions to sustainable development is a necessary step in incorporating the technology in the CSR strategy of the company.

3.2.2.6 World problems as driver for the product-process industries

Specifically, for the product-process industries, the results of the Delft skyline debates on world problems and innovation technology area are of interest. In the Delft Skyline debates, 75 scientists and engineers from several disciplines, majority from chemistry and chemical engineering from all over Europe, discussed world-scale problems and innovation areas, for solutions [15]. The results are summarized in the Tab. 3.5.

Tab. 3.5: World-scale problems and innovation areas [15].

Subject	Aspects
World-scale problems	
Shortage	Water, food, energy materials
Pollution	Air, water, soil
Human health	Diseases
Innovation areas	Food, health, living, transport

Tab. 3.5 (continued)

Subject	Aspects
Constraints	No waste, renewables, solar based
Technology areas	
Solar cells and electricity storage	Small scale, local
Fuel cells + manufacturing	Small scale, local
Molecular controlled reactions	Personalized medicines, tissues, and organs
Membranes	Safe affordable water
Process-intensified biomass conversions	Small scale, local

The purpose of this table is to stimulate thinking on contributing to solving these world-scale problems. Focusing innovation on these areas for both products and process innovation is likely to provide extra business in the future and will also contribute to SDGs.

3.3 Innovation classes and types

3.3.1 Innovation classes

There are many classifications for innovations, along similar definitions. Verloop [16] provides a classification, originally made by Gaynor, with the terms breakthrough, new, and incremental. McKinsey includes (lack of) knowledge of the market as an additional classification element [17]. We add as an additional parameter, new value chain, up front of the product (and process), as a new feed and a new supplier, which has additional risks. Combining all these elements led us to the innovation class definitions of Tab. 3.6.

Tab. 3.6: Innovation classification for combinations of markets, supply chains, products, and processes.

Innovation class	Market	Product	Supply chain	Process	Time to market (years)	Success (%)	Extra margin (%)
6 Breakthrough, radically new	New	New	New	New	8–19	15–20	0–60
5 Radically new	Existing	New	New	New			
4 New	Existing	Modified	Modified	New	6–15	30–40	0–10
4 New	Existing	Existing	New	New			
3 New	Existing	Existing	Existing	New			

Tab. 3.6 (continued)

Innovation class	Market	Product	Supply chain	Process	Time to market (years)	Success (%)	Extra margin (%)
3 New	Existing	Modified	Existing	Modified			
2 Novel	Existing	Existing	Existing	Modified	2–5	40–50	0–5
1 Conventional	Existing	Existing	Existing	Increased capacity			
0 Conventional	Existing	Existing	Existing	Existing			

The data of the last three columns of Tab. 3.6 are obtained from McKinsey [13]. They analyzed 118 chemical company innovation results. Their definition of time to market is from the formal project initiation, which is likely to be the same as start of the development stage to the time when the cumulative product sales equal the R&D cost. Their success rate is defined as the portion of projects in an innovation class that created positive returns on investment, expressed as net present value, using the cost of capital with no risk adjustment.

Here are short descriptions of each innovation class.

Innovation class 6: Breakthrough radically new involves a new market, a new product, a new supply chain, or a new process. The new market may be nonexistent or new to the company. For homogeneous new products, nearly always a market exists; so an even higher innovation class was not defined. With the new product, a radically new product is meant, which also involves a new supply chain, as new ingredients are needed and, of course, the process is new.

Examples are a Lab-on a chip product for medical personal diagnostics, and large household-sized batteries for "balancing" local solar energy production (Tesla "Powerwall").

Innovation class 5: Radically new involves an existing market known to the company, but the product, supply chain, and process are new. So, apart from the existing market knowledge, it has the same degree of novelty for the combination of new product, new supply chain, and new process. The time to market, success rate, and extra margin will be in between the classes 6 and 4.

Examples are:
– Hydrogen as transportation fuel (next to diesel and gasoline)
– Fischer-Tropsch-produced gasoline and diesel, with superior fuel efficiency and no sulfur component emissions
– Bio-based soft drink bottles, with improved barrier properties (PEF)
– Water-based paints (solvent-free)

Innovation class 4: New involves a modified product, a modified supply chain and a new process. A modified product means a product for nearly the same application, with a change in relative amounts of ingredients or one or two new ingredients. In product innovation, this is sometimes called a platform derivative. This class also includes an existing product for a new supply chain and a new process.

Examples are:

– Bioethanol containing gasoline (bioethanol process and supply chain are new)
– Bio-based polyethylene (monomer sourcing part is modified)
– Low-VOC water-based paints (improved post-reaction or VOC-removal processing)
– Ice cream with improved properties (caloric value, sensory properties) (flavor delivery, mouth feel) (improved product formulation and structure – through new processing)

Innovation class 3: It involves a radically new process for an existing product. A radically novel process means that novel principles for mass movement, reaction, separation, and heat exchange are applied. And, these principles are combined in novel ways in novel equipment. This type of innovation requires an enormous design effort, not only in creativity but also in developing new design methods, modeling, and simulation tools.

Example 3.1: Innovation class 3 in the chemical industry

An innovation class 3 example of the past is the fluid catalytic cracking process to convert refinery fractions into gasoline. The process, developed during World War II, is entirely based on gas-solid fluidization in the catalytic cracking and in the coke combustion, while the heat exchange is carried out by the solids flow. It allowed very large production capacities at very low investment and processing cost.

At that time, fluidization was not practiced in the process industries and for important elements of the process such as the standpipe between the combustor and the catalytic cracker riser, precise knowledge of the flow phenomena was not known. This standpipe has two functions: a) fast reliable catalyst transport from the combustor to the cracker by gravity and b) a seal for gasoline vapor so as not to flow to the combustor. Due to the enormous need for gasoline, the oil company (Exxon) took the risk and constructed the commercial process with enormous commercial-scale success. The process is still in operation in refineries [18].

A more recent example is reactive distillation. It combines reaction and distillation in a single column. Eastman chemical wanted to replace a conventional design with 11 unit operations by a single column, combining several extractive stripping and absorption with reaction operations. Knowledge of designing for these combined functions was minimal and it required several pilot plants before the basic principles were fully understood. The commercial-scale process was an enormous success, with reduced investment. The operation cost reduced by a factor of 5, while increasing the production safety, health, environment, and reliability [19].

Innovation class 2: Involves only as novelty a modified process. It contains no new processing functions in relation to the product manufacturing, but only fringe novelties, such as a heat integration with additional heat exchangers, improved process control; optimized operating conditions; and alternative utility sources; retrofitting a heat exchanger network. It requires on average 2–5 years for commercialization [17]. The average success rate of only 40–50% may come as surprise. Section 3.5.5 provides the reasons for this low success rate.

Innovation class 1: It involves a process with a higher capacity than previous commercial-scale processes already in operation. Examples are capacity increase by expansion or incremental debottlenecking, or a new plant with the same technology as the existing processes for the same product.

A word of warning is needed here. Often, a process is considered conventional when the applied unit operations are conventional, and the project is placed in this category. However, if those unit operations have not been applied for the particular product or streams, then that process is new and should be treated as new. It may still need a pilot plant to remove major uncertainties and reduce the risks to an acceptable level. Section 3.10 describes this risk of not considering a process as new.

Innovation class 0: It involves a carbon copy of an existing process in commercial operation. It may come as a surprise that projects in this class may sometimes fail. Section 3.5.5 of this book reports the reasons for the large percentage of failure even if the project involves no apparent novelty.

Overall remarks: Not all possible combinations are covered in the classification. The reader, however, can easily place his project in an innovation class using the degree of novelty for the parameters: market, product, supply chain, and process.

3.3.2 Innovation by serendipity

The *New Oxford Dictionary of English* defines serendipity as the occurrence and development of events, by chance, in a satisfactory or beneficial way, understanding the chance as any event that takes place in the absence of any obvious project (randomly or accidentally), which is not relevant to any present need, or in which the cause is unknown.

Innovations presented as examples of serendipity have an important characteristic: they were made by individuals able to "see bridges where others saw holes" and connect events creatively, based on the perception of a significant link. Walpore [20] observed "Serendipity is finding by accident and by sharpness of mind, something not looked for." So, it combines an accident and cleverness.

A recent example of serendipity is the discovery by my former Shell colleague Einte Drent, of a novel poly ketone polymer. He was trying to synthesize solvents from synthesis gas. He saw in one of his experiments that white flocks had been formed. Rather than discarding the result as a failure, he investigated the chemical composition of the flock and found that it was a poly ketone. This novel product was

further developed and appeared to be an engineering polymer with several properties superior to nylon [21].

Companies that want to be innovative foster a climate and they have employees
by which serendipity is stimulated. The results of serendipity innovations will, in the
end, fall in the categories of the previous section.

3.3.3 Social innovations

In some innovations, no new technology is involved and yet they can be called innovations. Some can be called social innovations. These are novel combinations by
stakeholders, such as described in Chapter 2, for system-level industrial symbiosis.
The case is cooperation between the city of Vermeulen, the Water Board, and Dow
Chemicals to take municipal waste water and use it as feedstock for boiler feed water
production.

These social innovations require a lot of communication and discussions, but involve low innovation project cost and low investment cost. The benefits are often
large economical, ecological, and social.

3.4 Innovation partners

3.4.1 In-house versus open innovation

Some companies perform all innovations inside their organization. This is called in-
house innovation. Others perform innovation with partners. This is called open innovation. The advantages and disadvantages are summarized in Tab. 3.7

Tab. 3.7: Advantages and disadvantages of in-house and open innovation.

Parameter	In-house innovation	Open innovation
Speed	Rapid	Slow; takes extra contracts and organization
IP	Intellectual property easily protected	IP protection needs negotiation and attention
Quality	Can be limited by narrow view	Can be high by outside expert knowledge
Cost	High	Low, by sharing cost between partners
When	Strong in-house knowledge base	Breakthrough concept knowledge is outside

When more than one company is involved in an innovation project, then that innovation project is called an open innovation project. If several projects with a common
theme are clustered and developed with the same partners, then this is may be called
an innovation platform. The advantages of the latter type of collaborations are that

the best partners for each knowledge element can be selected and the costs are shared. The disadvantages are:

a) Intellectual property contracts should be made and signed. Often, these take a considerable time, before the actual project can start.

b) Organizing and leading the project is more complex. Each partner has its own agenda and interest. To define a common goal and stay focused on this goal takes effort and requires a project leader experienced in open innovation.

3.4.2 Innovation partners

Potential innovation partners are depicted in Fig. 3.2, together with the main information flows.

Each potential partner type of an open innovation project is briefly described in the subsequent sections.

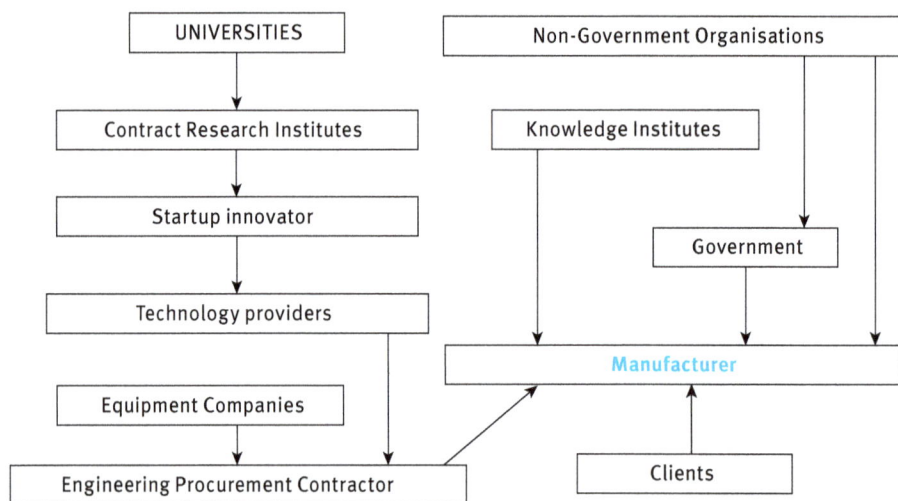

Fig. 3.2: Innovation stakeholders and connections.

3.4.3 Universities

Universities produce educated students and research results. University education programs focused on homogeneous product and process education and research are first of all technical universities and polytechnics. They have, in most cases, a chemical engineering department and/or a biotechnology department and/or a food technology department. Thirdly, general universities have, in some cases, also chemical engineering, biotechnology, and food processing departments. Mining departments may also have a section on ore processing.

The university education programs are, in general, for BSc and MSc students. Some universities also have programs for EngD students. For PhD research students, university departments often also offer education programs. Chapter 14 illustrates the content options for designing each of these education programs.

University research is focused on new elements of a technology and not on a complete integral solution, and on describing and publishing experimental results. The research is, in general, carried out by PhD students with no technical experience. The research papers and the PhD thesis, in most cases, has a technology readiness level (TRL) of 0 (TRL is defined in Section 3.6). Most of academic research stays at this bottom level.

Because of the enormous number of university research groups worldwide, the publications in scientific journals are readable by all. All the same, a large number of inventions from universities move up the TRLs and have become successful innovations.

Academic research also produces design methods and modeling methods. These methods facilitate the innovation velocity and facilitate optimized solutions. Modeling methods for reactive distillation and extractive distillation, for instance, have shortened the development time and have resulted in more optimum designs.

3.4.4 Contract research organizations

Contract Research Organizations (CROs), sometimes called knowledge institutes, carry out research for a customer. Often, these customers are small and medium enterprises with no or a small research and development department.

CROs also carry out research and development on their own budget and then later when the results are positive and protected by patents and other means, they sell the results to customers, often, via a license agreement.

CROs, in general, pick up ideas and research results from universities, and help small and medium enterprises (SMEs) to innovate. In this way, they play an important role in the general innovation level of a country. Because of this role, governments often subsidize these organizations and, often, a government customer, such as military, logistics and safety department is a customer of a CRO.

3.4.5 Start-up innovators

Start-up innovators are companies that start from nothing with an invention. The leader is a person who strongly believes that their product (technology) will change the world and she (let it be a woman) goes for it. Often, the product is ill-defined and she lacks marketing skills. By trying to sell it, she learns the hard way that she must modify her ways to make a product successful and learns how to position it in a market.

A spectacular example of a chemical product-process start-up is Synthon. It was started in 1991 by a Dutch chemist. Jacques Lemans. In 1993, the first generic product, dobutamine, a sympathomimetic drug used in the treatment of heart failure and cardiogenic shock, was launched, and quickly became a commercial success. Synthon is now a global company with over 1600 employees, producing generic drugs at affordable prices [22].

3.4.6 Technology providers

Technology providers provide unique knowledge about a specific technology to their customers so that the customer, an end-user, or an Engineering-Procurement-Contractor, can complete the design of a product or process with that knowledge. Often, technology providers also have an engineering department so that they can deliver the complete engineering package to purchase and construct the technology. Their unique knowledge is often applied to a specific design problem and then tested in their pilot plant facilities.

3.4.7 Equipment providers

Equipment companies manufacture and sell equipment. The design knowledge about the equipment remains for the larger part inside the company. They also have specialized manufacturing facilities for their equipment. Some equipment companies are very innovative with new equipment for existing and new problems.

3.4.8 Engineering procurement construction contractors

EPC contractors perform, in general, nearly all activities for the staged detailed engineering. They take the Front-End Loading design of the product manufacturer as the input. With that and the manufacturer's requirements, they make a detailed design, purchase the equipment and materials, and construct the process.

Their workforce consists of engineers of all disciplines and of a variety of experience levels. They are often specialized in process design for one industry branch, such as petrochemicals, food, pharmaceuticals, or electricity production.

For a product manufacturer, it is of paramount importance to select an EPC firm that has experience in the industry branch. For EPC activities, the motto is: The devil is in the details. If for instance, an EPC is chosen for a food production plant but has never done so, then it is likely that the process will be hard to clean for decontamination, the most important criterion for a food production facility.

3.4.9 Tolling manufacturers

Tolling manufacturers have processes that can be hired for producing an amount of product. Tolling manufacturers, for instance, have several small-scale processes suitable for reaction, distillation, and crystallization. Often, the process operation range is large and the construction material is such that many different media can be applied without causing strong corrosion. The toller's knowledge about what can be processed varies enormously. Some tollers leave the process operation, to some extent, to the operators of the hirer. Some tollers have enormous knowledge and experience and provide not only the process equipment but also their knowledge on how best to operate the process for the new product.

The advantages of involving a toll manufacturer are that: the product is quickly produced, no pilot plant has to be designed and constructed, and knowledge of the toller is obtained for the operation mode.

3.5 Portfolio innovation management

3.5.1 Objectives of portfolio management

Portfolio management is about managing all innovation projects to meet the strategic business objectives. This well-stated definition is obtained from Project Management Institute (PMI), as shown by Bakker [23]. The objective of portfolio management is to determine the optimal mix and sequence of projects in their innovation stages. It is therefore a dynamic decision-making activity, as some projects will stop at a stage-gate, while others proceed, needing more resources.

There are four major rules for portfolio management [23]:
Rule 1: All projects of a portfolio represent the planned investments by the company.
Rule 2: All projects are aligned to the company's strategic objectives.
Rule 3: A large portfolio will be divided into clusters of projects of similar characteristics.
Rule 4: All projects are quantifiable so that they can be measured, ranked, and prioritized.

Ad Rule 1: Often, companies also have small projects that are not part of the portfolio. These loose projects can be damaging to portfolio management, as these are not discussed in the portfolio management meetings and so are not properly managed. An investigation by Blichfeldt and Eskerod in 2008 of 128 companies showed that these small projects together often tie up considerable resources, originally allocated to the portfolio projects [23]. In this way, they damage the projects inside the managed portfolio.

This problem can be avoided by having a cluster called discovery or significant change, where all small embryonic projects are placed and an annual budget for that cluster is allocated. In this way, these projects no longer consume budget in an uncontrolled way. Management of such an ideation cluster can be further improved by having experienced researchers allocated part-time for their management, facilitating popping up of ideas during the year and providing budget (within the overall budget for that cluster).

Ad Rule 2: This rule can only be applied if the company has a clear innovation strategy.

Ad Rule 3: This clustering on similar characteristics is not easy. Several options are available, along business division lines or along technology type lines, but all these have the danger of being locked in thinking, meaning that novel ideas cannot be considered. It is better to first have a high-level innovation matrix clustering, as described in the next section. Further, clusters under each main heading of the innovation ambition matrix can then be formed. Section 3.5.2 proposes a very practical main clustering, from which further subclustering can be done.

Ad Rule 4: Metrics applicable to the whole portfolio and to individual innovation projects for ranking and prioritizing are very important. Wrong metrics will cause wrong outcomes. Metrics for companies can be developed or enriched using the generic criteria list of Chapter 4.

3.5.2 Ambition matrix for innovation portfolio management

For large established companies, the innovation ambition matrix described by Nagji can be very useful. It groups all innovation projects of a company into 3 clusters: core, adjacent, and transformational [24]. There are similar clusters described in literature, based on the terms in the innovation classes of Section 3.3., but the advantage of Nagji's clustering over others is that it has generic names and clear definitions, applicable to all product and process innovation-oriented companies.

3.5.2.1 Core cluster
In this cluster, existing products for existing customers and existing processes are further developed. The development of these minor changes of products will often be carried out in departments supporting existing products and processes, often close to the manufacturing facilities.

The projects of this cluster fall in innovation class 3, see Section 3.3. Their time to market is short. The success chances of introducing these modified products are high, and the expected internal rate of return is sufficient. Management is simple, as it will be in-house innovation by experienced employees, with known methods for product development.

Potential projects can be ranked and the best selected with a total budget the meets the set budget, with the criteria as follows. First, for Safety, Health, and Environment, clear questions regarding acceptability are to be answered yes or no. A no means that that project will not be part of the ranking procedure. Ranking of the remaining potential projects are mainly based on economics (foreseen additional sales, profit, and return on investment), technical feasibility, and social (market) acceptance. The criteria words, beginning with a capital letter in this paragraph, form the acronym SHEETS; described in Chapter 4 in detail.

3.5.2.2 Adjacent cluster

In this cluster, new businesses for adjacent customers and adjacent markets are developed. Here, it is important to recognize that new customers and new markets create additional risks. These risks can be mitigated by generating new knowledge about those customers and markets. Often, adjustments in products must be made to meet the requirements of those new customers. This, in turn, means adjustments to the manufacturing processes.

Projects of this cluster fall into innovation class 4, see Section 3.3. The time to market is much longer than for the core cluster. For chemical products, the expected internal rate of return is 13–18%.

Management of this cluster is critical to its success. It is very important for these projects to go through all the innovation stages, starting at the ideation stage, so that alternative options are generated and evaluated. If the ideation stage is neglected and a project is jump-started in a stage further up in the innovation funnel, then the best idea may not be pursued. When a new product enters the market, competitors may already have found a better way of designing and producing it. The ideation stage can however be quickly executed and most emphasis and budget will be on the succeeding stages.

Ranking of projects and selecting the best (in such a way that total budget stays within the total set budget for this cluster) is more difficult. Ranking criteria may be fixed by the company and experienced managers may use their judgement to perform this ranking. Tab. 3.8 is used inside a product-process innovation company for this purpose.

Tab. 3.8: Scoring table for economic attractiveness of novel products.

Criterion	Score 4	Score 3	Score 2	Score 1
Market size	Very large	Large	Moderate	Limited
Market growth	Very strong > 10%	Strong > 5%	Light 2%	None
Sales potential after 10 years	Very large	Large	Moderate	Small
Profitability after 10 years	>40% gross margin	>25% gross margin	>10% gross margin	<10% gross margin

For each score, the company has to first quantify what they mean by the terms. This will strongly depend on the company and its branch. Some companies consider 500 million euro very large, others consider 500 billion euro very large.

The scoring is carried out by experienced marketing employees. The total of all criteria can then be added up, so that each potential project gets a simple economic dimensionless value.

Attractiveness of opportunity, including risks, is used in the following criteria and scoring is suggested in Tab. 3.9. It is used inside a company with innovation on product-process combinations.

Tab. 3.9: Scoring table for potential innovation projects using risks criteria.

Criterion	Score 4	Score 3	Score 2	Score 1
Entry barrier – us	None	Overcome < 1 y	Overcome < 2 y	Overcome > 2 y
Entry barrier – competitors	Overcome > 2 years	Overcome < 2 years	Overcome < 1 year	None
Scale-up financial risks	Easy; can be outsourced	Moderate; cannot be outsourced	Hard; cannot be outsourced	Unclear; risks unknown
Our competitiveness	Very strong	Strong	Light advantage	No advantage seen
Own IP position	Strong for long time	Limited and short time	Unlikely	Unclear

The entry barrier is an important criterion about being legally allowed to produce and sell the product. For instance, for a novel food product, the entrance barrier is Food approval by a government body.

Scale-up financial risk means the risk involved in investing in production processes (for the new product). It must do with the novelty of the process. If the process consists of classic unit operations, can be completely defined such that an EPC contractor can design, and a pilot plant validation is not needed (see Section 3.11), then the score is 4.

The scoring is carried out by experienced managers and the individual score results as well as the total result is used for evaluating projects in the concept stage-gate to feasibility and for the next stage-gate.

3.5.2.3 Transformational cluster

In this cluster, breakthrough products are generated for new (or even, not yet existing) markets, targeting new customer needs. New manufacturing processes for these novel products are also generated.

Projects of this cluster fall into innovation classes 5 and 6. They have high risks and high rewards. The risks are reduced by following the stage-gate process.

For this cluster, it is important that sufficient number of new ideas are generated and are nurtured to reach sufficient definition, so that the best options can be selected for follow-up. Because new competences need to be acquired for these novel products for novel markets, it is also important to form new basic research groups. These can generate basic design data and knowledge so that the novel products can be embedded in the company. Chapter 6 describes in detail how for each innovation stage, basic design data are generated.

Road mapping to identify and link basic design data generation, skill development, supporting technologies, manufacturing technologies, and the products is very useful for this cluster. Goffin provides a generic picture of such road mapping [1]. Chapter 4 on concurrent product-process design will help to plan the design and research effort with the Delft Design Map and the fingerprint tables for innovation classes 5 and 6.

This transformational cluster may also be subdivided into technology platforms to serve specific (new) business divisions. DSM, for instance, has three technology platforms [25]:

- DSM biomedical: Innovative materials that deliver more advanced clinical procedures and improved patient outcomes
- DSM bio-based products and services: Advanced enzymes and yeast platforms, enabling advanced bioenergy and bio-based chemicals
- DSM advanced surfaces: Smart coatings and surface technologies to boost performance in the solar industry

For potential projects in the transformational cluster, strategic fit is the key criterion. Tab. 3.10 is used in an innovative product-process company that can be of help in evaluating these ideas and potential projects, notably in the discovery and concept stages.

Tab. 3.10: Strategic fit score for breakthrough ideas and projects.

Criterion	Score 4	Score 3	Score 2	Score 1
Input	Existing to us	Existing, new to us	New, proven far away	Emerging
Process technology	Existing to us	Existing, new to us	Proven elsewhere	Emerging
Product	Variants of existing to us	Existing, new to us	New, variant of existing	Radically new
Market	Existing to us	Existing, new to us	Immature, high entrance barrier, niche product	None, unknown

Clarification of some terms: "to us" means to the company involved, "proven far away" means proven in a world continent where the company has no manufacturing presence.

3.5.2.4 Budget distribution over clusters

The best distribution of innovation budget over the three clusters will depend on the business strategy of the company. Ngai provides a rule of thumb for the distribution of budget over the three clusters [24].

Core   10%
Adjacent   20%
Transformational 70%

The total corporate budget for innovation comprises, for most companies, the stages discovery, concept, and feasibility. The budget for individual projects in the stage development, detailed design, and start-up is decided by the relevant business section or a corporate decision.

3.5.3 Risk-adjusted value for innovation management

Risk-adjusted value for innovation management is still not a well-developed field. Academic research based on industrial practice is sparse. There is a variety of methods used by companies, but the most used "method" is judgement by management [23]. Here are some plausible guidelines for the innovation stages.

For the early stages of discovery concept and feasibility, the total spending is limited and the only risk is that all money spent is lost because all the projects do not pas the stage-gate to development. If this total spending can be carried by the company, the risk is very low. The total budget spent on these early innovation stages can be related to the typical spending of the industry in a particular branch. Most companies choose to go for the average value of branch. If they should catch up with their competitors, they may spend more. If they are ahead of the competition, then they may keep the spending the same as in the previous years, in particular, when the overall return of investment on innovation is high.

The temptation to cut innovation portfolio budgets for the early stages is always there when the going is tough. The immediate effect on the business performance is positive as the cost is reduced. In the long run, however, market share is lost to competitors, which then appears to be nearly impossible to get back, as innovations take time. The author experienced this when, in the eighties of previous century, research for a particular polymer product was stopped. Within 10 years, market share was lost and could never be recovered.

3.5.3.1 Risk-adjusted value development stage and beyond: crossing the valley of death

For the development stages, often major investments on prototyping tests and pilot plant test are needed. In general, the maximum budget for the whole development stage of the portfolio will be set by the top management. The individual projects are first evaluated using risks and the expected added value. Tab. 3.10 will be useful in this respect. All projects for the development stage can then be presented in the same way and ranked.

Special attention is paid to the so-called valley of death. When the projected cumulative profit versus time of an innovation project is made, it will first show an ever deeper negative value until the lower minimum is reached; losses will then become less as sales start to generate money and finally a net profit will be made. The negative part of this projected curve is called the valley of death.

The projected profit curve can only be made with some meaning for the first time, at the end of the feasibility stage, when the cost of the commercial-scale process, the prototype, and the pilot plant are known. The deepest point of the profit curve, the maximum projected loss, is especially the most critical part. If that projected loss is larger than what the company can bear or is willing to bear, then the project will stop or it is modified such that the valley of death is less deep.

The deepest point of the profit curve should also be compared to the long-run profit when more than one manufacturing plant will be built and the total sales become large. In Chapter 10, Fig. 10.1 provides a detailed description of the projected profit curve and the valley of death. Further reading on this topic is provided by Markham [26]. Cooper provides further reading material into innovation portfolio management [27–29].

3.5.4 Innovation management guidelines for small enterprises

3.5.4.1 Innovation guidelines for small enterprises

The main success factors for innovative small companies and institutes gathered from numerous literature sources have been summarized by Conquest [30] and are found in Tab. 3.11.

3.5.4.2 Guidelines for breakthrough innovation companies

If the innovation concerns breakthrough technology that can alter the company completely, then intuition of the top management plays a very important role. Verganti, who worked inside very innovative successful companies applying design-driven-innovation found no formal ranking method, but instead found that the only criterion was: do the CEOs find the idea sufficiently exciting and sustainable. If the CEO asked for a financial analysis of the value of the design, then that meant that the project would be stopped [31].

Tab. 3.11: Management success factors for innovation in small enterprises [30].

Aspect	Description
Leadership	
Visionary	Top management has innovation vision and shares it
Long-term commitment	Improvements are judged on contribution to long term
Risk acceptance	Top management accepts that not all innovations are successful
Strategy	
Positioning innovation	Improvements and renewal are important pillars of the strategy
Signals from environment	Our company can pick up signals, transform the strategy accordingly, and make clear what the impact needs to be on the present business process structure
Cooperation in chain	Our company actively shares strategic information and business drivers with chain the partners to innovate services, products, and processes
Steering	
Tasks and responsibilities	Innovation tasks and responsibilities are explicitly delegated
Innovation protection Rewards	Innovation initiatives are given room to grow Steering and rewards for innovation are explicitly embedded in the steering and reward structure
Culture	
Felt need to innovate	Employees' hearts and minds know the need to innovate
Learning climate	There is an open communication learning climate in which mistakes and problems are shared
Tolerance	Novel ideas are always appreciated, also when they are contrary to common wisdom
Knowledge and competence	
Diversity personnel	Our company has a large variety of people with regard to knowledge and competencies
Knowledge sharing	It is common to share knowledge on procedures and processes
Innovation ownership	People responsible for innovation have the knowledge and competence to create a collective atmosphere to sell innovation

It is also worth considering whether the CEO's intuition has been built by years of experience in successful breakthrough innovations. If that is the case, then follow-up is warranted. If, however, the CEO has no experience with successful breakthrough innovations, then failure is likely to happen. Kahneman, the Nobel Prize winner, describes in his book that intuition that has been formed by experience in a regular environment can, on average, be trusted [32]. A breakthrough innovative company may form such a regular environment.

3.5.5 Project failures and their causes

This section deals with project failure rates and their causes. As the causes are also related to portfolio management, the subject is treated here.

3.5.5.1 Innovation failure statistics

We specifically treat here process innovation failures – failures, where only the process is new (innovation class 3) and the product is conventional. The practice of the last 30 years of introducing novel processes is that more than 50% of those projects are disasters. A disaster is defined as that having more than 30% cost growth beyond the budget and more than 38% schedule slippage. These statistics have been obtained by Independent Project Analysis of over 12,000 projects in oil and gas, petrochemicals, fine chemicals, and pharmaceutical sectors [23]. This means that the risk of failure is large.

3.5.5.2 Causes of project failures

Here is list of causes of project failures in the stages ahead of detailed engineering, gathered by IPA and published in table 1.2 of [23]:
– Lack of adequate analysis of potential solutions
– Schedule is agenda driven rather than data driven
– Optimistic forecasting
– Overly aggressive appraisal strategy
– Lack of front-end loading (feasibility and development stage deliverables)
– High degree of complexity/innovation
– Unrealistic plan
– Unrealistic budget
– Lack of understanding of stakeholders' aspirations

In addition to this list, an experienced innovation manager of a product-process innovation company observed in his career, the following causes:
– Lack of strategic fit
– Lack of freedom to operate
– Lack of understanding of customer needs and profit

All causes can be avoided by sound project management.

3.5.5.3 Mega project failures

In addition to causes reported by Bakker for all types of project, Merrow specifically reports of failures of mega industrial process projects [33]. He found 7 major causes of failures, derived from data of mega industrial process projects in the IPA database. These mega projects also contain more novel technologies than the other projects in the IPA database; so the conclusions are very interesting regarding innovation. Here are the major causes of mega process failures.

1: Too little money is spent in innovation stages, often due to greedy lead sponsor of the budget.
2: Too ambitious a schedule, not taking into account time required for good industrial practices.
3: A good deal is not arranged between the budget provider and the project executor early in the project.
4: Too little effort is spent on the early innovation stages, ahead of detailed engineering.
5: The established budget for the innovation project is reduced by 20% by the CEO (lead sponsor).
6: Risk is transferred to the EPC contractor by a lump-sum contract.
7: Beating up (threatening to fire) project managers for an overrun-on cost.

The first four causes are mainly related to the early innovation stages ahead of detailed engineering, where novel technologies are developed.

The last three causes are mainly related to the detailed engineering and implementation stages.

This again leads to the conclusion that sound project management in each innovation stage is needed to increase the success rate of innovation projects.

3.5.5.4 Project failures in the detailed engineering stage

Here is the list of reasons for project failure from IPA for this stage, as reported by [23]:

- Lack of clearly defined scope
- Incomplete design and errors
- Wrong contracting strategy
- Wrong contractors
- Lack of adequate resources (skills and numbers)
- Personnel changes
- Personality clashes
- Inadequate risk management
- Lack of adequate change control

3.6 Project management of innovation

3.6.1 Objectives of innovation project management

Project management is about individual projects. It is needed to ensure that ideas result in a successful novel product and process market implementation. A project, in general, is a temporary organization, with a defined beginning and end, brought to life to deliver unique scope within given boundary conditions (schedule, cost, and quality). Management of these projects is about controlling the execution of the project so that it reaches its goal and stays within its constraints.

An innovation project is an endeavor in which human, financial, and material resources are organized in a novel way to undertake a unique scope of work of given specification, within the constraints of cost and time, to achieve beneficial change, defined by quantitative and qualitative objectives [23].

In the early innovation stages of ideation and concept, when the number of people and disciplines involved is limited to a few, this management is limited to ensuring alignment to business objectives and to ensure exploring and analyzing many potential solutions. In these early stages of discovery and concept, the project manager will only spend a fraction of his time on this management. Arranging meetings between project team members to monitor progress and to ensure that the project is focused on the goal and stays with its constraints is the main task.

In the next stages of feasibility and development, when more people, disciplines, and stakeholders are involved, this management becomes a full-time job. Critical to success factors for this project management are:

– Making sure what the deliverables for each stage-gate need to be
– Making sure that sufficient time and effort is spent to produce these deliverables and that no shortcuts are taken to jump over the stage-gates without the deliverables

3.6.2 Project management by stage-gate system

3.6.2.1 Product and process innovation stages: general aspects
A major method to innovate at the lowest cost and time has been the introduction of the stages and gates approach. The basic thinking is that in the early part of innovation, major risks of failure – on the nonoptimal concepts – are reduced quickly and at low cost by exploring the whole field of options, selecting the best, and then performing proof of principle experiments for the best option.

As illustrated in Fig. 3.1, at each gate between two stages, a decision is taken to continue or abandon the project. Some companies also include other options, such as put the project on hold, or recycle the project, that is, send the project team back to provide more or better information, or give a conditional go to continue, depending on specific future events to happen [29].

The purpose of the stages-and-gates method is to avoid spending large sums of money on wrong ideas for a long time and spending those large sums of money on good ideas and only in the last stages when the risks have already become smaller, because more reliable information is available.

Cooper initiated a stage-gate method for product innovation in 1990. Many product innovation companies quickly adopted his method. His method has also been validated in the end by investigating many companies, including healthcare, chemicals, and polymer companies. Companies that had a formal stages-and-gates method in place and those that also had committed gate keepers did far better than companies with no formal stage-and-gate method. [28]. A reference book on the stages-and-gates method is by Cooper [27].

This stage-and-gate approach to innovation was applied by many process-focused companies, in particular, oil and gas companies, even before the product-innovation-oriented stages of Cooper. They however used different names for the stages. Bakker provides an overview of the different names for the same or similar stages [23].

In this book, we provide a uniform nomenclature for all product- and process-focused industry branches with the major activities for each stage, as shown in Tab. 3.12. It is partly based on the product innovation stage definitions of Cooper and partly based on process-focused companies, and EPC companies [23]. We condensed these descriptions into a single stage-gate description to fit the concurrent innovation of products and processes.

Table 3.12 is structured from top to bottom on general (people) management, business, design, and experimentation. This table will be helpful in getting an overview of the steps, from idea to implementation, and for a mutual understanding of the employees of R&D, engineering, and business departments in their use of terms, and in understanding their differences in attitude.

Table 3.13 indicates for each stage which disciplines are involved and to what extent. In combination with Tab. 3.12 and Belbin team roles, as described in Appendix A3, teams can be composed for each stage.

This stage-gate approach to industrial innovation ensures that unfeasible ideas are quickly discarded with minimal effort and that good ideas are pursued. When the stage-gate approach is combined with the systematic design approach of this book, good ideas are further enriched so that both the company and society greatly benefit from the novel products and processes.

3.6.2.2 End-of-life stage

We do not show in the overview the end-of-life stage for product and process as for that stage, no generic design methods are available. In the design of products and processes in the earlier stages, considerations for end-of-life will have to be considered, so that product and process recycles are made easy. This foresight aspect of design for

Tab. 3.12: Key items in the stages and gates for product-process design-driven innovation.

Items	Innovation stages					
	Discovery	Concept	Feasibility	Development	Detailed engineering	Implementation
General management						
Purpose	Idea generation	Screening	Feasibility concept	Validation	Engineering procurement construction	Launch Start-up
Required attitude	Optimism	Optimism	Realism	Realism	Pessimism	Pessimism
Project management	Nurture part time	Research part time	Project part time	Project part time	Project full time	Project + Operation
Open innovation	Low	Academic	Knowledge institute	T. provider toll manufacturer.	EPC	–
Funding	Game changer	Annual R&D budget	Annual R&D budget	Specific project	Specific project	Business
Business						
Risk of failure (%)	100	100	50	30	10	10
Invested (%)	0.01	0.1	2	10	94	100
Capex estimate uncertainty (%)		200	50	30, 10	0	
Market items		Demand opportunity	Size estimate	Size determination		Launch
Business question	What starting?	Full range explored?	Best selected?		Everything in place?	What did we learn?
Business case			Initial	Confirmed value, risks		
Stage-gate decision	Strategy fit	Business case	Business case	Business case	Top management	
Duration (years)	1–2	1–5	1–2	0.5–1	1–2	0.2–0.5

Tab. 3.12 (continued)

Items	Innovation stages					
	Discovery	Concept	Feasibility	Development	Detailed engineering	Implementation
Design						
Product design	Sketch	Concept	Prototype	Complete	Detailed	Description
Process design	Sketch	Concept	Unit operations	Main equipment	Detailed	Start-up plan
Main deliverable	Sketch concept	Concept economic potential		Basis data design, test results	Design books	Plan, post start-up review
Project engineering			FEL-1	FEL-2, FEL-3	Procurement package	
Experimental	Proof of principle	Proof of concept	Product prototype	Pilot plant	–	Start-up

Tab. 3.13: Typical company functions involved in product and process innovation.

	Discovery and concept	Feasibility and development	Detailed engineering	Product launch and process start-up
Science	90	10	5	5
Engineering	5	75	75	5
Operation		5	5	75
Marketing		5	5	5
Business	5	5	10	10

As percentage per stage.

end-of-life is treated in the design chapters of this book and in life cycle assessment part of Chapter 12.

3.6.3 Project entries to stages by the technology readiness level method

Technology transfer from a university, institute, or technology provider to a company, such as an end user company is not easy. The readiness level for commercial-scale implementation should be established so that what still needs to be done is clear.

TRL assessment is very useful in this respect. It is about quickly assessing a novel technology for its innovation stage. It is particularly used for novel technologies resulting from academic research or technologies provided by start-up companies. End users then use the result of that assessment to decide to pursue the technology (partly) in-house or to leave it for further development outside the company.

There are several TRL tables available from NASA [34], ESA [35], DOE [36], and EC [37]. However, these are defined by government organizations and do not use terms familiar to the industry. They also do not consider the (mass) manufacturing of the technology, but focus only on the product technology. To fill this gap, a manufacturing readiness level (MRL) table for novel manufacturing technologies (for existing products) has been additionally defined [38]. Recently, the European Association of Research Technology Organisations (EARTO) developed a TRL-level description for both novel products and processes [39]. From these tables, we derived a TRL table for novel homogeneous products and for their novel manufacturing processes using terms familiar to these industries, shown in Tab. 3.14. We added level "0" so that ideas that have not reached TRL 1 can be categorized.

The table is particularly useful to assess novel technologies from universities and from technology providers. It is also useful in open innovation projects with various partners to help establish consensus on the technology status and what needs to be done for commercialization of various process options [40].

Tab. 3.14: Technology readiness level definitions for novel products and processes.

TRL level	Description	Innovation stage
0	Idea of novel product or process stated	Discovery
1	Experimental proof of key novel element of product or process, called proof of principle	Discovery
2	Concept design of product and process	Concept
3	Experimental proof of product concept elements and process elements	Concept
4	Experimental validation of integrated product (prototype) and integrated process (lab scale)	Feasibility
5	Industrial professional's technoeconomics assessment of product and process	Development
6	Novel product prototype tested by industry Novel process technology demonstrated by industrial pilot plant	Development
7	Novel product launched in market – first commercial-scale demonstration of process plant in operation	Implement
8	Full launch of product to market; learning points of demo plant incorporated in commercial process design	Implement
9	Commercial process in operation and evaluation report available	Implement

A warning must be made here. The TRL table suggests that novel technologies at TRL 1–4 are outside the industrial realm. Also, in the industry, ideas are researched from TRL 1 to 4.

3.6.4 Team formation

3.6.4.1 Behavior characteristics of team members and teams required over the stages

Team members should be chosen for their skills as also whether they can work together. The Belbin group formation method is very successful in forming groups that cooperate well.

3.6.4.2 Group design with Belbin team roles

Project groups are mostly designed based on disciplines required. However, if groups must cooperate closely for the desired results, then they should get along with each other. In addition to discipline design, the Belbin team roles can be used to design the group or team. Appendix 1 shows how to apply this method.

3.6.4.3 Desired behavior in different stages

The desired behavior of people in the various innovation stages is also important. Optimism should prevail in the early stages of discovery and concept, as most ideas will not make it to commercial implementation and the lack of knowledge should not hamper people to believe in the success of the project. In the feasibility and development stages, realism should prevail, as now it is important to show and validate that the new product and process indeed works. In the detailed engineering and start-up stage, pessimism should prevail, as now all details of the design and the start-up are right. Pessimistic people are very good in finding detailed concerns as to why the novel product and process will not work. All these concerns should be turned into specific modifications and actions to get the product and process working.

3.6.5 Intellectual property creation and protection

Intellectual property is something that is new, useful in commercial practice, and is documented and protected from use by others. Intellectual property can be created by a design, as a design always has new elements, and is nearly always useful when exploited in practice. It is always documented. Protection from commercial use by others is then the last step. There are many ways to protect intellectual property. One way is to keep it secret. For small companies, this may be doable. For larger companies, this will be hard. A second way is to file patents. Now, the information shown in the patent is made public but cannot be used by the competitors for business purposes. A third way is to file copy rights for essential documents containing the innovation. A fourth way is to file a design right. A fifth way is to file for a trade mark.

The main error made on this topic is that the novelty is presented orally or in writing, in the open. Open means any public meeting place of any publication in any form. Even bilateral discussions with no formal nondisclosure agreement are public. Once the novelty is public, no patent protection is possible anymore.

Literature on intellectual property generation and protection is vast. Verloop argues that intellectual property is a form of intellectual capital, and should be used defensively, that is, preventing other companies from commercially exploiting the innovation and offensively, by exploiting it in various ways [16].

3.7 Discovery stage

The purpose of the discovery stage is to obtain ideas and to define the ideas to such a level that it becomes clear what they are about so that at the end of the discovery stage, it is clear whether they fit to the company's strategy and can pass on to the concept stage or not (see also Fig. 3.1).

3.7.1 Managing projects in discovery stage

The discovery stage can be compared to hatching of an egg. The only essential conditions for success, provided by the hen, are protection and warmth. The actual development inside the egg is invisible and autonomous.

The same holds for ideas generation and development in the discovery stage. Companies highly dependent on a continuous stream of new ideas to feed their innovation funnel, protect researchers in their early idea development. Some, like IBM, do it by letting researchers spend 10% of their time on pursuing their idea, without requiring a formal budget and reporting.

Other companies have a formal organization such as a significant changes organization, where researchers can ask for a budget to pursue their idea any time of the year, with a minimal description of the idea. A member of the game changer team will then help to describe the idea such that it fits in the company's strategy and will provide a budget to perform the most critical test to demonstrate the feasibility of the idea. Often, this is a proof of principle experiment for a novel product or process step.

If the business case looks sufficiently attractive but the idea involves many innovation aspects, then the idea may also be analyzed for the specific additional innovations that will be required. Once the idea has been proven in some way and the business case has been sufficiently developed, the researcher will propose the idea for taking up in the annual research budget round, see also the table, generic criteria for stage-gate decisions.

3.7.2 Creativity methods

In the discovery stage, creativity is needed to generate many and good ideas. The literature on creativity methods for innovation is vast. A good introduction to further information is by Nijstad and Paulus [41] and Hermann and Felfe [42].

Brainstorming is a good method to generate unusual associations and new perspectives to the problem at hand. The most important rules are:
a) No critical remarks on ideas
b) Associate and enlarge ideas

Brain writing is another useful method. It works well for introvert people, as they can write their ideas down. Process engineers in general are introvert; hence this method is mentioned. The complete method can be found in Appendix A3.

3.7.3 Discovery stage-gate decision evaluation criteria

What in general needs to be done in the discovery stage is expressed here, under elements of the reporting for the stage-gate decision.

3.7.3.1 Idea description
If the idea is a product, then a functional description of the purpose of the product is needed. Options for product composition to fulfil the product functionality may be given. If the idea concerns a process, then a block flow diagram with feedstock and product streams is required.

3.7.3.2 Proof of principle
The result of a proof of principle experiment for the main critical novelty of product or process is needed. The challenge for the researcher is to define an experimental setup that is simple, quick to construct, and also such that the test result is convincing to others that the novel principle works. In general, it requires several attempts before the proof of principle is successful.

3.7.4 Business case

A business case is the business justification or opportunity assessment of a project [43]. It is very clear from this definition that it specifically belongs to a specific company. The business case cannot be established by others outside a company. In other words, the company owns the business case. Even better, a specified person inside the company owns the business case. This is not the project leader.

The business case of an idea in the discovery stage may be developed by answering the following questions [43]:
1) Does it create value for the company?
2) Does it fit to the strategy of the company?
3) Does the company need the idea?
4) Can the company (together with others) develop the idea?

For a project to pass the discovery stage-gate, these four questions of the business case may be used. If positively answered, then the preliminary business case for the idea is established.

3.7.4.1 Creating value
The question of creating value for the company, for radically novel products may be hard to answer. It may be answered by stating, yes, we believe in its capacity to create value for company and society.

A simple economic potential estimate may be provided on the potential economic contribution per year of the novel product or process, when commercialized. It may be done by just estimating a sales price per ton, subtracting from it, the feedstock cost per ton of product, and estimating the total sales in ton/year.

3.7.4.2 Strategic fit
The fit of the novel product or process with the company strategy's will be indicated here. If it does not fit to the strategy and a strategy change of the company is not foreseeable, then the idea should not be pursued. However, some champions of an idea cannot be easily stopped and because of their stubbornness and creativity, they may find ways to pursue the idea.

3.7.4.3 Necessity of the idea
Even if the novel product and/or process fits the company's strategy, the question whether the idea is needed remains to be answered. If yes, it will be pursued. If not, it can be stored and retrieved when circumstances change and when the idea is needed. This however requires an idea storage and retrieval system that most companies will not have.

3.7.4.4 Idea development doable (with others)

The question, can the idea be developed by the company with help of others, is also often hard to answer. It requires knowledge about the technology providers and academic research groups that are available to solve some of the foreseen knowledge gaps.

3.8 Concept stage

The purpose of the concept stage is to obtain a full picture of the goal of the novel concept, the knowledge needed, the knowledge available, and the most critical success factors that are to be investigated in the concept stage so that at the end of this stage, a decision can be taken, based on all criteria, to pursue the concept or not (see also Fig. 3.1).

For a novel product, this involves developing a customer value proposition, technical requirements for product and process definition, identification of critical to quality variables, and experiments to prove the concept.

3.8.1 Customer value proposition in concept stage

For a novel product concept, it means that, first, a customer value proposition is made. This is a marketing statement that describes why a customer should buy a product. For a business-to-business product, it consists of the sum total of benefits a vendor promises that a customer will receive in return for the customer's associated payment [44]. Table 3.15 contains its main elements.

Tab. 3.15: Customer value proposition elements.

Customer value proposition elements	Description
Product type	Single product, product family, or product platform
Technical value proposition	Technical advantages over the existing product
Product technology	Product components
Manufacturing process technology	Process options for manufacturing of new product
Materials technology	Input materials for the new product

This customer value proposition requires a strong interaction between the product developers and the business and marketing people of the company. Design thinking and design models help these conversations enormously by having drawings and pictures of the product design, by having both a functional product description (what it does) and several physiochemical alternatives fulfilling those functions. By also making basic process concept designs with material balances, the first indications of feedstock cost per

alternative are available. Following a checklist, derived from human senses, may be useful to make sure that all relevant aspects of the product requirements are covered:
- Feel for relevant body parts
- Visual appearance
- Smell
- Taste
- Sound

3.8.2 Technical product requirements in concept stage

The primary consumer perception requirements should be translated to technical requirements. Other functional goals and requirements from the customer value proposition should also be translated into technical requirements. This translation may involve several steps before a list of quantified technical specifications of the product is obtained. Quality function deployment and house of quality methods may be used for this translation of customer needs into technical requirements. The summary of the technical product requirements is called critical-to-quality requirements.

3.8.3 Process concept generation

Process concept generation involves defining the reactor system, recycle structure, and the separation system. The new method of process concept generation comprises designing a functional block-flow diagram with its material inputs and outputs for the chosen product concept. If several product concepts have been generated, then at least one functional block-flow diagram for each product alternative will be made. Major energy inputs and outputs will also be determined. Finally, feasible unit operations will be selected for each functional (task) block, or even better, functions will be combined and feasible functional integrated concept solutions will be defined.

   The dominant method of the past, designing with unit operations, does not deliver breakthrough innovation anymore. Process design by task integration is now emerging as the dominant design method, as the advantages in investment and operating cost, lower environmental impact and inherent safe design have been proven for hundreds of commercial-scale implementations [8]. The design method is described in detail in Chapters 6 and 7.

3.8.3.1 Proof of process concept
From the process concept design, all key issues are derived and a lab-bench-scale research program is defined and executed to remove these key issues and establish information for a better process concept design.

3.8.3.2 Selection of best process alternative
The best process alternative will be selected as the base-case design, based on SHEETS criteria (see Chapter 4), with, often, a high weight factor for lowest feedstock cost and lowest investment cost. Selection methods are provided in Chapter 7.

3.8.3.3 Process concept design
For the selected best alternative, a process concept design will be made, containing:
– Feedstock specifications
– Product specifications and quality performance requirements
– Catalyst type and amount
– Byproduct stream specifications
– Stream compositions for each stream
– Heat balances for each process unit
– Selected unit operations with pressures, temperature, and residence time
– Selected equipment types and main sizing aspects
– Choice of construction materials
– Main control strategy (indicated for sensitive process aspects)
– Auxiliaries
– Utilities needed

3.8.3.4 Basic design data generation
Basic design data will be generated for the concept design. Chapter 6 provides details of what data are to be generated and how this can be done efficiently.

3.8.4 Concept stage-gate evaluation

Business and research management will be informed about the results of the concept stage and about the planning and cost of the feasibility stage. For all criteria, information will be provided such that a decision to pass the stage-gate or stop the project can be made.

3.9 Feasibility stage

The purpose of the feasibility stage (see Fig. 3.1) is to verify the feasibility of the product concept and the process concept for the success of all critical aspects.

3.9.1 Concurrent product-process design and testing

Technical product feasibility is proven by experimentally testing the integrated product. Often, this integrated product is called prototype. Here, unexpected effects of integration may cause the product to fail. Often, the failure can be repaired by analysis of the cause of the failure. Because the tests are carried out at laboratory scale, modifications are fast and at low cost.

Through the direct customer and end-consumer feedback, both the product design and the optimal process design can be fixed. The product design includes not only the optimal composition and structure, but also the tolerable composition levels (including impurities). Also, the option to have multiple material suppliers is validated, which will help later in the negotiation of raw materials purchasing and supply conditions. Finally, the tolerable key product attributes and their relationship with composition and processing should be very clearly reported for testing in the next stage: the manufacturing stage.

3.9.2 Microplant in feasibility stage

The best process design concept is integrally tested experimentally at laboratory scale. This test unit is best called a microplant.

A properly designed microplant contains all major process units in the same lineup and recycle structure as the commercial-scale design. It contains all reactions and separations and it can operate in the commercial design window of process conditions.

Its purpose is to study the reactions and separations under the design conditions and with recycles. Reaction conversion rates and byproduct formation under recycle conditions are determined to improve reaction kinetics. Separation performance results are used to validate and improve physical equilibrium models for the separation steps.

Including the recycle flows in the microplant is also very useful. Heavy boiling species, for instance, can buildup in it, affecting the reactions and the separations. Knowing this is of enormous value for improving the commercial-scale design and reducing its risk of failure.

Because of its small scale, the construction and operation costs are low. Often, is it made from glass due to which, problems such as foaming and fouling are also quickly localized and the process is quickly modified to solve them. It can be used to test various process design options [45].

A downscaled microplant is a very cost-effective element to reduce risks and provide data for optimization of the commercial-scale design. It is the smallest scale an integrated version of the commercial-scale design concept, including recycle flows. The typical production capacity range is 1–100 g/h and the equipment diameter is typically 1–5 mm [45].

The emergence of microscale equipment with well-defined residence time distribution (r.t.d.) and uniform temperatures has helped enormously to boost the use of microscale units.

For new pharmaceutical products, these microplants increase the speed of process development and thereby reduce the time to market. In some cases, the laboratory-scale process is run for several weeks to quickly provide the product for clinical testing.

In fact, a whole new industry branch of very small-scale devices has emerged [46], with over 274 companies providing these devices for the pharmaceutical industry [47]. A new term, microfluidics, is often used for these devices. The term refers to devices, systems, and methods for the manipulation of very small fluid flows, as small as microliter/h, nanoliter/h, picoliter/h, and even femtoliter/h flows [48]. Other terms used are lab-on-a-chip and flow chemistry.

Also for fine chemicals, the application of microflow systems is increasing. In this industry branch, most of the production is carried out in a batch-operated mechanically stirred reactor. Cooling is provided by having a solvent that evaporates from the reaction liquid, which in turn is cooled by an overhead condenser. The cooled liquid solvent is returned to the reaction liquid.

This reactor and its operation have large scale-up uncertainties, which are solved by an experimental program on the batch reactor. By-product formation is also often far more than the amount under ideal conditions due to the far longer residence time, than required, for the reaction itself. Also, the solvent, after its use, becomes a waste by-product as well as the cleaning liquid in between batches.

Boodhoo describes the following microprocess technologies in some detail [49]: micromixers, microheat exchangers, and microreactors. Of the latter class, multi-channel reactors and spinning disk reactors are described in detail. Günther and Jensen [50] provide an overview of multiphase microfluidics for chemicals production at laboratory scale, and Kenning et al. [51] cover microfluidic unit operations: extraction, absorption/desorption, distillation, as well as particle or droplets separation.

If successful, the process design is completed with process instrumentation and control and the product packaging method is designed. The degree of design definition for an economic assessment is typically 50%. Meaning that 50% of the investment is based on the process design and 50% is based on correlations for cost of plot preparation, concrete structure, and outside battery limits.

The remaining uncertainties regarding process integration and scale-up will be addressed in a pilot plant in the development stage. The design and costing of the pilot plant and the cold-flow model, however, belong to the feasibility stage, as it may well be that the cost of the pilot plant is so high that the project stops at the feasibility stage-gate.

The feasibility of the product and the process design are assessed at the stage gate using all criteria of Chapters 10–13, focused on validation of its feasibility for all those criteria. Details on the feasibility stage gate evaluation will be found in the evaluation chapters.

3.9.3 Scale-up strategy and information

The first question to answer is what scale-up strategy is to be followed. For novel products with a limited patent protection and high added value, such as for pharmaceuticals, a short R&D time is very important. Equipment costs are in general a minor item of the overall cost, so easily scalable equipment requiring only one validation step are preferred. Millimeter flow reactors, for which the brute force method can be applied, have a large advantage here.

For bulk products, for which reliable production to satisfy the customers and to minimize the start-up cost of the commercial-scale process is very important, the scale-up risk needs to be low. The innovation class table with its risk data of Section 3.3 can be used to assess the risk involved.

Table 3.16 provides for some, often applied, process equipment available scale-up methods.

Tab. 3.16: Scale-up methods available for various process equipment.

Equipment	Residence time distribution	Micromixing	Mass transfer
Trickle bed	B.F., Cor. + dilution	B.F	B.F., Cor. + wetting
L, G packed bed	B.F., Cor.	B.F.	B.F., Cor.
Static mixer L, G.	B.F., Cor., Mod.	B.F., Mod. Cor	B.F., Cor., mod.
Static mixer G/L	B.F., Cor., Mod.	B.F.	B.F., Cor. mod.
Tray column	Cor		
Plate column	Cor		
G/L bubble column	Cor., Mod.	Unknown	Cor.
Cross flow bub. col.	Cor., Mod.	Unknown	Cor.
Mech st. tank L, G	Cor., Mod	Mod.	–
Mech st. tank G/L	L. Corr. G unknown	Unknown	Unknown

Scale-up methods: brute force (B.F.), correlation based (Cor.), modeling-based (Mod.).

For reaction, the r.t.d. is always relevant. For two-phase reactors, both the r.t.d. and the mass transfer are always relevant. If the reactions are fast and multiple reactions are involved, often micromixing is also relevant.

For crystallization, micromixing, r.t.d., and mass transfer are relevant.

For gas/liquid phase separations, such as in distillation and absorption, both the mass transfer and r.t.d. are important.

A special warning must be given for the scale-up of two-phase mechanically stirred reactors. These reactors have a highly nonuniform turbulent intensity profile. Near the impeller, high shear and energy dissipation occurs. There, the dispersion is made very fine. Further away, the turbulent field is less intense and there, coalescence of the dispersed bubbles or droplets can occur. At scale-up, the areas of high and low turbulent intensity change and often, the ratio of high and low intensity changes. This makes scale-up notoriously uncertain. Emulsion polymerizations carried out in mechanically stirred tank reactors, where the dispersion quality also determines the final product quality, scale-up is therefore still a high-risk affair. Choosing other equipment such as static mixers would avoid this scale-up problem.

3.9.4 Scale-up information equipment

For all process equipment, scale-up information will have to be provided, so that the commercial-scale design base and its associated risk are known for each process equipment. In some cases, the risk may be reduced by mitigation measures.

3.9.5 Business case feasibility stage

In the feasibility stage, the business case is much further developed. Also, here, the business case values of all projects are then established and so the value of the whole portfolio of innovation projects [23].

3.10 Development stage

The purpose of the development stage is to reduce the risk of introducing the novel product and process to acceptable levels (see also Fig. 3.1). This means that at the end of the development stage, the chances of a successful product launch are high and that the commercial-scale process will meet the product requirements and process capacity and that it can be started up within the averaged industry start-up time.

3.10.1 Product testing in development stage

For product performance at the client, different product versions are produced at a processing scale that is large enough to perform accurate scale-up of the process and/or is large enough to produce sufficient amount for testing the product in real conditions at the direct customer and/or end user of the final product with the new ingredient/component. Customer evaluations of the product should reveal the appreciation for and the value of the product for them. Modifications of the product are still possible and needed, not only to verify the product functions but also to maximize the business opportunity by tuning the value added parts of the product and reduce the costs (raw materials, processing) of producing the product.

Often, a mini plant is run to produce different product versions, based on raw materials (different suppliers, quality levels, process settings, etc.), and to validate the product in the (industrial) customers processes and in the final consumers' application. For this purpose, sufficient material for testing should be produced, and this often also determines the size of the mini – or pilot plant. If this scale is not large enough and a pilot plant will be necessary, then that plant will be constructed and operated in the development stage.

Also, critical logistics processes, such as storage, filling, packaging, transportation, unloading, etc. at the direct customer end may be identified and verified with the product batches. Again, if major investment cost is involved, then that investment will be part of the development stage.

3.10.2 Process testing in development stage

3.10.2.1 Mini plant purpose and design

The mini plant, ideally, is also a downscaled version of the commercial-scale design, including all recycles, but now it also contains the same type of reactors and separators, with the same number of mass transfer units, r.t.d.s, and the same construction materials as in the commercial scale. It typically has a production capacity of 0.1 kg/h [45].

With this mini plant, further validation of the design concept and the process models are established and problems are quickly identified and solved. Potential interactions between the construction materials and the components such as catalysis by construction material and corrosion of construction material can now also be observed, as these plants operate continuously and for a long period (several months). Additionally, corrosion test material samples are placed in all streams to optimize the construction material choices for the commercial scale. Automatic process control and data storage are purchased from a reliable supplier of distributed control systems.

In some cases, a mini plant avoids the need for a large pilot plant, saving large investment. The pilot plant section shows when a larger pilot plant is still needed.

The following design documents for a mini plant are required [45]:
– A process flow sheet
– An engineering flow diagram
– All equipment types and sizes
– Objectives of instrumentation
– Specifications for insulation of equipment and piping
– List of measuring points
– Safety concept and logic: how sensors are linked to actuators and the safety procedure

3.10.2.2 Pilot plant purposes and design

The first question to answer is: Is a pilot plant needed? The pilot plant can be needed for:
– A: Producing a large amount of product for quality and market testing
– B: Identifying scale-up problems

- C: Validating process models and providing input for mass and energy balance updates
- D: Training operating personnel of the future commercial plant
- E: Improving the estimates of catalyst lifetimes
- F: Testing construction materials for corrosion
- G: Validating process control and operating procedures

A pilot plant is also very useful to get reliable stream compositions. This increases the stream knowledge factor F of the start-up time correlation, from nearly 0 to 1 (see the section on start-up). The difference in start up time of those two F values is on average 3.2 months. The amount of money lost by not having a specification product for that period, in general, easily outweighs the cost of a pilot plant.

Ad A: Pilot plant for producing large amount of product for testing

Often, a large amount of product is needed for performance tests at one or more clients. This testing is often desired by the client even if the product is promised to be the same and only the process has been modified. This holds even for intermediate bulk products. Trace amounts, not detected by analysis, can still affect the product performance at the process of the client; for instance, when the new trace components degrade the catalyst in the process of the client.

If the product is novel, then, often, large amounts are needed for all kinds of market launch tests [45].

Ad B: Pilot plant to identify scale-up process problems

A pilot plant is needed to identify scale-up process challenges if [52]:
- More than one new process step is involved
- The process step contains a new step and has a recycle stream over more than one step
- The process contains a novel solids processing step
- The feedstock is a crude source, whose composition knowledge is incomplete

If the pilot plant is needed but is still omitted, then start-ups, in most cases, lead to a failure, which means that the process as such cannot be operated and/or meet the design specifications on product quality and capacity [54 Merrow, 1988]. Additional budget of over 30% of the investment is often needed to solve the problems [23 Bakker 2014]. These costs far exceed the pilot plant cost, which are, in most cases, less than 10% of the commercial-scale investment cost.

Ad C: Pilot plant to validate process simulation models
Process simulation models are now the main source of detailed information of process streams composition knowledge everywhere in the process, including the final product output stream. This information is used to select the construction materials, select and size equipment, and define utility requirements. The start-up time correlation parameter F also indicates the importance of this knowledge, as shown in the start-up section of this chapter. It is therefore important that the simulation models are reliable. Extensive validation of these models can be carried out by measurements of stream compositions, temperature, and pressure, and comparing them with model predictions.

Ad D: Training operating personnel of the future commercial plant
A pilot plant can also be used to train operating personnel of the future commercial plant. Nowadays, this training is mostly carried out using dynamic simulation models, but additional training with a real pilot plant helps personnel to get acquainted with, for instance, taking samples for analysis and seeing what the intermediate stream samples and product samples look like and how miss-operation also shows up in changing the colors of those samples.

Ad E: Improving the estimates of catalyst lifetimes
Long-term pilot runs under actual process conditions will show the catalyst lifetime and how the process stream compositions change with deteriorating catalysts. This type of testing can also be done in a mini plant.

Ad F: Testing construction materials for corrosion
By placing many material test samples in many pilot plant streams and running the pilot plant for a long time will provide accurate corrosion rate information, which can be used to improve the final materials selections for the commercial-scale plant. Many problems in commercial-scale plant operations are still connected to materials corrosion [52]. This corrosion testing can also be done in a mini plant, see the section above on mini plant.

Ad G: Validating process control and operating procedures
In process designs with strong dynamic interactions between various process parts, process control may be difficult to achieve. Testing the chosen process control design and tuning the process control in the pilot plant can be very important. Also, operating procedures can be tested by process operators.

This validation is particularly important for batch processes, which by nature have a dynamic behavior, Validation of process control design and operating procedures are there a necessity

3.10.2.3 Pilot plant as downscaled version of commercial-scale design

If a pilot plant is needed, then it should always be a downscaled version of the commercial-scale plant, with all unit operations and all recycle flows. Also, the commercial-scale process conditions such as temperatures, pressures, and residence times should be inside the pilot plant operating window. Only in this way will the tests be representative of the previously optimized commercial-scale plant.

If the pilot plan is not designed as a downscaled version of the commercial scale, then this means that the commercial-scale plant must contain the same process type units and recycle as the pilot plant. In some cases, this means that multiple units must be selected in parallel for the commercial scale because the selected equipment for the pilot plant did not appear to be scalable. It will be clear that this way of scale-up means higher investment cost of the commercial-scale process, so it is highly undesired, although it is still a practice in some companies. If the commercial-scale design also differs from the pilot plant in process type and recycle structure, then the risk of failure increases enormously. According to Merrow, this is one of the major causes of commercial-scale failure, meaning that the start-up time is 40% longer than budgeted for (see start-up time prediction) and 40% additional cost to remedy the failure [33].

3.10.3 Pilot plant engineering, procurement, and construction (EPC) company's choice

The detailed design and construction of the pilot plant should be done with the same rigor as the commercial scale plant. The selection of the EPC company for the pilot plant is therefore important. A company dedicated to pilot plant EPC is strongly preferred. If an EPC company is chosen that handles both commercial-scale projects and pilot plants, then it is likely that the pilot plant will get a far lower priority and lower qualified personnel, as the size of pilot plant project, in money terms, is a fraction of a commercial-scale project. This can in turn mean that the agreed timing is not adhered to and that the pilot plant will contain errors. For this reason, an EPC company that is only concerned with pilot plants is the strongly preferred choice.

3.10.3.1 Equipment scale-up effects determination

Commercial-scale equipment will be larger than preceding test equipment in the development stage. In general, this means that the hydrodynamics in the equipment will be different. The following transport phenomena will change with the change in hydrodynamics and, in turn, can affect the performance of the equipment:

- r.t.d.
- Micromixing
- Mass transfer
- Dispersion of a second phase
- Heat transfer
- Momentum transfer

For most standard pieces of equipment, these transport phenomena are known as a function of geometric sizes, fluid velocities, and fluid properties; often, in the form of correlations. The relations between the phenomena and the performance are also known and, often, expressed in models.

Especially for many conventional reactor equipment and gas-liquid contact equipment, correlations for the transport phenomena are available in textbooks such as Perry's *Chemical Engineers' Handbook* [54].

A word of warning is needed here. Often, the mass transfer correlations assume implicitly that the r.t.d. of the fluids is plug flow, while deviations from plug flow may occur. Often, these mass transfer correlations have been obtained under the conditions of plug flow deviation. At scale-up, these deviations from plug flow can become larger and then the commercial-scale performance is worse than predicted using the mass transfer correlation.

For tray-type gas-liquid columns, often the r.t.d. of the liquid is assumed to be completely back-mixed for the gas phase. For very large diameter trays, with one of a few down comers for the liquid phase, this assumption is sometimes not valid. Technology providers of large distillation trays make available this knowledge for their clients.

If the relation between the equipment size and the phenomena are not known, then that relation must be determined. There are two established methods for determining the relation:

- Method A: The phenomena are determined in the same piece of equipment, but used in a different process.
- Method B: A large scale of that equipment is constructed and the relevant phenomena are studied in that piece of equipment. To keep the cost low, often simple fluids such as air and water are used, rather than the actual fluids. Such a test facility is called a mock-up or cold-flow model.

3.10.3.2 Mock-up design and testing for hydrodynamic scale-up effects

Nowadays, computational fluid dynamics models are often made for equipment with complex geometries and complex hydrodynamic flow behavior, affecting the performance upon scale-up. These models are then used to design and optimize the equipment geometries of the commercial scale. These models need therefore validation. This validation is carried out in the so-called mock-ups where inert fluids such as water and nitrogen are used and mixing, r.t.d., and mass transfer are measured and compared with the model predictions.

3.10.3.3 Development stage-gate evaluation

The purpose of development stage-gate evaluation is to decide to go to the detailed design stage or not. As the detailed design stage is very costly, this is an important decision. The economic evaluation of the commercial-scale design for this stage-gate is therefore very important.

Formal evaluation methods found in the evaluation chapters can be used as input for the gate decision. It is advised to consider all criteria for the decision, not just the economics. The gate-keepers for the decision should be experienced and have the authority to take the go/no-go decision.

3.11 Detailed design stage

3.11.1 Detailed product design

The product has already been designed in the development stage. However, product pricing adaptation to latest developments in the market, details of packing, logistics transport planning, and, if needed, additional buildings are now to be defined.

3.11.2 Detailed process engineering

The detailed process engineering stage comprises detailed process design, procurement of all process equipment and construction of the process (see also Fig. 3.1). These three steps are in general executed by EPC contractors. For a detailed description of the activities of this stage and how the manufacturing company cooperates with the EPC contractor, the reader is referred to other books, such as by Bakker et al. [23].

The starting point for the detailed design is the commercial-scale concept design provided by the manufacturing company. If only a preliminary concept process design is delivered, then the EPC contractor will make a concept design. In general, a process design is first made with sufficient definition to allow an investment estimate of ±30%, on which the manufacturing customer can decide to continue the project or not, followed by a design allowing a ±10% accurate estimate. These two levels of definition and accuracy required differ a little from company to company. Clarification between the customer and the EPC contractor is always needed. Also, a complete evaluation for all other criteria for now and future trends on safety, health, environment, social acceptance, legal items of the local country (license to operate) is carried out as part of concept design. The customer then takes a decision to go for detailed engineering, procurement of equipment and construction or not.

Nomenclature of the process concept designs varies enormously between companies and clarification should always be asked for. Process concept design with 30% accuracy in capital investment is also called FEL-1. The next concept design with 20% accuracy is called FEL-2, and the third process concept design with 10% accuracy is called FEL-3 [23].

Detailed engineering is an enormous effort, with many engineering specialists involved. Here, only some main elements are mentioned.

1. A complete piping and instrumentation diagrams (P&ID) is made for every part of the process. It shows all piping connections to all equipment and all controls and instrumentation. This is an enormously important piece of information for the process construction and for a safe and proper operation. To get a feel of its importance, consider that, on average, piping concerns 20% of the total process investment cost. A large part of that cost is making specifications for each pipe on construction material, tracing or not, sizing, and precise 3-D location.
2. A complete process description is made for each unit operation, containing dimensions, capacities, temperatures, pressures, construction material, chemical and physical knowledge, and data.
3. A 3-D plant layout plan is made showing all the equipment and piping.
4. A detailed commissioning and start-up instruction manual is made.
5. Equipment detailed specifications for purchasing are defined.
6. Detailed drawings for special equipment, to be custom made, are made.

3.11.3 Choice of EPC contractor

The choice of the EPC contractor for process design and construction is very important. The selection should be made by the project leader of the manufacturing company. Any involvement by business leadership in the selection of the individual contractors is inappropriate. Any involvement of the lead sponsor's purchasing organization in the selection of contractors is likely to be catastrophic. Even more important is the selection of the right contractor team, a team that the manufacturing company team can work with creatively [33].

The EPC contractor should have experience in the relevant process industry branch. For the food industry branch, for instance, that contractor will know how to prevent contamination, how to ensure complete cleaning of equipment, and how to avoid undesired microorganism growth.

For a breakthrough novel process unit within the overall process, additional contractors to the main contractor may be involved, with specific knowledge of designing for the materials flowing through the process unit and with knowledge of designing for similar stream compositions. That contractor will, for instance, design, in additional measures, to prevent or removing local fouling, or select, at critical locations, an instrument that is robust to fouling.

3.11.4 Demonstration plant

For a radically novel process that, in the end, will be built for very large capacities and with a large investment, a demonstration process is first designed and operated, in most cases. This demonstration process is, in general, 1–10% in capacity of the final commercial scale. It may be 10–30% of the full-scale investment cost.

The main reasons for having this demonstration plant are to reduce the real and perceived risks. The risk can be technical and/or business (uncertain new market development)

The technical risk of a novel process (and product) is reduced by simply lowering the investment cost. The higher investment cost per ton of product for this demonstration scale may also be earned back later by incorporating the learning points of the demonstration plant in the final full capacity plant. The savings by these learning points can be small, but can also be up to 40% of the final investment, depending how much redundancy has been built in by the engineers in the demonstration plant [52]. The redundancy measures can be operating safety margins, oversized equipment, or even spare equipment.

The business risk can be due to new market development, which is uncertain. Sometimes, this demonstration plant is needed to convince investors with no experience in the process industry (inside or outside the company) that reliable processing is feasible and the promised return on investment can be made.

It is very important that up-front it is made certain that the demonstration plant will make a profit. It is also important that the lower return on investment of the demonstration plant is clearly reported to the business department of the company. If either is not carried out, the commercial-scale plant will not become a reality.

An example of a demonstration plant prior to the final full-size plant is the Shell GTL plant in Bintulu. It has capacity is 14,700 bbl/day of liquid products. It started up around 1990. Its capacity is a factor of 100 smaller than the commercial-scale PEARL plant in Qatar of 140,000 bbl/day of liquid products. The latter plant started up in 2011 [55].

3.12 Process start-up and product launch

A product market introduction and, in particular, a consumer product introduction requires an enormous amount of planning and preparation. It is therefore taken as a separate project phase, as illustrated in Fig. 3.1. Here are some main elements for consumer product introduction. A business-to-business product introduction will contain similar elements, but to a lesser extent.

3.12.1 Panel for product testing

The consumer product will be first tested by a panel consisting of company people and later by a panel of outside potential customers of the product. If it is a food product, the panel will be asked to mention the look, smell, mouth feel, and taste. If, for instance, it is a laundry soap, the potential users will be asked to read the instruction and use it in a washing machine. All reactions of the panel will be monitored and the product and instruction may undergo a final adaptation, based on the panel's feedback.

Here is an example of such a panel test for a new, more concentrated laundry soap. Because of the more concentrated soap, far less soap is needed to be used per laundry batch. The external panel members, however, could not believe that so little soap would work and still added (nearly) the same amount as before.

3.12.2 Product launch planning

A detailed planning will be made by the producer for all elements of the product launch. Market intelligence will be used for the product launch planning, such as how much, when, and where to supply the product for the first time. A prelaunch will nearly always be planned in which the product will be launched only in a certain location (a town or a region). The purpose of the prelaunch is to validate market expectations and to detect imperfections in logistics. After the successful prelaunch, the full product introduction will start. Contract preparation and signing in time with all relevant stakeholders is of course an essential element in the planning.

3.12.3 Matching the timing of marketing and manufacturing

It is of utmost importance that planning is carried out by close contact between marketing and manufacturing to fit the manufacturing production availability and capacity to market.

In some cases, the pilot plant will be used to make sufficient product for panel testing or even for prelaunch. Up front, in the design of the pilot plant, years earlier than the product prelaunch, special attention will also be paid to product packaging. All these will be considered by having close contact between marketing, development, and manufacturing.

For consumer product-oriented companies, this regular exchange of information and matching marketing with development manufacturing is a common practice. For business-to-business manufacturing-oriented companies that move to consumer products, this will be new, and structured communication between marketing, development, and manufacturing are often absent, causing risk of mismatching product prelaunch and manufacturing.

3.12.4 Information to supply chain and customers

All relevant stakeholder in the supply chain (transport, storage, shop, etc.) and customers will be informed about the prelaunch and the launch of the new product by means of proven and novel media. The product description for each stakeholder is an important aspect. Fine-tuning of the product introduction price is an important element of the product introduction to the market too.

3.12.5 Process implementation

In the implementation stage of a novel process for an existing business-to-business product, the start up for the first time should be viewed as an experiment that is not allowed to fail. If it fails, then the client does not receive the product and, in turn, cannot produce its own product. The client will then look for other suppliers, as security of supply is enormously important to stay in business.

This means that precautionary measures must be taken. The measures are about the process design, construction, preparation, start-up, and regular operation.

3.12.6 Recognition of new commercial implementations require special preparation

The first step in any implementation is to recognize that something new is to be implemented. Most often, failures are due to not recognizing that the process is new and perceiving that the process is the same as before. An analysis of projects containing revamps, expansions, green fields highly innovative, and lowly innovative revealed that the most successful projects, in time and cost, were innovative projects. However, the worst projects were also innovative, but in the latter cases, the technical difficulty was grossly underestimated. The innovative projects were treated as standard technology and in many cases, adequate testing of the process was absent as also were sound basic data packages [56].

Heuristic: A process is novel if it has not been in operation with that feed, product, catalyst, piece of equipment, or those conditions, at commercial scale.

This heuristic rule is often violated. A case I have experienced was a crystallization process step, applied for decades in the beverage industry, introduced at a bulk chemical plant. Because the process step was considered commercially proven, smooth operation was expected. The process step was interrupted every few weeks, requiring cleaning before regular operation could be continued. After questioning, it appeared that in the beverage industry, cleaning was carried out every week as part of the normal operating procedure. The bulk chemical process, however, was expected to run continuously for four years.

Other cases I have experienced, is the use of a new batch of catalyst, which is perceived as the same as before. The same procedure for start-up is followed and then all kinds of problems appear because, in reality, the catalyst behaves differently from the previous catalyst. What can happen if novelty is not recognized is dramatically illustrated in example 3.2.

Example 3.2: Shell Moerdijk – new catalyst charge leads to explosion
A dramatic explosion occurred at the Styrene process at Shell, Moerdijk, Netherlands in 2014, because a new catalyst was charged to a reactor. Fortunately, nobody was injured, because nobody was present in the plant. The damage, however, was so large that it took 1.5 years to rebuild the process.

The catalyst was perceived as an improved version of the previously used catalyst. Performance testing under production conditions had been carried out at lab-scale. However, nobody had considered the consequences of the difference in chemical composition of the catalyst during preconditioning and start-up, and no formal new start-up and operation risk assessment was carried out. The new catalyst contained 5% Chromate VI instead of the 0.2% in the old catalyst.

Preconditioning of the catalyst in the commercial reactor occurred with ethylbenzene, as always had been used before. Ethylbenzene was considered an inert flushing liquid just to heat up the catalyst bed. In reality, the chromate reacted exothermically with the ethylbenzene, causing a runaway reaction and resulting in an explosion that destroyed large parts of the process. As always with a calamity like this, there were additional causes resulting in the explosion. The spray installation for the trickle-bed reactor did not function optimally for distributing the ethylbenzene, by which local hot spots in the catalyst bed occurred, which could not be noticed by the operators. At these hot spots the copper oxide in the catalyst started to react with the ethylbenzene, causing more heat to be produced, which caused evaporation of the ethylbenzene and thereby an enormous pressure increase. In the design, this high pressure was not considered and a pressure relieve valve was not part of the design. In addition, the operators of the process at that time had never carried out a start-up of this reactor. They did not notice differences such as an unusual pressure profile, a strongly fluctuating ethylbenzene flow, and a controller alarm. Manual operation, rather than automatic control operation, also contributed to the explosion [57]

3.12.7 First commercial-scale process start-up

3.12.7.1 Start-up preparation
From literature on industrial process start-ups, Harmsen extracted four critical success factors for the design stages and six for the start-up stage [58]. The design success factors are:
1. Project is identified as new.
2. An integrated pilot plan, as downscaled version of the commercial-scale process has been used.
3. R&D, detailed design, and construction personnel have meeting to ensure that all knowledge is integrated in the design and construction.
4. Scale-up knowledge for unit operations is available.

The six start-up preparation measures are:
1. Potential problem analysis
2. Complete start-up team
3. Operators trained in start-up and normal operation
4. Precaution measures
5. Start-up plan
6. Documentation

With these preparation measures, he analyzed 10 Shell Chemicals' start-up cases. It appeared that for cases where the process design success factors were fulfilled and where all preparation measures were taken, the start-up time was a factor 5–10 shorter than the start-up time predicted, with the start-up time correlation [58].

3.12.7.2 Start-up manual

An important element of successful implementation is having an operation manual and a start-up manual. Process start-ups still fail, often due to a lack of proper start-up procedure and lack of operator training [52]. The start-up manual includes the precommission period, when all individual equipment are cleansed and tested using inert gas and liquids. Also, process controls are tested as much as possible for the limits of the process conditions.

Independent Project Analysis has made a statistical correlation for the start-up time of a new process, derived from their large database of start-ups of commercial-scale processes. Most of them concerned containing solids processing units but there were also of others [53]. The start-up time correlation is given in Tab. 3.17.

Tab. 3.17: Start-up time correlation, derived from industrial cases [53].

$t_{start-up} = 3.3 + 3.7\,N - 3.2\,F + S$		
Parameter	**Dimension**	**Description**
$t_{start-up}$	Months	Start-up time from moment all feeds are started till design targets are met
N	–	Number of new process steps
F	–	Fraction of mass and heat flows that are known $F = 0$, if mass and heat flows are not known $F = 1$, if mass and heat flows are all precisely known
S	Months	Solids processing parameter $S = 0$, if no solids are processed $S = 0.7$, if refined solids such as plastics are produced $S = 10.8$, if raw solids are fed to the process

A process step is new if it has not been applied before at commercial scale for that specific input or output. Here, a process step is defined as a section of the process, such as a reaction section or a purification section, with a defined input and output. A section has at least one unit operation, which may include several unit operations.

This start-up time correlation can be used to plan the start-up and the related marketing and sales planning. It can also be used to justify a certain development effort, such as having a pilot plant get the value of F from near zero to near 1. The difference in start-up times of those two conditions is on average 3.2 months. The amount of money lost by not having on-specification product for that period, in general, easily outweighs the cost of a pilot plant.

Further reading on how to make innovation projects successful is provided by Savelsbergh et al. [59].

References and further reading

[1]   Goffin K, Mitchell R. Innovation Management, Strategy and Implementation Using the Pentathlon Framework, 2nd edn, Basingstroke: palgrave macmillan, 2010.

[2]   Cooper RG. Winning at New Products: Creating Value through Innovation, Hachette UK: 19 Sept, 2017.

[3]   Arora A. Implications for Energy Innovation from the Chemical Industry, NBER, 2010.

[4]   Valencia RC. The Future of the Chemical Industry by 2050, Hoboken: J. Wiley, 2013.

[5]   Seider WD, et.al Product & Process Design Principles, 3rd edn, Hoboken: John Wiley & Sons, USA, 2010.

[6]   Almeida-Rivera C, Bongers MM, Zondervan E. A Structured Approach for Product-Driven Process Synthesis in Foods Manufacture, Computer Aided Chemical Engineering, 39, 2017, 417–441.

[7]   Little AD, Pathway to Innovation; A Global Benchmark study April, 2010, Available from: http://www. adlittle.com/downloads/tx_adlreports/ADL_InnoEx_Report_2010.pdf. Last resourced 15 November 2010

[8]   Harmsen GJ, Dutch Chemical Technology, Top of the world, lecture for AIChE, section BENELUX, the Hague, oral presentation, 2006.

[9]   Jaber MY. Learning Curves: Theory, Models, and Applications, Michigan: CRC Press, 2017.

[10]   Swanson RM. A vision for crystalline silicon photovoltaics, Progress in Photovoltaics: Research and Applications, 14, 2006, 443–453.

[11]   WBCSD, CEO Guide to the Circular Economy, 2017. Available from: http://www.wbcsd.org/Overview/ Resources Last resourced 10 January 2018

[12]   Chakravorti B. What Businesses Need to Know about Sustainable Development Goals, HBR, Nov 20, 2015.

[13]   Cheng B, et.al. Corporate Social Responsibility and Access to Finance, Strategic Management Journal, 2017, Last resourced 10 January 2018. Available from, https://dash.harvard.edu/handle/1/ 9887635.

[14]   Thomson Reuters, ASSET4 ESG scores on credit views key ESG scores on over 3,200 global companies, 2017, Available from: https://customers.reuters.com/community/fixedincome/material/ ASSET4ESGSCORES.pdf

[15]   Gorak A. Delft Skyline Debate, Chemical Engineering and Processing, 51(Special Issue), p1–149, 2012.

[16]   Verloop J. Insight in Innovation, Amsterdam: Elsevier, 2004.

[17]   Mehdi Miremadi M, et.al., Chemical innovation for the ages, 2017. Available from: http://www.mckin
       sey.com/industries/chemicals/our-insights/chemical-innovation-an-investment-for-the-ages. Last
       resourced 10 January 2018
[18]   Palucka T. The Wizard of Octane: Eugene Houdry never trained as a chemist but he made the
       greatest advance in the history of petroleum chemistry, American Heritage of Invention &
       Technology, 20(3), 2005, 36–45.
[19]   Harmsen GJ. Reactive Distillation: The frontrunner of Industrial Process Intensification: A full review
       of commercial applications, research, scale-up, design and operation, Chemical Engineering &
       Processing: Process Intensification, 46(9), 2007, 774–780.
[20]   Walpole H, letter to Horace Mann, 28 January 1754. Available from: http://www.thefreedictionary.
       com/serendipity, sourced 17 Sept. 2017
[21]   Drent E, et.al. Polyketones. In Encyclopedia of Polymer Science and Technology, Vol. 3, John Wiley
       and Sons, 2001.
[22]   Synthon, 2017. Available from: http://www.synthon.com/Corporate. Last resourced 10 January 2018
[23]   Bakker HLM, et al. Management of Engineering Projects – People are Key, Nijkerk: NAP, 2014.
[24]   Nagji B, Tuff G. Managing Your Innovation Portfolio, HBR, 90(5), 2012, 67–74.
[25]   Meesters G DSM technology platforms, 2017. Available from: https://www.dsm.com/corporate/
       about.html Last resourced 25 January 2017
[26]   Markham SK, Traversing the Valley of Death: A practical guide for corporate innovation leaders,
       2014. Available from www.TraversingTheValleyOfDeath.com. Last resourced 10 January 2018
[27]   Cooper RG. Where Are All the Breakthrough New Products?: Using Portfolio Management to Boost
       Innovation, Research-Technology Management, 56(5), pp25–33, 2013.
[28]   Cooper RG. Stage-gate systems: A new tool for managing new products, Business Horizons, 33(3,
       May–June), 1990, 44–54.
[29]   Cooper RG, Edgett SJ. Best practices in the idea-to-launch process and its governance, Research
       Technology Management, 55(March-April), 2012, 43–54.
[30]   ConQuastor. Innovatie Business Process Management, Utrecht: ConQuastor BV, 2013.
[31]   Verganti R. Design-Driven Innovation: Changing the Rules of Competition by Radically Innovating
       What Things Mean, Boston USA: Harvard Business Press, 2009.
[32]   Kahneman D. Thinking, Fast and Slow, London: Penguin Books, 2012.
[33]   Merrow ED. Industrial Megaprojects, Concepts, Strategies, and Practices for Success, Hoboken: John
       Wiley & Sons, 2011.
[34]   Mankins JC, Technology Readiness Levels: A White Paper, NASA, Office of Space Access and
       Technology, Advanced Concepts Office, 1995. Available from: http://fellowships.teiemt.gr/wp-
       content/uploads/2016/01/trl.pdf Last resourced 15 January 2018
[35]   ESA, Technology Readiness Level (TRL) The ESA Science Technology Development Route." European
       Space Agency, Future Missions Office, Technology Preparation Section. "Technology readiness levels
       (TRL)" (PDF). European Commission, G. Technology readiness levels (TRL), HORIZON 2020 – WORK
       PROGRAMME 2014–2015 General Annexes, Extract from Part 19 – Commission Decision C(2014)4995.
[36]   DOE, "Technology Readiness Assessment Guide (DOE G 413.3–4)." United States Department of
       Energy, Office of Management. Sep 15, 2011.
[37]   EC2014 "Technology readiness levels (TRL)" (PDF). European Commission, G. Technology readiness
       levels (TRL), HORIZON 2020 – WORK PROGRAMME 2014–2015 General Annexes, Extract from
       Part 19 – Commission Decision C (2014) 4995.
[38]   Dodd Graettinger CP, et al., Using the Technology Readiness Levels Scale to Support Technology
       Management in the DOD's ATD/STO Environments: A Findings and Recommendations Report
       Conducted for Army CECOM (CMU/SEI-2002-SR-027). Carnegie Mellon Software Engineering
       Institute, 2002.

[39] EARTO, 2014, The TRL Scale as a Research & Innovation Policy Tool, EARTO Recommendations, 2014. Available from: http://www.earto.eu/fileadmin/content/03_Publications/The_TRL_Scale_as_a_R_I_Pol icy_Tool_-_EARTO_Recommendations_-_Final.pdf Last resourced 15 January 2018

[40] Harmsen J. Novel sustainable industrial processes: From idea to commercial scale implementation, Green Processing and Synthesis, 3(3), 2014, 189–193.

[41] Nijstad BA, Paulus PB. Group Creativity: Innovation through Collaboration, New York: Osford University Press, 2003, 326–338.

[42] Herrmann D, Felfe J. Effects of Leadership Style, Creativity Technique and Personal Initiative on Employee Creativity, British Journal of Management, 25, 2014, 209–227.

[43] Rademaker E. 101 Gouden Tips voor de Business Case, Teteringen: Van Ierland Uitgeverij, 2007.

[44] Anderson JC, et.al. Customer Value Propositions in Business Markets, Harvard Business Review, 84(3), 2006, 90–99.

[45] Vogel GH. Process Development from the Initial Idea to the Chemical Production Plant, Berlin: Wiley-VCH, 2005.

[46] Yole, Micro fluidic applications, 2015. Available from: http://www.yole.fr/iso_album/microfluidicappli cations_emergingappli_yole__june2015.jpg Last resourced 15 January 2018

[47] FluidicMEMS, microfluidics lab on a chip, 2016. Available from: http://fluidicmems.com/list-of-microfluidics-lab-on-a-chip-and-biomems-companies Last resourced 15 January 2018

[48] Sekhon BS, et.al. Microfluidics technology for drug discovery and development – An overview, International Journal of Pharmtech Research, 2(1), 2010, 804–809.

[49] Boodhoo KVK, Harvey AP. eds. Process Intensification for Green Chemistry, Chichester: John Wiley & Sons, 2013.

[50] Günther A, Jensen KF. Multiphase microfluidics: From flow characteristics to chemical and materials synthesis, Lab Chip, 6, 2006, 1487–1503.

[51] Kenig EY, et al. Micro-separation of fluid systems: A state-of-the-art review, Separation and Purification Technology, 120, 2013, 245–264.

[52] Harmsen J. Industrial Process Scale-up A Practical Innovation Guide from Idea to Commercial Implementation, 2nd ed. Amsterdam: Elsevier, 2019.

[53] Merrow EW. Estimating start-up times for solids-processing plants, Chemical Engineering, 24, 1988, 89–92.

[54] Perry RH. Perry's Chemical Engineers' Handbook, 8th edn, New York: McGraw-Hill, 2008.

[55] Helvoort T, et.al. Gas to Liquids Historical Development of GTL Technology in Shell, Amsterdam: Shell Global Solutions Int, 2014.

[56] De Groen T, et.al. 2x2 Your Choice of Projects, Twice as Cost Effective, Twice as Fast, a Guide for Key Decision Makers in the Process Industry, Leidenschendam: napDACE, 2002.

[57] Biesboer F, Explosie Shell illustreert falend risicomanagement: onvoorzien scenario, de ingenieur 8 aug. 2016, p.50–53.

[58] Harmsen GJ Kritische Succesfactoren bij het ontwerpen en opstarten van chemische fabrieken, NPT Processtechnologie, pp15–17, 1996.

[59] Savelsbergh C, Martijn Jong M, Storm P, Project success: creating excellent projects. How can I make my project successful? Maastricht: AMI 2011. Resourced from: http://ami-consultancy.com/wp-content/uploads/2016/04/Project-Success-creating-excellent-projects.pdf Last resourced 15 January 2018

4 Designing for innovation

4.1 Introduction

This chapter provides theory about design for innovation, an integrated method for breakthrough concurrent product-process design, planning for design and experimentation for innovation, setting criteria for design, and shows roles of modeling in design.

This chapter is meant for industrial product and process designers and post-MSc engineering doctorate (EngD) students to generate novel product-process solutions. The thinking modes and methods that are provided facilitate breakthrough concurrent product-process designs.

For graduate students, step-by-step design methods for novel products and processes are described in Chapters 6 and 7. Modeling and optimization are discussed in Chapters 8 and 9. Evaluating designs is covered in Chapters 10–12.

4.2 Design thinking

4.2.1 Design for innovation theory

Design as a verb is a problem-solving thinking activity. It starts with assigning a specific problem and ends with presenting a solution to the client [1]. The design approach to a problem is to lay out possible solutions and try to improve upon them. The approach is mainly alternating between two thinking modes: a creativity mode and an analysis mode. These two steps are repeated until a satisfactory solution is obtained. The focus is thus on getting a specific solution to a specific problem.

Dorst [2] shows design thinking (and other types of thinking modes) in a scheme, which is summarized in the Tab. 4.1.

Tab. 4.1: Types of thinking to solve the unknown (based on [2]).

Type of thinking	Value to be created	What (object(s))	How (working principles of object(s))	Result	Challenge
Deduction	–	Known	Known	*Unknown*	Justification of result
Induction	–	Known	*Unknown*	Known	Discovery of working principles
Abduction-1	Known	*Unknown*	Known	Known	Conventional problem solving or conceptual process design
Abduction-2	Known	*Unknown*	*Unknown*	Known	Conceptual product and process design

https://doi.org/10.1515/9783110782127-004

Abduction-1 and abduction-2 are the types of thinking carried out in the field of engineering design. Understanding these differences help companies organize their innovation and prepare an organization environment, so that both designers and scientific researchers can cooperate. It also helps students first educated in scientific research and subsequently educated in engineering design understand the large differences in the modes of thinking they must acquire.

In abduction-1, which may be called normal abduction or conventional problem solving, the engineering designer generates intermediate design solutions of the object or objects (the "what") that will create value based on existing knowledge about the working principles ("how") of the various objects. Using the knowledge, the result of these designs can be simulated to predict the value created. The "what" (object(s)) is adjusted until the analyzed result from the simulation meets the desired outcome. The simulations can take different forms: from simple sketches to visualizing the working principle to complex mathematical modelling and prototype testing. There are many iterations between solution generation, simulation, analysis, and evaluation [2]. Conventional process design follows abduction-1 reasoning. The product is specified. Most feeds are specified. The chemical reaction stoichiometry for the conversion of the feed to the product is known. Working principles for reaction and separations (the "how") can be selected from textbooks. Unit operations can be selected from textbooks, while sizing is determined using available relations. Chapters 6 and 7 systematically deal with conventional design for novel products and processes.

In abduction-2 both the "what" (the desired end solution) and the "how" (the working principles of the end solution) are unknown. The "what" and the "how" are then obtained by alternating between "what" and "how" to simulate intermediate solutions. Often the intermediate solutions also involve reframing the solution space by thinking first which problem is hardest to solve, redefining the framework of that problem, and then continuing to find intermediate solutions. The reframing of the solution space may also be done several times. This type of design thinking is particularly fruitful for solving complex problems for which no off-the-shelf solutions exist and which do not lend themselves to solutions generated by scientific research because they cannot be assigned to a specific discipline.

In concurrent product-process design, abduction-2 design thinking is applied, as the product, its related manufacturing process, and the end-of-life recycling process are unknown at the start. Section 4.3 provides methods to help the designers in generating combined solutions.

4.2.1.1 Design and risks

An additional view on what characterizes design is from Petroski [3]. He views engineering design as a succession of hypotheses that such and such an arrangement of parts will perform a desired function without fail. Petroski stresses that a design result is always novel and thereby always will have a risk of failure. Petroski also remarks that implicitly the designer assumes that the design will not fail if it is used as intended.

The risks and ways to reduce them associated with these two aspects of design are treated in this book explicitly by advocating testing novel designs and writing startup and operation manuals for commercial scale implementation.

4.2.1.2 Design links with society and nature

The International Council of Societies of Industrial Design (ICSID) defines design (includes services, processes, and system) as: "Design is a creative activity whose aim is to establish the multi-faceted qualities of the objects, processes, services, and their systems in whole life cycles. Therefore, design is the central factor for cultural and economic exchange" [4].

The exchange between society and designers is treated extensively in this book in Chapter 2 and in Chapter 4 showing the relations between modal aspects of reality, including society and nature on the one hand and on setting design criteria derived from these aspects on the other hand.

4.2.1.3 Further reading on design

The field of design thinking is expanding rapidly. Good entries are Dorst [2], Verganti [4], Verkerk et al. [5], Lawson [6], and De Bont [7]. The latter treats designing all kinds of industrial products, and services in complex contexts.

4.2.2 Design knowledge types

Knowledge of designing can be categorized in many ways. The authors found the categorization of Vincenti fitting with their view on knowledge of designing. Vincenti distinguishes the following categories as presented by Verkerk et al. [5] (relevant chapters in our book are mentioned for each category):

(a) Knowledge of fundamental design concepts: elements a design is made of
     This knowledge is explicitly discussed in Section 7.2 on synthesizing product and process designs, and Chapters 8 and 9 on modeling.
(b) Knowledge of criteria: the kinds of demands a design should meet
     This subject is covered in Section 4.5 of this chapter, and in Chapters 6, 10, 11, and 12.
(c) Knowledge of quantitative data
     In our book, these are called basic design data, and are discussed in Chapter 6.
(d) Knowledge of theoretical methods and models
     Chapters 8 and 9 deal with modeling for products and processes in detail.

(e) Knowledge of practical considerations
 This aspect is dealt with in Sections 4.6 and 4.7.
(f) Knowledge of design approaches
 This main topic is explicitly discussed in Chapters 4–12.

Experienced designers have gained large amounts of knowledge of all these categories. Novice designers do not have a notion of these types of knowledge. Entry courses on process design treat the knowledge types (a), (b), (c), and (d) explicitly. Knowledge types (e) and (f) are obtained when designers execute design projects.

4.2.3 Differences between design thinking and scientific research

The main differences between design thinking and scientific research thinking are found in Tab. 4.2. The first four items have been obtained from Verganti [4]. The others have been obtained from Lawson [6].

Tab. 4.2: Differences between creative design and scientific research (based on [4, 6]).

Parameter	Creative design	Scientific research
Number of ideas	Create numerous ideas	One vision or hypothesis
Peer values	Beginner approach to problems	Knowledgeable scholar approach
Focus	Variety and divergence	Convergence on one paradigm
Work culture	Neutral: if it works it is good	Personal vision
Goal	Specific solution	General theory
Stopping criterion	Good enough	Best
Method	Many (anything that helps)	Hypothesis and experiment
Reliability	Judgment and calculation	Experimental validation
Essence	Prescriptive	Descriptive

4.3 Design methodologies

4.3.1 General design methods for products, processes, and systems

Research into design methodologies (or methods) to improve the quality of the design and design process in all aspects relevant for stakeholders only started in the 1960s. Products and processes in many fields (housing, transport, infrastructure, agriculture, etc.) were designed, developed, realized, and improved before and over many centuries, but were done in an evolutionary way. The execution of these design processes was lengthy and was characterized by a trial-and-error approach and did not always lead to the desired performance. As soon as more theoretical knowledge and understanding was obtained within the science and engineering disciplines, more design options could be created, analyzed, and tested (using small scale prototypes, mathematical models) before realizing new designs or improving existing designs.

With the aim to have design processes leading to better designs faster – and thereby increasing the rate of innovation – formal design methodologies were developed and implemented since the mid-twentieth century. The following general design methods are worth mentioning because of their wide application in various fields. These design methodologies also formed the basis for more specific design methods tuned for specific engineering fields and will be described in more detail including references in Sections 5.2 and 5.3:

(a) Quality function deployment (QFD)
(b) VDI 2221 and VDI 2206 guidelines ("Verein Deutscher Ingenieure (VDI)": German engineers association)
(c) Total design (Pugh)
(d) Product design and development (Ulrich, Eppinger)
(e) Circular design:
    a.   Cradle to cradle (Braungart, McDonough – 2002)
    b.   Blue economy model (Pauli)
    c.   Circular design (Ellen MacArthur Foundation)

In Section 4.3.2 the design methods applied in the chemical engineering innovation field are listed.

4.3.2 Design methods for chemical products and processes

The development and implementation of design methodologies for chemical products and processes started at the end of the 1960s. This development history and the work leading to the Delft Design Map (DDM) methodology (described in Section 4.4) are depicted in Appendix A4 and comprise the following methods and developers:

(a) Heat-exchanger networks design (Linnhoff and Fowler)
(b) Conceptual design of chemical processes (Douglas)
(c) Computer-based optimization (Kocis and Grossman)
(d) Task-driven process design (Siirola)
(e) Process design principles (Seider and Lewin)
(f) Chemical product design (Cussler and Moggridge)
(g) Design and development of biological, chemical, food and pharmaceutical products (Wesselingh, Kiil, and Vigild)
(h) Product and process design principles (Seider, Seader, Lewin, Widagdo)

(i) Delft template for conceptual process design (Grievink and Swinkels)
(j) Product-driven process synthesis (Almeida, Bongers, Zondervan)
(k) DDM for product and process design (Swinkels, Section 4.4.2 of this book)

4.4 Designing for innovative products and related processes

4.4.1 Introduction

This section provides methods for designing both the novel product and the related novel process effectively and efficiently. It covers all innovation classes and for all innovation stages (see Sections 3.3 and 3.6). It has been used by post-MSc engineering doctorate (EngD) trainees at Delft University of Technology (TU Delft) and in industry. It facilitates teamwork and project and time planning.

4.4.2 Product-process design method: Delft Design Map (DDM)

The presented design method was first defined and developed by Grievink and Swinkels in the period 2002–2014 [8]. In this period, it was also further refined and tested by case work in a consumer product company by Almeida-Rivera et al. [9]. It combines all important elements: the value chain (with sub-systems supply chain and market), product, process, design cycle steps, and innovation stages.

To aid the designer, all these elements have been systematically placed in DDM, which is given in Tab. 4.3.

The first column in this table lists the design steps from the engineering design cycle [8] and [9]. These "design (cycle) steps" range from "1 scope" to "7 report."

The top rows of this table list 12 "design levels" shown horizontally in the DDM (third row, Tab. 4.3). The design levels are grouped into substages (framing, supply chain imbedding, process technology, process engineering, and final design):

Framing:
– Project framing (PF)

Supply chain imbedding:
– Consumer wants – product concept (CW-PC)
– Product concept – property function (PC-PF)
– Input-output structure (I-O)
– Subprocesses (SP)

Tab. 4.3: Delft Design Map structure (based on [8, 9]).

Stage-Gate™:		Concept			
Substages	**Framing**	**Supply chain imbedding**			
Design levels ⇒ **Design steps** ⇓	**(PF)** **Project framing**	**(CW-PC)** **Consumer wants – product concept** (product performance – quality function)	**(PC-PF)** **Product concept – (product) property function**	**(I-O)** **Input-output structure** Products, feeds, wastes, utilities, process functions	**(SP)** **Subprocesses** Subprocesses, action blocks, technology, I-O structures
1 **Scope** (of design)	6 ⇓	6 ⇓	6 ⇓	6 ⇓	6 ⇓
2 **Knowledge** (about objects)	6.5 ⇓	6.5 ⇓	6.5 ⇓	6.5 ⇓	6.5 ⇓
3 **Synthesize** (alternatives, structure, scale)	7 ⇓	5.2, 5.4, 5.5, 7.2.2, 8.1–8.3 ⇓	5.2, 5.4, 5.5, 7.2.2, 8.1–8.3 ⇓	7.2.3 ⇓	7.2.3 ⇓
4 **Analyze** (physical behavior)	7.3 ⇓	7.3, 8.4–8.6 ⇓	7.3, 8.4–8.6 ⇓	7.3, 9 ⇓	7.3, 9 ⇓
5 **Evaluate and select** (evaluate performance vs specifications, and select most promising design alternatives)	7.4, 7.5 ⇓	4.5, 7.4, 10–12, 7.5 ⇓	7.4, 8.7, 10–12, 7.5 ⇓	7.4, 10–12, 7.5 ⇓	7.4, 10–12, 7.5–7.6 ⇓
6 **Report** (decisions and results)	13	13	13	13	13

Process technology:
– Task network (TN)
– Unit network (UN)
– Process integration (PI)

Process engineering:
– Equipment design (ED)
– Operability integration (OI)
– Flow sheet sensitivity and optimization (FO)

	Feasibility			Development			Final
	Process technology			Process engineering			
(TN) Task network Allocation, targeting, mechanism, rates of tasks, integration in compartments	**(UN)** Unit network Purification, conversions, separations, assembly, stabilization	**(PI)** Process integration Energy, solvents, water, emissions, production time	**(ED)** Equipment design Structure, size, materials, reliability	**(OI)** Operability integration Safety, monitoring, control, flexibility, availability	**(FO)** Flow sheet sensitivity and optimization		**(Final)** Final design Product and process (PFD)
6 ⇓	6 ⇓	6 ⇓	6 ⇓	6 ⇓	6 ⇓		
6.5 ⇓	6.5 ⇓	6.5 ⇓	6.5 ⇓	6.5 ⇓	6.5 ⇓		Design data, heuristics, and models
7.2.3 ⇓	7.2.3 ⇓	5.6,5.7,5.8, 7.5 ⇓	7.2.3 ⇓	7 ⇓	9 ⇓		Flow sheet and control structure
7.3, 9 ⇓	7.3, 9 ⇓	5.8, 9 ⇓	9 ⇓	9 ⇓	9 ⇓		Mass and energy yield factors
7.4, 10–12, 7.5 ⇓	7.4, 10–12, 7.5 ⇓	7.4, 10–12, 7.5–7.6 ⇓	7.4, 10–12, 7.5 ⇓	7.4, 10–12, 7.5 ⇓	7.4, 10–12, 7.5–7.6 ⇓		SHEETS evaluation and selected design alternatives, 3.12
13	13	13	13	13	13		Go/no go?

Final:
–   Final design (final)

The design levels have a clearly ordered sequence that is synchronized with the innovation stages (DDM, first row) defined in the Stage-Gate™ framework [10].

Each cell in the resulting table represents a set of focused design activities to be carried out. A cell is called a "Design Activity Space (DAS)." Each activity space is confined by the design level and design step. The required activities in each space are thereby relatively limited and straightforward. Advice for the activities to be executed in each design activity space is indicated with a section reference number in this book.

4.4.3 Explaining the Delft Design Map (DDM) for product-process design

The DDM table's purpose is threefold. First, the DDM table is used to quickly find information on what design activity to perform and how. The second purpose of the table is facilitating the planning of the work to be done. The third purpose is to help categorize and check the design result, and whether all activities have been carried out.

In this paragraph, the engineering "design (cycle) steps" and the "design levels" are described in further detail. Recall that the design (cycle) steps are carried out for each design level and that each combination of a design step and a design level forms a cell in the DDM, called a design activity space (DAS). The numbers in each cell of the DDM table refers to a chapter or section of this book.

4.4.3.1 Design (cycle) steps description

In the bullet list below, the design (cycle) steps are described with a few examples of design aspects and activities that are relevant for selected design levels. The design levels will be elaborated in more detail after the bullet list. Chapter 6 covers items (1) and (2), Chapter 7 covers items (3–6), and Chapter 13 covers item (7) reporting.

(1) Set the design scope involving:
  – Design goal
  – Constraints
(2) Acquire and assess knowledge about potential building blocks (depending on design level):
  – Target market size and dynamics
  – Competitor activities/patents
  – Raw material supply specifications
  – Pure component properties (physical, safety, health data) for components and combined streams
  – Thermodynamic equilibria, kinetics
(3) Synthesize alternative designs (networks/structures) to meet the design scope by:
  – selecting suitable building blocks/objects at this design level
  – varying order and connectivity of these building blocks
  It is important to aim for solutions quantity first before moving to step (4) Analyze, to avoid the too-early elimination of design options that are also valuable in combination with other options. Creativity methods and tools discussed by Tassoul [11], "TRIZ" methodology [12], and synthesis heuristics (Section 4.7) are sources that help creating many design options.
(4) Analyze the physical-chemical behavior of generated alternative designs (networks):
  – Estimates, computations and/or model simulations connecting resource flows to building blocks
  – Expert analysis by sketches and drawings of product and process block flow diagrams

The aim of this step is to quantify the behavior of design options at hand at the design level. Aim to arrive at quantitative data like yield estimates, mass and heat balances, separation efficiency, utilities used, etc., as far as possible. These quantitative data are the input into the next design step (5) evaluate and select, where the quantitative data are translated into performance and compared with the performance goals set in step (1) design goal setting, and the most promising design options are selected for the next design level.

(5) Evaluate and select:

Evaluate the performance of alternative solutions against the SHEETS (see Section 4.6.3) sustainability goals and constraints:

S: Safety

H: Health

E: Environment

E: Economy

T: Technology

S: Social

For the supply chain-related design levels, the estimates will be less accurate than for the estimates at the equipment design level; still, it is crucial to make every attempt to quantify the performance of the various design options for each design level, as it will make clear where design options perform well or do not at all and where the design focus should be drawn to in the subsequent design levels.

Select a limited number of preferred solutions from the alternatives for the next design level:

– Eliminate/weed-out unfeasible options.
– Retain promising parts/interesting aspects of eliminated options for implementation in other options.

(6) Report all design outcomes, drawings, mathematical models, analysis and evaluation results, key decisions, underlying rationales, and numerical data for transfer to the next design level.

– Write, store, and share the activity reports for each design activity for each design step in each design level (see also Chapter 13).

4.4.3.2 Executing design activities in all 12 design levels

In this section design activities for the design levels in the framing, supply chain imbedding, process technology, process engineering, and final substages of the DDM will be shown.

1. Design-level PF: project framing (substage: framing)

Important design activities in the PF-Project Framing design level can be summarized as follows:

- Frame the product and/or process design task within the overall business innovation project ("BOSCARD": background, objectives, scope, constraints, assumptions, risks, deliverables):
  - Background of business context
    - Marketing and supply chain aspects and opportunities
      - 6P model (proposition, promotion, place, packaging, price, product)
    - Design driver (company profile, market pull, technology push)
    - Design type (new product family and process platform, platform derivative, manufacturing improvement, platform development)
  - Objectives
  - Scope
  - Constraints
- Describe the foreseen design approach
  - Design methodology
  - Assumptions
- Project organization
  - Team members and roles
  - Risks
  - Deliverables and milestones
  - Project and time planning
- Result: *project brief*

2. Design-level CW–PC: consumer wants – product concept (substage: supply chain imbedding)

- Identify direct product users
- Identify stakeholders in product/process life cycle
- List (qualitative) stakeholder/consumer needs
- Translate (qualitative) needs into product performance attributes (quality function) and their quantitative metrics
  - For example, by using QFD: house of quality (HoQ) methodology (see Section 8.3)
  - Rank product performance attributes
  - Performance attributes (quantitative) specifications
    - Targets and acceptable ranges
  - Make relationships explicit between different product attributes (using HoQ method)
  - Identify the key (product) design challenges
- Result: *HoQ (1): product performance (quality) attributes* **mapped onto** *consumer wants*

3. Design-level PC–PF: product concept – (product) property function (substage: supply chain imbedding)
– Create/synthesize product concepts (with existing or novel product application processes, aiming to fulfil the product (quality) performance metrics)
  – Find thermodynamic phases and physical scales to compose different product structures.
  – Find chemical/biological compositions of these various phases.
  – Analyze the behavior of the structured products using mathematical modeling and/or experimentation.
  – Evaluate product design options against performance metrics.
– For each product design option also review:
  – Tolerances on variability of product structure and compositions per grade
  – Expected economic life span of product(s) and evolution of its price range
  – Overview of suitable alternative feeds to make the product(s)
  – Recovery of materials from spent or end-of-live product as secondary feed for manufacturing or in a circular economy
– Result: *HoQ (2): product property function*
  – Product composition, structure, physical properties – combined with the product application process, **mapped onto the product performance (quality) attributes**

4. Design-level I-O: input-output structure (substage: supply chain imbedding)
– Imbedding of the process in a manufacturing supply chain and choice of future site
– Specification of products and cogeneration options for chemical products and energy
– Selection of chemical product grades to be manufactured and available feeds
– Demarcate the process from its environment by setting boundaries ("battery limits": inside battery limits; (ISBL) outside battery limits (OSBL) (see also Section 10.4.1)
– Specification of production modes (campaigns) and operation mode(s)
– Estimation of mass balances based on conversion yield factors
– Identification of possible waste streams
– Estimation of expected economic and physical life span of process plant
– Duties and SHEETS specifications for process plant, external permits, and licenses
– Result: *input-output diagram* **(including feeds, (by)-products, solid/liquid/vapor (S/L/V) waste streams, technologies**

5. Design-level SP: subprocesses allocation (substage: supply chain imbedding)
– Decompose the main process into subprocesses (see **NOTE**) with high-level manufacture process functions – for supply chain mapping.
NOTE: These subprocesses should be envisaged as large supply chain blocks (not as a split of a single product plant) to be located on different locations, and/or run by separate legal entities, and/or using a single technology and/or sell their products to more

*customers. It is not meant as the split of a single production plant into reaction, separa-
tion, and polishing steps. Each subprocess will be further designed, with its distinctive
tasks network (TN) and unit network (UN) designs. For maximum process intensification opportunities at the TN and UN design levels, the subprocesses should be defined
as broad as is realistically possible.*

- Choose the use of key technologies per sub-processing function (licensing).
- Decide on the making, buying, or selling of intermediates or product components.
- Design the connectivity and recycle structure between subprocesses.
- Decide on (re-)use of standard processing equipment for short process life span.
- Or: design subprocesses with innovative processing operations for a long economic plant life.
- Result: *Block diagram* **with selected subprocesses and processing actions
  (and different feeds, (by)-products, S/L/V waste streams; including preliminary mass balances (based on yield factors) and evaluation regarding gross
  margin, transportation cost estimates, discharging or recycling waste
  stream components, etc.**

6. Design-level TN: task network (substage: process technology)

- Task network: allocation, targeting, integration
  - Decomposition of subprocesses into a network of elementary tasks (*allocation*)
*A task achieves a change in the state of matter from an inlet to a target outlet state,
thus demanding a processing duty (targeting). Per task, a (bio)chemical and physical
mechanism (e.g., reactions and phase transfer) and operating conditions are selected.*
- Balancing of rate processes and finding space-time requirements
- Identification of compatible tasks suitable for integration in spatial compartments
- Sizing of compartments, with due consideration of various production modes
- Results: **task network(s), represented in a** *task block diagram*

7. Design-level UN: unit network (substage: process technology)

- Create spatial compartment and structures (e.g., ideally mixed volume compartment,
  volume compartments with surface structures for direct or indirect contact between
  streams (in plug flow, indirect, or mixed-flow contacting) and operating modes (e.g.,
  batch, fed-batch, or continuous) for the (task-integrated) spatial compartments.
- Integrate compartments in operational units achieving the main processing functions (e.g., conversion and separation).
- Adjust connections and recycle structures between operational units.
- Result: **process unit network at meso-level, represented in a** *unit block diagram*

8. Design-level PI: process integration (substage: process technology)
- Efficiently matching internally consumed and generated physical resources by creating a network of exchanges between process units, involving mass (solvents), and various forms of energy (thermal, mechanical, and electrical power).
- Obtain any remaining unbalanced duties from the utility systems.
- Consider the operability aspects of the resource exchanges networks.
- Deal with ceilings on integrated emissions from the plants (e.g., SO_2 and CO_2).
- Account for multiproduct integration aspects with respect to unit run times and storage.
- Set targets for the system's availability as built-up from availability of units.
- Result: *Resource exchange networks* **at the macrolevel of the entire subprocess**

9. Design-level ED: equipment design (substage: process engineering)
- Select and size *existing* type of equipment, which can host process unit operations.
- Or: design better performing *new* equipment for task integrated operational units.
  - Spatial structures and space-time requirements fix the main equipment dimensions.
- Choose construction materials to safely accommodate processed materials and conditions.
- Set targets for "RAM" (= reliability, availability, maintainability) for equipment items.
- Fix multiplicity and overdesign of identical equipment (for size, safety, and reliability).
- Identify (inequality) constraints confining feasible window of operation of equipment.
- Determine process variables in equipment to be monitored and controlled with sensors and actuators.
- Result: *equipment specification sheets*

10. Design-level OI: operability integration (substage: process engineering)
- Apply safety analysis, evaluation, and design adaptations (e.g., DOW Fire & Explosion Index).
- Perform flexibility analysis and feasible windows of plant operation.
- Specify key features of an operator – process information interface with time scales.
- Select instrumentation for process performance and product quality monitoring.
- Develop models to convert measured plant data to process management information.
- Design a plant-wide stabilizing control structure per operating mode.
- Check on feasibility of mode switches, startup, and shutdown.
- Determine "RAM" features and run-time per production mode.
- [Optional: design an optimizing control structure for product quality and throughput.]
- Result: *Process flow diagram* **optimized regarding process control**

11. Design-level FO: flow sheet sensitivity and optimization (substage: process engineering)
- Select the key performance indicators (KPIs) of the process plant.
- Identify sensitivities of KPIs to main design variables of process units. Perform an uncertainty analysis of the design and make necessary adaptations.
- Do a Pareto trade-off optimization of some KPIs for final designs.
- Result: **optimized** *process flow diagram (PFD)*

12. Design-level final: final design (substage: final)
- Consolidate and compile:
  - All design data, heuristics and models
  - Product composition and structure
  - Process flow diagram, control structure, equipment design
  - Mass and energy balances, including yield factors
  - SHEETS performance metrics
  - SWOT (strengths, weaknesses, opportunities, and threats) of the design and risk analysis and mitigation suggestions
- Evaluate:
  - Design project results
  - Design project process
- Recommendations for next steps:
  - Go/no go? To Manufacturing and Product Introduction Stage-Gate^{TM} phases.
- Write and file final report.
- Close design project.

4.4.4 How to plan and execute design activities using the Delft Design Map

The general way of working with the DDM is based on the presumption that the design results in the preceding design level $(N - 1)$ are available for transfer to design level (N). Here, the design is further expanded for which the steps of the engineering design cycle are executed (again). The design results so obtained in design level (N) are passed on to design level $(N + 1)$.

For a complete novel product and process design the team would indeed start at the first design level and carry out the design activities as indicated by the engineering design steps (see Fig. 4.1); then move to the next design level and repeat the cycle of design steps again. This can be repeated until all design levels have been covered. This is indicated with the "arrows" in Tab. 4.3. No design process is strictly sequential, so knowledge and understanding generated in the later design levels can be "ploughed back" into earlier design levels to improve the design. The circular setup of the design steps around the design aims to suggest that steps can be worked on simultaneously and iteratively by the team members, rather than in a strictly sequential manner. As long as the information is reported, classified, and shared between the team members, the design results can be captured.

Another way of working with the DDM is to first assess the design case at hand, and then use the DDM to assess what design levels and design step combinations (design activity spaces: cells in the DDM) are crucial at the start of the project; execute the design activities and use the results to plan further navigation through the DDM, especially when a complete novel product and process design is not required. The DDM can help the team making a navigation plan and allocate resources accordingly. Yet another way is to use one of the predefined "fingerprint DDM tables" relating to a design case's innovation class. This will be further explained in Section 4.4.5.

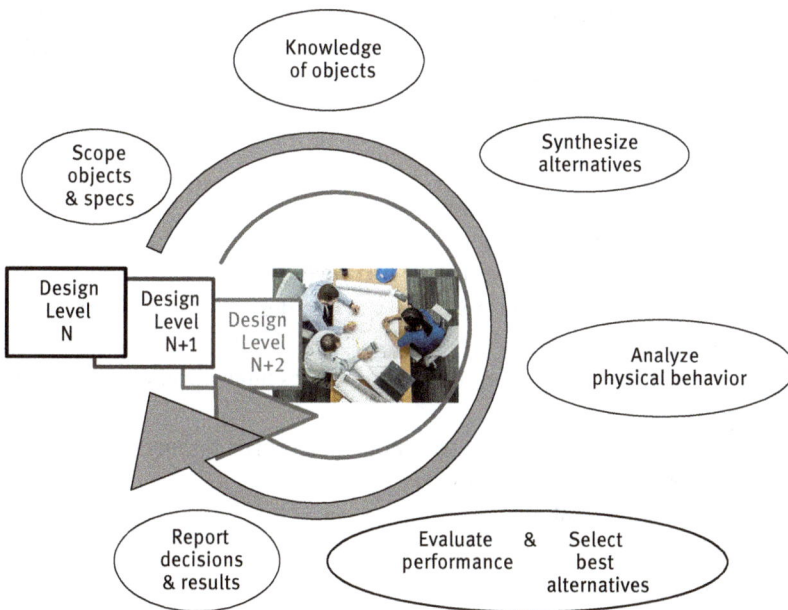

Fig. 4.1: Design steps of the generic engineering design cycle. Photo: Creative Commons.

Almeida-Rivera et al. [9] describe the strengths and weaknesses of the strictly sequential, the simultaneous, and the "anticipating sequential" approaches. The strictly sequential approach is the most intuitive and therefore currently most adopted, but it also the one that needs multiple iteration cycles. Attempts are being made for simultaneous designs of products and processes, but this requires the capability of formulating mathematical descriptions of the product physical and chemical functions in use and the manufacturing process to produce the required product structure/composition. For some cases, this modelling is feasible, and Section 4.7 provides a method for this modelling. However, for most product-process cases modelling solves only a part of the design problem. When product-process design for optimal product recycling is included, an extra dimension is added to the design problem, making it a larger challenge to model and solve the design problem.

As this field has only recently begun to develop, the current best bet is the "anticipating sequential approach" that can use mathematical optimization at restricted design levels but follows the sequential approach to a large extent.

The key aspect to stress the anticipating sequential approach as a hybrid to concurrent design of product and process: you do not necessarily move through the design levels in a sequential way and without recycling. Multiple design levels can be started, especially when novel product/manufacturing processes are targeted, and new knowledge and insights need to be acquired that potentially and most probably affect each other mutually.

The design levels are merely a demarcation of product and process aspects that need to be addressed, but the design team needs to maintain overview of the impacts of the different design levels. Moving back and forth between design levels for analysis, evaluation, and certainly also synthesis steps is key for the overall best result, and the DDM caters to this very well [9].

4.4.4.1 Practical benefits of working with the Delft design map

The practical benefits of using the DDM are:
– Sure-footed and faster design processes due to a clear design structure
– More inventive power by greater awareness of the design levels, design steps, and design options
– Better options for supply chain integration and process intensification at multiple scales in process
– Comprehensive coverage of process integration of all shared resources
– Applicability to processes for structured products
– Better integration of operability aspects into design
– Proper documentation of design decisions and rationales
– Easier reuse of a design for retrofitting and revamps by such documentation

4.4.5 Design planning with the Delft Design Map for various innovation classes

Chapter 3 provides a table with innovation classes for products and processes with their markets and supply chains, including several examples. In this paragraph, we are linking each innovation class to the design focus that is required. For several innovation classes, so-called fingerprint tables are defined using the DDM (Tab. 4.4 and Tabs. 4.5–4.10). The design activities are classified as "key design focus" (K), "optional design focus" (o), "default design activity" (D), and "check implications of design decisions" (C). With these fingerprint tables the designer can quickly see where the main design efforts should be focused.

The fingerprint tables have been derived from observing about 200 PDEng/EngD product/process innovation design projects carried out for industrial partners in the past decade by Delft University of Technology (TU Delft).

4.4.5.1 Fingerprints of Delft Design Map for design tasks planning

Tab. 4.4: Legend belonging to fingerprint Tabs. 4.5–4.10.

Legend	Design levels	
K: Key design focus o: Optional design focus D: Default design activity C: Check implications of design changes _: No activity	PF: project framing CW/PC: product concept/consumer wants PC/PF: product concept/property function I-O: input-output structure SP: subprocesses TN: task network	UN: unit network PI: process integration ED: equipment design OI: operability integration FO: flow sheet sensitivity and optimization Final: final design

Tab. 4.5: Fingerprint for innovation class 6: "radically new."
(New market, completely new product, new supply chain, and new process).

Stage-Gate™:		Concept					Feasibility		Development			
K: Key design focus o: optional design focus D: Default design activity C: Check implications of design changes _: No activity	**Framing**	**Supply chain imbedding**					**Process technology**		**Process engineering**			**Final**
Design levels ⇒ Design steps/ activities ⇓	(PF)	(CW-PC)	(PC-PF)	(I-O)	(SP)	(TN)	(UN)	(PI)	(ED)	(OI)	(FO)	(Final)
1 Scope	K	D	D	D	D	D	D	D	D	D	D	D
2 Knowledge	D	K	K	K	K	K	K	D	D	D	D	D
3 Synthesize	K	K	K	K	K	K	K	K	K	K	o	K
4 Analyze	K	K	K	K	K	K	K	K	K	K	o	K
5 Evaluate and select	K	K	K	K	K	K	K	K	K	K	o	K
6 Report	D	D	D	D	D	D	D	D	D	D	D	D

Tab. 4.6: Fingerprint for innovation class 5: "radically new."
(Existing market, new product (new platform), new supply chain, and new process).

Stage-Gate™:		Concept				Feasibility			Development			Final
K: Key design focus o: optional design focus D: Default design activity C: Check implications of design changes _: No activity	**Framing**	**Supply chain imbedding**				**Process technology**			**Process engineering**			**Final**
Design levels ⇒ Design steps/ activities ⇓	(PF)	(CW-PC)	(PC-PF)	(I-O)	(SP)	(TN)	(UN)	(PI)	(ED)	(OI)	(FO)	(Final)
1 Scope	K	D	D	D	D	D	D	D	D	D	D	D
2 Knowledge	D	K	K	K	K	K	K	D	D	D	D	D
3 Synthesize	K	K	K	K	K	K	K	K	K	K	o	K
4 Analyze	K	K	K	K	K	K	K	K	K	K	o	K
5 Evaluate and select	K	K	K	K	K	K	K	K	K	K	o	K
6 Report	D	D	D	D	D	D	D	D	D	D	D	D

Tab. 4.7: Fingerprint for innovation class 4: "new."
(Existing market, modified product (platform derivative), modified supply chain, and new process).

Stage-Gate™:		Concept				Feasibility			Development			Final
K: Key design focus o: Optional design focus D: Default design activity C: Check implications of design changes _: No activity	**Framing**	**Supply chain imbedding**				**Process technology**			**Process engineering**			**Final**
Design levels ⇒ Design steps/ activities ⇓	(PF)	(CW-PC)	(PC-PF)	(I-O)	(SP)	(TN)	(UN)	(PI)	(ED)	(OI)	(FO)	(Final)
1 Scope	K	D	D	D	D	D	D	D	D	D	D	D
2 Knowledge	D	D	D	K	K	K	K	D	D	D	D	D
3 Synthesize	K	_	_	K	K	K	K	K	K	K	o	K
4 Analyze	K	C	C	K	K	K	K	K	K	K	o	K
5 Evaluate and select	K	C	C	K	K	K	K	K	K	K	o	K
6 Report	D	D	D	D	D	D	D	D	D	D	D	D

Tab. 4.8: Fingerprint for innovation class 3: "new."
(Existing market, existing product, existing supply chain, and new process).

Stage-Gate™:		Concept				Feasibility			Development			Final
K: Key design focus o: Optional design focus D: Default design activity C: Check implications of design changes _: No activity	**Framing**	**Supply chain imbedding**				**Process technology**			**Process engineering**			**Final**
Design levels ⇒ **Design steps/** **activities ⇓**	(PF)	(CW-PC)	(PC- PF)	(I-O)	(SP)	(TN)	(UN)	(PI)	(ED)	(OI)	(FO)	(Final)
1 Scope	K	D	D	D	D	D	D	D	D	D	D	D
2 Knowledge	D	D	D	D	D	K	K	D	D	D	D	D
3 Synthesize	K	_	_	C	C	K	K	C	K	K	o	K
4 Analyze	K	C	C	C	C	K	K	C	K	K	o	K
5 Evaluate and select	K	C	C	C	C	K	K	C	K	K	o	K
6 Report	D	D	D	D	D	D	D	D	D	D	D	D

Tab. 4.9: Fingerprint for innovation class 2: "novel."
(Existing market, existing product, existing supply chain, and modified process).

Stage-Gate™:		Concept				Feasibility			Development			Final
K: Key design focus o: Optional design focus D: Default design activity C: Check implications of design changes _: No activity	**Framing**	**Supply chain imbedding**				**Process technology**			**Process engineering**			**Final**
Design levels ⇒ **Design steps/** **activities ⇓**	(PF)	(CW-PC)	(PC- PF)	(I-O)	(SP)	(TN)	(UN)	(PI)	(ED)	(OI)	(FO)	(Final)
1 Scope	K	D	D	D	D	D	D	D	D	D	D	D
2 Knowledge	D	D	D	D	D	D	D	D	D	D	K	D
3 Synthesize	K	_	_	_	_	_	_	_	_	_	K	K
4 Analyze	K	C	C	C	C	C	C	C	C	C	K	K
5 Evaluate and select	K	C	C	C	C	C	C	C	C	C	K	K
6 Report	D	D	D	D	D	D	D	D	D	D	D	D

Tab. 4.10: Fingerprint for innovation class 1: "conventional."
(Existing market, existing product, existing supply chain, and capacity increase process).

Stage-Gate™:			Concept			Feasibility			Development			
K: Key design focus o: Optional design focus D: Default design activity C: Check implications of design changes _: No activity	**Framing**		**Supply chain imbedding**			**Process technology**			**Process engineering**			**Final**
Design levels ⇒ **Design steps/ activities** ⇓	**(PF)**	**(CW-PC)**	**(PC-PF)**	**(I-O)**	**(SP)**	**(TN)**	**(UN)**	**(PI)**	**(ED)**	**(OI)**	**(FO)**	**(Final)**
1 Scope	K	D	D	D	D	D	D	D	D	D	D	D
2 Knowledge	D	D	D	K	K	D	D	D	D	D	D	D
3 Synthesize	K	–	–	K	K	–	–	–	–	–	–	K
4 Analyze	K	C	C	K	K	C	C	C	C	C	C	K
5 Evaluate and select	K	C	C	K	K	C	C	C	C	C	C	K
6 Report	D	D	D	D	D	D	D	D	D	D	D	D

This DDM has been developed over decades with strong interaction between developments in the global field of product and process design, developments in industry, and developments in teaching at TU Delft. Its complete historic development is found in Appendix A4.

Exercises on DDM and on the other topics discussed in this section can be found in Section 4.10.

4.4.6 Examples of the Delft Design Map (DDM) results

4.4.6.1 Input-output (I-O) examples

Example 4.1
In Fig. 4.2, an example of an Input-Output (I-O) diagram is shown for a production process of fuels, syngas, heat and power) from biomass and other co-feeds (e.g., natural gas and coal) [8]. In Fig. 4.3 an Input-Output (I-O) diagram example is shown for the production of bio-butadiene from bioethanol. Fig. 4.2 provides very broad overview information that will need agreement between stakeholders and designers, whereas Fig. 4.3 focuses on the quantified mass balance for the I-O diagram (obtained by using stoichiometry and conversion factors for reactions, separation factors, and estimating the largest heat effects of reaction and separation).

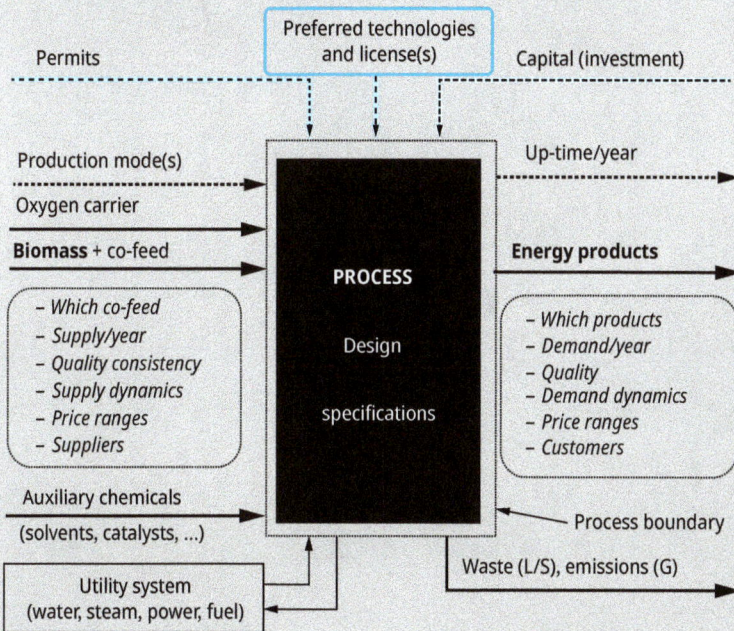

Fig. 4.2: Input-output (I-O) diagram example for the production of energy products (fuels, syngas, and heat power) from biomass and other co-feeds (e.g., natural gas and coal) [8] (obtained permission from Wiley & Sons).

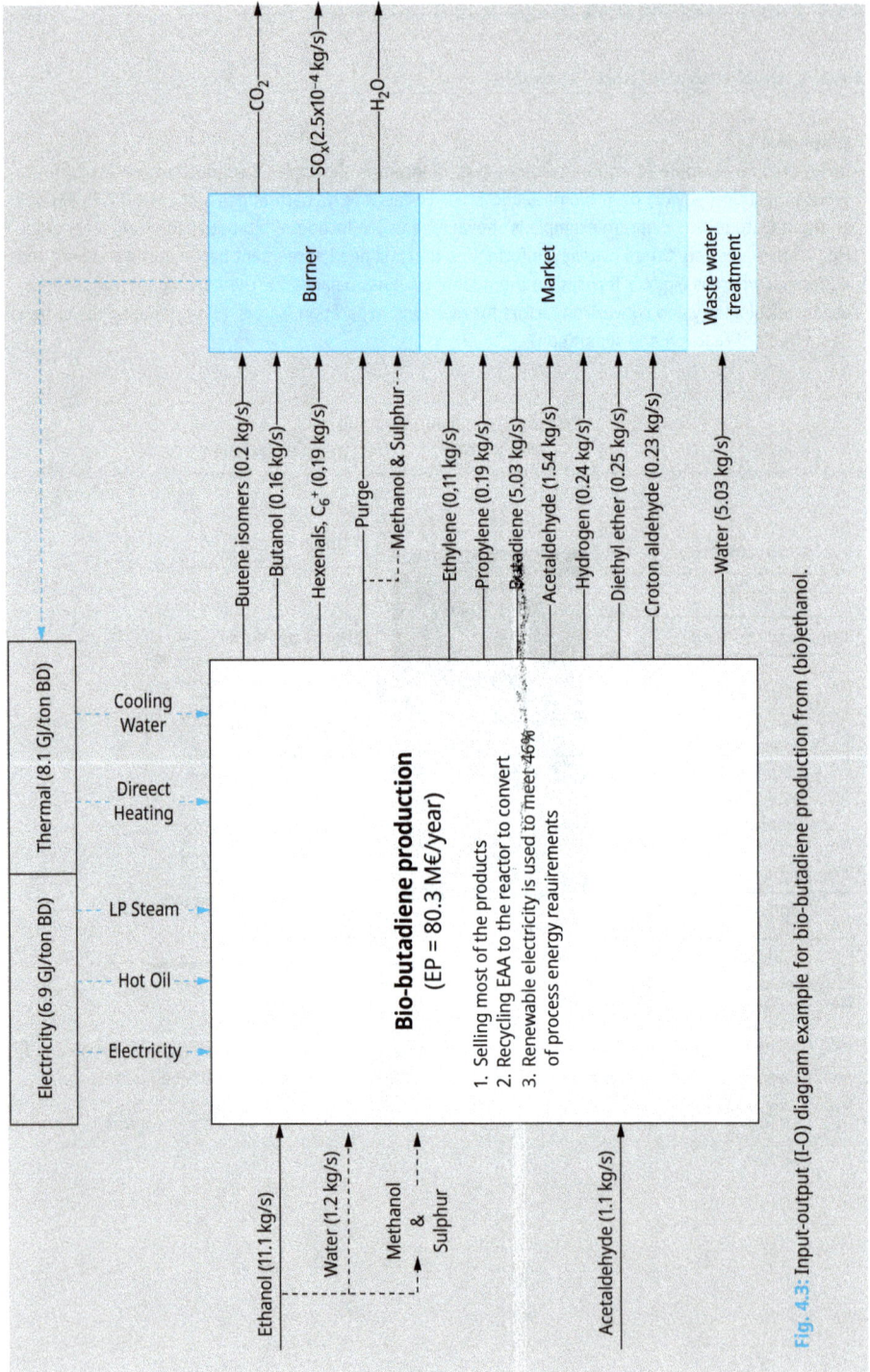

Electricity (6.9 GJ/ton BD)   Thermal (8.1 GJ/ton BD)

Bio-butadiene production
(EP = 80.3 M€/year)

1. Selling most of the products
2. Recycling EAA to the reactor to convert
3. Renewable electricity is used to meet 46%
   of process energy reauirements

Cooling Water
Direect Heating
LP Steam
Hot Oil
Electricity

Ethanol (11.1 kg/s)
Water (1.2 kg/s)
Methanol & Sulphur
Acetaldehyde (1.1 kg/s)

Butene isomers (0.2 kg/s)
Butanol (0.16 kg/s)
Hexenals, C$_6$+ (0,19 kg/s)
Purge
Methanol & Sulphur
Ethylene (0,11 kg/s)
Propylene (0.19 kg/s)
Butadiene (5.03 kg/s)
Acetaldehyde (1.54 kg/s)
Hydrogen (0.24 kg/s)
Diethyl ether (0.25 kg/s)
Croton aldehyde (0.23 kg/s)
Water (5.03 kg/s)

Burner
Market
Waste water treatment

CO$_2$
SO$_x$ (2.5x10^{-4} kg/s)
H$_2$O

Fig. 4.3: Input-output (I-O) diagram example for bio-butadiene production from (bio)ethanol.

4.4.6.2 Subprocess (SP) examples

Example 4.2

In Fig. 4.4, the example (from [8]) of the overall I-O diagram for a biomass to energy process is split into three subprocesses 1–3 representing the syngas (carbon monoxide/hydrogen mixture) generation from a co-feed (e.g., natural gas, subprocess 1), the syngas production from biomass (subprocess 2) and the production of fuels, heat, and power (subprocess 3) from the combined syngas feeds of subprocesses 1 and 2. This split is made because these different subprocesses could be owned by different companies (legal entities), or are based on different locations/sites, or both. These aspects make it impossible to identify and integrate tasks from these different subprocesses, because it is infeasible because of location or because the companies want to keep the boundaries of their subprocesses and are not willing to integrating these subprocesses further, but would rather optimize/redesign these within their own company.

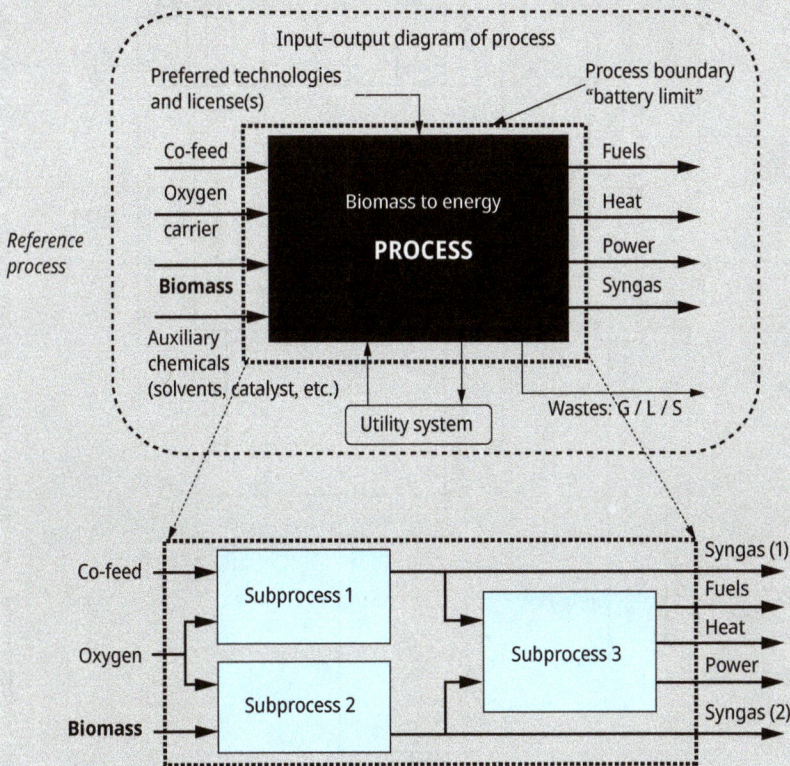

Fig. 4.4: Example of splitting an overall (input-output) process or supply chain into subprocesses 1–3 [8] (obtained permission from Wiley & Sons).

Example 4.3

Figure 4.5 shows the subprocesses diagram for the bio-butadiene production process from bioethanol in six subprocesses.

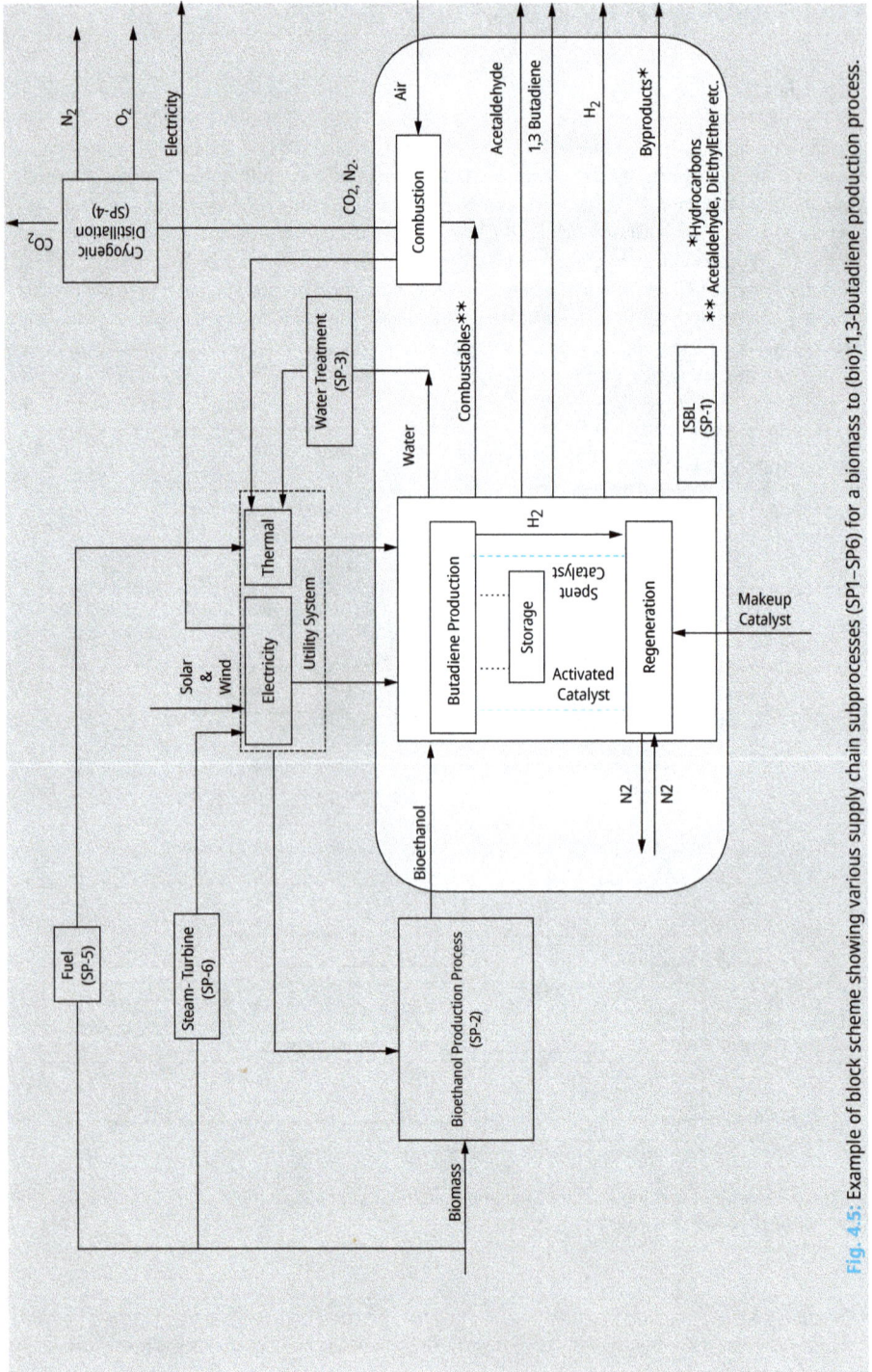

Fig. 4.5: Example of block scheme showing various supply chain subprocesses (SP1–SP6) for a biomass to (bio)-1,3-butadiene production process.

4.5 Planning for design and experimentation in innovation

Design and experimentation both play important roles in innovation. The major methods for planning both activities are provided in Chapter 3. The DDM, however, can also be used for planning both design and experimentation. In all cases where knowledge cannot be obtained from written sources and where analysis cannot be done by calculations, experimentation will be needed. Validating designs will also be done by using experimentation.

There is a general efficient approach to planning design and experimentation, namely, by making a design first and listing all assumptions needed for that design. For all assumptions that are critical to success and for which no written data can be obtained, an experimental plan is made. If the execution of that experimental plan is expensive and takes a long time, then the design may be altered in such a way that the assumption is not needed any more, and the related experimental program is then avoided. In this way, only the most essential experiments are carried out.

This working method is particularly fruitful if the concept designer and the experimentalist maintain strong interactions. As soon as the experimentalist foresees expensive long-lasting experiments the designer can modify the design and ask for an estimate of time and effort for the major assumptions to be tested, again.

4.5.1 Design sequence ranking method

Information and knowledge from many different knowledge fields are affecting the design. For strong interactions between many fields of information, a decision ranking method can be of value. This was proven for an industrial case with eight different fields of information by Korevaar in 2004 [13]. Each field is judged by its influence on the performance of the design by the team members with a simple score: strong, weak, or no. Each field is also scored on active or passive related to the other fields. With these scores, the fields are classified as:
(I)   High performance influence and highly active
(II) High performance and passive
(III) Low performance and active
(IV) Low performance and passive

The field (I) experiments are performed first. If their result is negative the whole design is stopped and a new design is made.

4.6 Embedding design by criteria and context setting

4.6.1 Introduction to the purpose of criteria and context setting

In designing a product, a process, or any artefact, the designer tries to take all relevant aspects into account in such a way that the new artefact fits well in reality. Only then the new artefact will be successful. At the start of the design, this can be done by deriving design criteria from a description of reality that is relevant to the design. This subject is discussed in Section 4.6.2.

4.6.2 Comprehensive list of modal aspects for defining criteria from modal aspects of reality

The most comprehensive description of reality – in terms that are easy to understand and that are relevant to design problems – is provided by Verkerk [5]. He provides a list with 15 modal aspects of reality, shown in Tab. 4.11.

   This table also contains suggestions on design criteria type that can be defined for each modal aspect for a design (second column). Guidelines for this procedure are provided for each modal aspect hereafter.

Tab. 4.11: Modal aspects and design criteria.

Modal aspect	Criteria type	Examples
Belief	Agreement with belief	Halal, kosher
Moral	Moral values	Safe, health, sustainable
Juridical	Legal	Within the laws
Esthetical	Esthetic	Attractive appearance
Economic	Economic	Profitable, efficient
Social	Social	Socially accepted
Linguistic	Communication	Clear reporting
Shaping	Shape	Correct 3-D shape
Logical	Logical	Correct model calculations
Sensitive	Sense	Tasty product
Biotic	Biological	No pathogen contamination
Physical/chemical	Properties and principles	Technically feasible
Kinematical	Flow rates	Heat and mass flows correct
Spatial	Geometry	Within plot area
Numerical	Numerical	Mass balances closed

Modal aspects from Verkerk [5].

Regarding the modal aspect "belief," the generation of design criteria needs elaborate consideration. The governing beliefs in the countries in which the product and the process will be present will have to be considered for acceptance. The obvious criterion for food products and processes for Muslim belief, for instance, is whether it is halal. However, this criterion also holds for nonfood products where the feed materials for the manufacturing process are from animal origins. Even the location choice for the manufacturing process can be a subject of belief aspects. A process for liquefied natural gas was planned for a location in the northern part of Australia. Late in the project, when its location was publicized, it appeared that the location was considered holy ground by the aboriginal inhabitants. The location had to be changed due to their protests. Consequently, the design had to be modified too, due to the different conditions at the new location.

Moral values play a strong role in safety, health, environment, social, and sustainable development criteria. Most companies have developed specific design criteria for these aspects, and report on these in their annual sustainability reports.

Social acceptance of new products and processes are, of course, of enormous importance. If a product is not accepted, there will be no market for it. Social acceptance is very difficult to predict. In most cases it needs testing.

The linguistic model aspect is about human communication. A design is information to be communicated to others. Therefore, the linguistic model aspect is very important. Clear reporting, presenting, and obtaining feedback are vital for any design.

The shaping aspect of design is about the three-dimensional shape of product and process. The shape of a product is important for its use, transport, and storage. The shape and outline of the process must fit within the given plot limits.

The logical aspect is about reasoning, modeling, and calculations regarding the design. Deriving the correct decisions based on logical arguments and reasoning is important.

The sense aspect is particularly important for product design. In most design cases, for each sense at least one criterion has to be defined.

The biotic aspect is particularly important for biological and food products and processes. Contamination by pathogenic microorganisms, for instance, must be avoided. Also, for process design biotic criteria are important, for example, in preventing microorganism-induced corrosion.

Physical and chemical aspects are vital in all product and process designs. The technical feasibility of a product or a process will always depend on the physical and chemical properties of the materials selected.

The kinematic aspect is about movement. It is particularly important for flows inside the process. If the fluid velocity becomes higher than the speed of sound, unwanted effects may occur.

The spatial aspect is the space taken up by the product and process, and by the material transport related to the product and process design.

The numerical aspect is involved in all computations made for the design. The calculations should be based on correct mathematical equations. Mass and heat balances, for instance, must be correct.

4.6.3 SHEETS criteria list

For students to learn scoping a design, a limited set of criteria should be used. Often such a limited set is also used within companies. This limited set can be seen as the minimum criteria set for each innovation project and its design. In this book the set is called SHEETS. It concerns the following criterion types:
– Safety
– Health
– Environmental
– Economic
– Technical
– Social

This list also incorporates the triple P (3P) definition of sustainable development of people (social), planet (environmental), and prosperity (economics). Chapters 10–12 provide detailed descriptions of these SHEETS criteria and how to use them for evaluating designs.

4.7 Role of modeling and simulation in concurrent design

Models and simulations play various roles in design. In the discovery and concept stage it is a quick way of determining the feasibility of an idea or concept. The famous inventor Tesla could imagine or simulate in his mind the working of a novel concept, like the dynamo. Only when he could envisage it working, would he construct a prototype. Some product developers work in a similar way. They say things like, "I envisage this to be working." For novel process concepts, experienced engineers too can imagine the working of a complex process. Based on the authors' knowledge, a few guidelines can be given to facilitate this type of imagination and simulation. One guideline is, consult an experienced developer on your idea and discuss. A second one is to take time to visualize the working of your idea using drawings, schematics, or mock-ups as presented next.

A more common way of modelling a concept is to produce a drawing, construct a mathematical model, and (mathematically) simulate the working of the product or process. In this way problems with the concept surface soon and modifications can be made to overcome the problems. This is an important part in the analysis of preliminary design solutions.

In the radical concurrent product-process design, modeling can also play this role. As the synthesis of a product and its process, design concepts are created and analyzed in a concurrent manner; the mathematical modeling role is more complex. A workable design flow diagram involving modeling and simulation of product and process has been generated by Grievink, Swinkels, and Almeida. A simplified version is shown in Fig. 4.6 (see also Section 8.5).

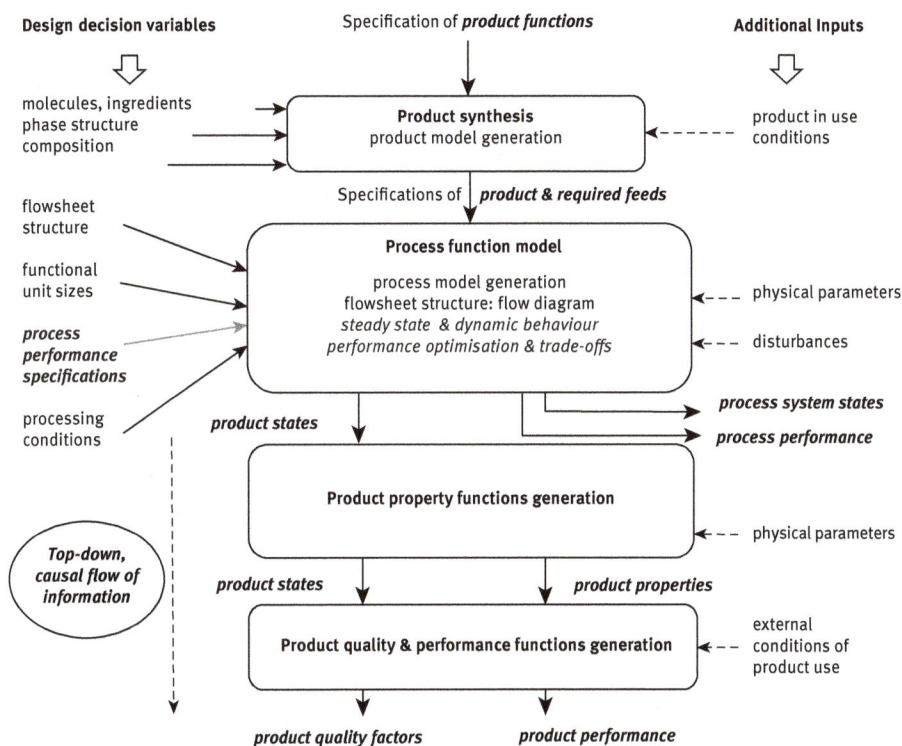

Fig. 4.6: Concurrent product-process-concept design flow diagram.

Specifications of product functions are input into the product synthesis step. A product model is built and its function is simulated based on external use conditions and design decisions. The design decision variables are molecule identities through which phase structure and composition are identified and manipulated. By model simulations the behavior of the product concept in its use/application is predicted. By trial and error, the product synthesis and model analysis steps are carried out until a satisfactory product concept is obtained.

The resulting product composition is then taken as the specification for the product the process concept design should deliver. A process concept design, based on various process functions is then synthesized and analyzed for its behavior. Again, this is done by simulating with a functional model: a process model. After several iterations of process design modifications and simulations an acceptable process design is obtained. The design results are the process system states, physical conditions of the process functions, flows between the function blocks, the process performance, and the product state.

In the third block (product property function), all (physical and chemical) product properties are predicted by a model that contains all relevant interactions between the constituting materials and the product structure.

In the fourth block (product quality and performance functions), the product quality and product performance in use is predicted. The product state, (physical) product properties, and the external conditions for using the product are the inputs for predicting the product quality factors and product performance.

Chapter 8 deals with product modeling, simulation, and optimization in detail. Chapter 9 deals with process modeling. The connectivity between product and process modeling and simulation will be covered in Section 8.5.

4.8 Exploiting experience in design (design heuristics)

4.8.1 Strength of design heuristics

Designing using heuristics is enormously fast. Particularly in the concept stage, this facilitates making several designs quickly and selecting the best. The resulting concept design will also be more reliable than those made by using theory only.

Experience can be exploited for design in the following ways:
(a) By using rules of thumb (design heuristics)
(b) By asking experienced people for advice and feedback
(c) By going back in memory on previous cases and derive knowledge to use for the case at hand

(a) Design heuristics is coded knowledge obtained from many experienced designers. Many heuristics for process concept design are provided by Seider et al. [14] and Douglas [15]. For process design in all innovation stages, Koolen is also a very good resource [16].

Heuristics for product design are scarce. Most of this type of knowledge is available within product innovation companies, and strongly context-bound to product design in foods, cosmetics, pharmaceuticals, etc. Martin provides heuristics for each of these product types [17].

(b) Obtain advice and feedback from experienced people. They have a lot of knowledge from experience and insight, but this is not easily transferred to others. This knowledge is called tacit knowledge. By stating the design solution, or a design problem, and asking them to go back in their own memory, giving them time (can be minutes, can be days) on similar cases, and probing them for stories during conversations or reflection on decisions. Through this process, a lot of useful information can be obtained.

An example from Harmsen's own experience is presented here.

Example 4.4

I (Harmsen) had made a trickle-flow fixed bed reactor design and showed it to an experienced engineer. He said: "What is the bulk crushing strength of the catalyst and is it below the maximum pressure drop that could occur when the bed is in operation, and when dust fouling has happened on the bed?" I looked up the bulk crushing strength of the catalyst particles, and it appeared to be even below the pressure drop of the empty bed. I needed to redesign the reactor and had to select a larger catalyst particle size to further reduce the pressure drop.

(c) Consult your own experiences. By asking yourself the same question as you would ask experienced people, you will find that you have more knowledge than you imagined you had.

If while designing using heuristics, a conflict of rules occurs, then rethink the whole design in such a way that the conflict disappears. Philosophically this is the same as solving the conflict of a thesis and an antithesis, by moving the whole problem to a different level or different context by which synthesis is possible.

4.8.2 Weaknesses in using design heuristics

Heuristics are based on truths related to a historically fixed context (sometimes called orthodoxies). Some heuristics, for instance, are based on the context that electricity is an expensive energy form. It has happened recently that electricity becomes very cheap or even trades at negative day ahead prices (albeit for certain periods during the day) due to excess wind and solar energy produced, and solar cells becoming very cheap, etc.; then these heuristics have become invalid.

4.8.3 Tapping into experience

Experiences shape our intuition. The intuition is used for practical purposes including design purposes. It is a very fast and a very effective form of thinking for all kinds of design activities. This is the reason you see so many experienced employees in design departments.

The intuitive form of thinking has one disadvantage: it blocks the individual pieces of memory about specific historic cases. The individual learning points of historic cases, however, can also be of interest to the design at hand. Kahneman [18] provides a simple method of drilling through the intuition barrier. His method is: Ask the experienced person (may be yourself), to go back into his own past and "envision" the cases that were a success and those that were failures. The person will fall silent for a minute or more, as he goes back in his memory. Then ask the question, "What is the learning point relevant to the design case at hand?" With this method, drilling through the intuition barrier is easy. Kahneman calls this the outside view, as it gives us views beyond the barrier of intuition.

Heuristics for breakthrough products and their processes are not publicly available. Heuristics for product design and for process design are provided in Chapter 7.

4.9 Industrial example of design driving innovation

Example 4.5

The prime example of a design driven breakthrough process innovation is the countercurrent extractive reactive distillation to produce methyl acetate by Eastman Chemical. Agreda et al. [19] describe this innovation in detail with all steps from concept generation to commercial scale implementation.

The implemented design is a one-column process containing two countercurrent extractions, a reactive distillation, and distillation sections at the top and bottom. It is dynamically controlled allowing for stoichiometric feeds of the two chemicals. At that time there was no equivalent concept of a complex combination of phenomena in a single column. Even simple reactive distillation was just being researched at that time of conception.

The following steps were taken in this design driven research, development, and implementation.

Step 1: A conventional design with classic unit operations would consist of two reactors and eight distillation columns. This design was tested at pilot plant scale. Due to its complexity, it was deemed desirable to search for a more economical process, requiring far less unit operations, hence, a challenging design goal was assigned.

Step 2: Many ideas were considered and computer simulations were used to test the ideas. After close examination of the problems of azeotropic distillations and the chemical equilibrium reaction, the concept was generated to use countercurrent extractions to avoid azeotropic distillations and to use reactive distillation to avoid reaction equilibrium limitations and to carry this all out in a single column. The top and bottom part of the column were designed as classic distillations to purify the product streams. An economic analysis revealed that the process was sufficiently economically attractive to pursue, and laboratory testing was recommended.

Step 3: Laboratory experiments were carried out to generate kinetics and physical properties. The feasibility of the concept was also tested in a laboratory setup. These tests confirmed the simulation calculations. The liquid holdup for reaction appeared to be a very critical factor for success. New economic analysis showed favorable results and large-scale testing was recommended.

Step 4: A bench-scale test glass column was built with a diameter 0.1 m and a height of 9 m. With this column in operation, corrosion tests were carried out as well as detailed measurements of component concentration profiles over the height of the column.

Step 5: A commercial scale design was made using the data from the bench-scale tests. Steady-state and dynamic computer simulation programs were developed to generate data for the design of the column. Again, an economic analysis was made. The results reinforced the expectation that it could be a viable commercial process. A downscaled pilot plant was then designed, constructed, and tested. Its dimensions were a diameter of 0.2 m and a height of 30 m. Special precaution was taken to ensure adiabatic operation. A dynamic control was also installed to allow stoichiometric feed ratio as intended for the commercial scale. Again, corrosion tests were also carried out. The results proved that steady-state production of methyl acetate of the required purity could be made over a significant period of time. Again, an economic analysis was made of the commercial scale process.

Step 6: the decision was taken to make a detailed design, to construct, and to operate the commercial scale process. The commercial scale process contained an additional methanol recovery column to allow for a lower temperature bottom flow to reduce potential corrosion and to minimize chemical losses during upsets of the operation. Additional measures concerning sufficient liquid holdup on the reaction trays were taken.

Step 7: Operation procedures were written with special attention to startup. Plant commissioning started with inspection and water checkout. Many problems were identified and corrected at this stage.

The plant started up in May 1983. Many problems including froth-related problems had to be solved in the startup phase. However, no production losses were incurred. Agreda contributes this to the pilot plant testing. The plant now operates routinely at rates above 100% of the design. The design production rate is 22,700 kg/h. It is about five times lower in investment cost and energy cost compared to the conventional process setup with 11 unit operations [20].

It is clear from his description that at the beginning it was not known how the process would work, nor what the process concept would be. Suitable simulation tools were not even available but were developed during the research and development stages. Also, a novel control method had to be developed. This innovation would not have happened if a design approach had not been followed from the start and through all subsequent stages to commercial implementation. It illustrates the power of design-driven design. The case also shows the power of the stage-gate approach and having the commercial scale solution always in mind, and from there, designing whatever research and development activities are needed.

This design-driven success led Siirola, the principal design engineer of Eastman Chemical, to develop a process synthesis method based on tasks (functions) rather than classic unit operations [20]. It is a nice illustration where first, design solutions are generated and then, through reflection, a general design method is derived. Or as Henderson put it: "Science owes more to the steam engine, than the steam engine owes to science" [21].

4.10 Exercises

Exercise 4.1: Designers and designing
State whether the following statements are true or false:
(1) Novice designers have a good grasp of design approaches when starting out in a project.
(2) A design result is always novel and always has a risk of failure.
(3) A designer does not make any implicit assumptions while designing.

Exercise 4.2: Design thinking or scientific research thinking
Categorize the following scenarios as either Design Thinking (DT) or Scientific Research Thinking (SRT):
(1) Product is good enough to be released.
(2) Elaborate explanation on why something works.
(3) General theory.

(4) Numerous ideas are to be tested.
(5) Focus and converge to a single paradigm.
(6) Use different methods as long as it helps.
(7) Use judgment and empirical calculations to assess reliability.
(8) "If it works it is good."

Exercise 4.3: Design thinking types

Categorize the following activities into one of the four types of thinking: deduction, induction, abduction-1, abduction-2:
(1) Reduce energy use in an ammonia production plant by applying heat integration.
(2) Determine whether a battery electric vehicle (BEV) is cheaper to run than an internal combustion engine (ICE).
(3) Develop high temperature heat pump fluids with low global warming potential.
(4) Design a low temperature laundry powder detergent.
(5) How do the batteries of battery electric vehicles deteriorate?
(6) Implement electrical heating equipment in process plants for reducing CO_2 emissions.
(7) Develop novel hydrogen storage materials based on chemical bonding.

Exercise 4.4: Delft design map (DDM)

State whether the following statements are true or false:
(1) The DDM is an easy lookup guide for steps to be carried out while designing.
(2) The DDM enables easy communication between design team members and stakeholders.
(3) The DDM is a visual aid for presenting final design reports.
(4) The DDM ensures that a fully sequential design process is adopted without iterations.
(5) The DDM benefits both novice designers and more experienced designers.

Exercise 4.5: DDM design levels, design steps, activity spaces

Appoint the design activity space (DAS) in which each of the following items/knowledge/activities will be addressed/generated/executed, needed/handled; indicate this with a combination of design level (PF-final) and design step (scope – report): e.g., PC-PF/analyze):
(1) Perform a project team analysis (core qualities analysis, Belbin)
(2) Make a project and time plan for the design project.
(3) Determine a gross margin estimate (product sales minus operating costs).
(4) Define the value proposition of novel product to consumer and other stakeholders.
(5) Make a stakeholder's overview and define important consumer needs, and translate these into product specifications and ranges.
(6) Calculate or estimate the carbon footprint (raw materials use, utilities use).
(7) Design a heat exchanger network for the production plant.

(8) Compute the net ingredient costs for a heterogeneous Fischer-Tropsch catalyst extrudate product.
(9) Gather knowledge about available commercial raw material feeds streams.
(10) Specify the raw material feed streams assumed to be available for the production process.
(11) Define various location options for the manufacturing process for green hydrogen using renewable energy.
(12) Create supply chain network alternatives for integrating the manufacturing of "green high strength steel for automotive application" with the production of feeds (upstream), utilities generation, waste treatment, and further downstream integration with steel product customization.
(13) Create supply chain network alternatives for integrating end-of-life steel recycling into high strength steel manufacturing.
(14) Gather information on the use of the QFD –HoQ design tools.
(15) Carry out a safety and health analysis of the manufacturing processing options.
(16) Specify the duty and different mechanisms to accomplish the tasks forming the manufacturing process.
(17) Gather information for the pure component properties table.
(18) Draw the process flow diagram (PFD) and compute the process stream summary table.
(19) Compute various distillation sequence options.
(20) Investigate different reaction/separation task integration options.

Exercise 4.6: DDM fingerprints for design project planning
For each of the following product-process designs, fill out the empty DDM fingerprint Tab. 4.12:
(1) Modify the design of a bioethanol plant by changing the pretreatment step from acid hydrolysis of the cellulosic biomass to steam explosion.
(2) Electrify a distillation column in an existing manufacturing plant.
(3) Identify products based on short cellulose waste fibers from a paper manufacturing plant and design the process(es) to produce these new products.
(4) Convert a spray-drying production process (dissolving/dispersing ingredients in water followed by spray-drying) for laundry detergent to a novel process using granulation technology to reduce carbon footprint.
(5) Redesign the Fischer-Tropsch process for linear hydrocarbons and waxes from synthesis gas for an order of magnitude higher activity catalyst.

Tab. 4.12: Empty DDM fingerprint table.

Stage-gate™:			Concept				Feasibility			Development			
K: Key design focus	**Framing**		**Supply chain imbedding**				**Process technology**			**Process engineering**			**Final**
o: Optional design focus													
D: Default design activity													
C: Check implications Design changes													
_: No activity													
Design levels ⇒ **Design steps/ activities** ⇓	**(PF)**	**(CW-PC)**	**(PC-PF)**	**(I-O)**	**(SP)**	**(TN)**	**(UN)**	**(PI)**	**(ED)**	**(OI)**	**(FO)**	**(Final)**
1 **Scope**												
2 **Knowledge**												
3 **Synthesize**												
4 **Analyze**												
5 **Evaluate and select**												
6 **Report**												

Abbreviations

6P	Proposition, promotion, place, packaging, price, product
BOSCARD	Background, objectives, scope, constraints, assumptions, risks, deliverables
C	Check implications of design changes
CW-PC	Consumer wants – product concept
D	Default design focus
DAS	Design activity space
DDM	Delft design map
ED	Equipment design
EngD	Engineering doctorate
FO	Flow sheet sensitivity and optimization
HoQ	House of quality
I-O	Input-output
ICSID	International Council of Societies of Industrial Design
K	Key design focus
KPI	Key performance indicators
o	Optional design focus
OI	Operability integration
OSBL	Outside battery limits
PC-PF	Product concept – (product) property function

PDEng	Professional doctorate in engineering
PI	Process integration
PFD	Process flow diagram
RAM	Reliability, availability, maintainability
QFD	Quality function deployment
SHEETS	Safety, Health, Environment, Economy, Technology, Social (aspects)
S/L/V	Solids/liquids/vapors
SP	Subprocesses
TN	Task network
Triple P	People, planet, prosperity
TRIZ	Theory of inventive problem solving – (Russian acronym)
VDI	Verein Deutscher Ingenieure
UN	Unit network

References and further reading

[1] Dorst K Understanding Design, 175 Reflections on Being a Designer, Amsterdam, Netherlands: BIS Publishers, 2006.

[2] Dorst K Frame Innovation: Create New Thinking by Design, Cambridge MA, USA: MIT Press, 2015.

[3] Petroski H To Engineer Is Human: The Role of Failure in Successful Design, New York NY, USA: Vintage books, 1992.

[4] Verganti R Design-Driven Innovation: Changing the Rules of Competition by Radically Innovating What Things Mean, Boston MA, USA: Harvard Business Press, 2009, 4.

[5] Verkerk MJ, et al. Philosophy of Technology, New York, USA: Routledge, 2016.

[6] Lawson B How Designers Think – The Design Process Demystified, 4th edn, Oxford USA: Elsevier, 2006.

[7] De Bont C Advanced Design Methods for Successful Innovation, Den Haag, Netherlands: Design United, 2013.

[8] Grievink J, Swinkels PLJ, Van Ommen JR Basics of Process Design. In: De Jong W, Van Ommen JR (eds). Biomass as a Sustainable Energy Source for the Future: Fundamentals of Conversion Processes. 1st edn, Hoboken NJ, USA, Wiley & Sons Inc., 2014, 184–229.

[9] Almeida-Rivera C, Bongers P, Zondervan E A structured approach for product-driven process synthesis in foods manufacture. In: Martin M, Eden MR, Chemmangattuvalappil NG (eds). ComputerAided Chemical Engineering Vol 39: Tools for Chemical Product Design – From Consumer Products to Biomedicine. Amsterdam, Netherlands: Elsevier BV, 2017, 417–441.

[10] Seider WD, et al. Product & Process Design Principles, 3rd edn, Hoboken NJ, USA: John Wiley & Sons Inc., 2010.

[11] Tassoul M Creative Facilitation. Delft, Netherlands: VSSD, 2009.

[12] TRIZ – a powerful methodology for creative problem solving, 2017. Resourced from: http://www.mindtools.com, https://www.mindtools.com/pages/article/newCT_92.htm.

[13] Korevaar G Sustainable Chemical Processes and Products: New Design Methodology and Design Tools, Delft, Netherlands: Eburon, 2004.

[14] Seider WD, et al. Product & Process Design Principles, 3rd edn, Hoboken NJ, USA: John Wiley &Sons Inc.,2010.

[15] Douglas JM Conceptual Design of Chemical Processes. New York NY, USA: McGraw-Hill, 1989.

[16] Koolen J Design of Simple and Robust Process Plants, Weinheim, Germany: Wiley-VCH, 2001.

[17]   Martin M, Eden MR, Chemmangattuvalappil NG, eds. Computer Aided Chemical Engineering Vol 39: Tools for Chemical Product Design – From Consumer Products to Biomedicine. Amsterdam, Netherlands: Elsevier BV, 2017.

[18]   Kahneman D. Thinking, Fast and Slow, London, UK: Penguin Books, 2012.

[19]   Agreda VH, Partin LR, Heise WH High-purity methyl acetate via reactive distillation, Chemical Engineering Progress, 1990, 86(2), 40–46.

[20]   Siirola JJ Industrial applications of chemical process synthesis, In: Anderson JL, Bischoff KB (eds). Advances in Chemical Engineering Vol 23: Process Synthesis. Academic Press, 1996, 1–92.

[21]   Henderson LJ. Quote in Moore, WJ. Physical Chemistry, 4th edn, London, UK: Longmans Green and Co., 1962.

Part B: **Design generation**

5 General and sustainable design approaches

5.1 Introduction to general and sustainable design approaches

This chapter is about design approaches. Its purpose is for designers and systems engineers to quickly find literature about a method suitable for their design case. Section 5.2 is about general design approaches. Section 5.3 is about designing from a sustainable perspective. Section 5.4 is about designing with a specific purpose.

5.2 General design approaches

5.2.1 Product design and development

A comprehensive product design and development method is provided in the textbook by Ulrich and Eppinger [1]. The method is explained step-by-step. It treats concept design as well as more detailed design. It also treats the experimental development of the new product by prototyping. Furthermore, it combines product design with design for manufacturing. Its applicability is wide.

5.2.2 Product quality function deployment

In the quality function deployment (QFD) design approach, first of all, the voice of the customer is defined by marketing analysis and competitor analysis of how their product fulfils the customer needs. Then this information is transformed into new product function requirements followed by designing the manufacturing process. Finally, process controls are designed to ensure product quality and the manufacturing process implements the process commercially [2 Akao]. The method can be applied to any type of product and process design. Section 7.2.2 on (bio)chemical product and process design is based on this approach.

A practical textbook by Akao is made available [2]. For QFD, an ISO quality method 6355-1:2015 is available [3].

5.2.3 Total (product) design (Pugh)

In the total design method developed by Pugh and published in 1991 [4], a matrix is provided by which many alternative product designs can be evaluated using several criteria based on consumer needs and requirements. It represents a very structured

https://doi.org/10.1515/9783110782127-005

method but requires considerable effort and it is not clear whether the optimum design has been reached. Guler provides a product development model, which requires far less effort and can be used by small design teams [5].

5.2.4 Resilient mechanical product design by VDI guidelines

Hedrich explains how resilience can be designed into a new mechanical product [6]. The method is described in VDI 2221 guidelines of the German Engineering Association design [7, 8]. The method focusses on obtaining a resilient product, thus controlling uncertainty in its product design, its manufacturing, and its use [6]. A textbook describes the method in detail [9].

For mechatronic system designs, the VDI 2206 method has been developed over the years. The latest version is provided by Graessler [10]. Mechatronics engineering, also called mechatronics, is an interdisciplinary branch of engineering that focuses on the integration (systems engineering) of mechanical engineering, electrical engineering, and electronic engineering leading to a specific product device. Figure 5.1 shows all steps and their connections to create and implement a new mechatronics product. A comprehensive textbook that provides detailed information about each step of the VDI 2206 method is available [11].

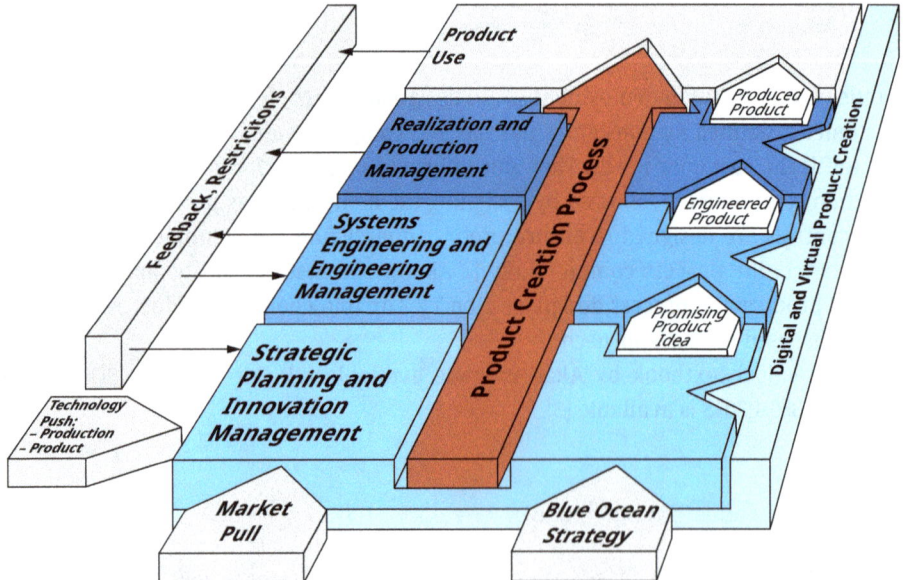

Fig. 5.1: Mechatronics product creation process [10].

5.3 Sustainable product and process design approaches

5.3.1 Introduction

This section introduces sustainable design methods for products and processes. Section 5.3.2 describes the cradle-to-cradle product method that focusses on no environmental pollution over the closed cycle.

5.3.2 Cradle-to-cradle product design

5.3.2.1 Design for cradle-to-cradle
Braungart's cradle-to-cradle product design method for eco-effectiveness is based on three essential elements [12, 13]:
A: Design for recycling of all product components staying within the technology cycle (see Section 5.4).
B: Design for end-of-cycle products obtained from biological origin and ending in the biosphere, such that all decomposed components ending in the environment after their use act as nutrients for plants to grow.
C: Use solar as sole primary energy source.

This method is used for novel consumer product design and focusses on protecting the environment.

5.3.3 Product and process design for circular economy (Ellen MacArthur Foundation)

The widely accepted definition of circular economy is the one by the Ellen MacArthur Foundation: "a circular economy is one that is restorative and regenerative by design, and which aims to keep products, components and materials at their highest utility and value at all times, distinguishing between technical and biological cycles" [14]. Figure 2.4 of Chapter 2 shows the two closed mass cycles and their mostly applied process steps. This figure can be used by designers for inspiration to design for closed mass cycles for the entire life cycles of product and process.

In the circular economy model no accumulation of mass should occur. Presently carbon dioxide is accumulated in the earth's atmosphere. The source of that is fossil carbon.

This means that fossil sources that end up in the atmosphere cannot be used. The sole source of energy should come from the sun, directly by radiation, or indirectly, for instance, from wind.

Another accumulation is plastic ending up in oceans. The obvious solution is complete recycle of plastics to be reused.

Cramer has developed a practical set of 10 guidelines for circular product design. She calls it the 10R Ladder of circularity [15]:

Refuse   Do not use native material but use refused material
Reduce   Use less material
Renew   Use waste material in new ways
Reuse   Use waste material again
Repair   Repair the device or material
Refurbish   Revive the material
Remanufacture   Make a new product from used materials
Repurpose   Reuse product for other function
Recycle   Recycle material for use
Recover   Recover some value from used material such as energy

The ladder should be started at the top. If a certain step has been tested and cannot be further applied the lower step should be tested.

As the used material often has a zero or even negative price, while the new product has a higher market price profits can often be foreseen. In general start-up companies exploit this potential earning power]. Cramer estimates that in the Netherlands alone, 7 billion €/year can be made by circular economy, just considering water, energy, and materials [15].

As the method has economy in its title it is appropriate to add that circular economy cycles should also be economical. In free market societies it means that each economic entity in a value chain should make a profit. The author found out that for all design cases explored by EngD trainees at TU Delft this appeared to be achievable; so closing material cycles also makes economic sense.

5.3.4 Design for industrial ecology

Industrial ecology is the study of the flows of materials and energy in industrial and consumer activities, of the effects of these flows on the environment, and of the influences of economic, political, regulatory, and social factors on the flow, use, and transformation of resources. Core elements are [16]:

– the biological analogy
– the use of system perspectives
– the role of technology change
– the role of companies
– dematerialization and eco-efficiency
– forward-looking research and practice

Unfortunately, this handbook provides little information of use for designers. To my knowledge, there is no other book on designing for industrial ecology available.

However, some elements of industrial ecology can be applied in design. Ecological systems are resilient because of many connections of the food chain from biomass, via small animals, big animals, and finally via microorganisms, all components are degraded to elements and small molecules that are used to start the cycle again. The energy that drives the whole cycle is sunlight that is stored in biomass and then that energy drives the animals. Industrial ecology could copy this by starting with molecules from biomass or forming olefins from hydrocarbons using sunlight as primary energy source for the endothermic reaction. Subsequent conversions to useful products are then exothermic, so do not need energy.

5.3.5 Design for industrial symbiosis

A subset of industrial ecology is industrial symbiosis. It is about transferring the biology analogy of the ecosystem to the industrial system. Chertow investigated 22 industrial symbiosis cases and concluded that environmentally and economically desirable symbiotic exchanges are all around us and now we must shift our gaze to find and foster them [17].

Table 5.1 provides an overview of the main elements of industrial symbiosis.

Tab. 5.1: Industrial symbiosis elements derived from nature.

Nature ecology system elements	Derived industrial symbiosis elements
Food chain and closed material recycles	Closed material cycles
Many different species connected in network	Industry and society parts connected in network
Primary energy from the Sun	Solar energy-based
Energy stored and transported via food	Energy in intermediate materials

The designer can use the Dow Chemical method of finding and implementing industrial symbiosis synergies [18]. It involves the following steps:

Step 1: Perform a high-level feasibility study of potential cases on environmental, social, and economic benefits and select cases with positive outcomes on these three criteria.

Step 2: Research material properties and potential applications for those cases.

Step 3: Locate potential suppliers and users of streams and define potential connections.

Step 4: Obtain interest from partners in the project both on technical and economic points of view.

Step 5: Finalize project.

Step 6: Monitor and improve project.

Magan describes an actual industrial symbiosis case (which he calls a by-product synergy case) and from which he derives the following steps [19].

Step 1: Organize facilitated meetings between potential partners of an industrial site and explain the concept and the potential benefits. The meeting should be chaired by an independent person to avoid perceived bias for a certain partner.

Step 2: Form working groups of engineers from potential partners to analyze options and prioritize best cases.

Step 3: Overcome barriers, such as legal, technical, economic, communication, competing priorities, and geographic and perception challenges by maintaining communication and cooperation within the working groups.

Step 4: Make action plans.

In Magan's experience, steps 1–4 can be carried out within a year. After that first year, a quarterly meeting should be held to implement action plans and bring in new companies [19].

Synonyms used for industrial symbiosis are:

By-product synergy

Eco-industrial development

Eco-industrial parks

Industrial ecology

Industrial ecosystems

5.3.6 Product design for biomimicry

The term *biomimicry* appeared for the first time in 1981 and was popularized by scientist and author Janine Benyus in her book *Biomimicry: Innovation Inspired by Nature* [20]. Biomimicry is defined by her as a "new science that studies nature's models and then imitates or takes inspiration from these designs to solve human problems." Since then, an array of articles and books has been appearing in which the idea is further developed. Here are the basic ways to apply biomimicry.

The simplest way is to take a natural material with a very special property and use its specific property such as strength, hardness, or chemical reactivity and design a product with the same material for an application. The industrial process for making the material may also be copied from the way nature makes it.

A second way is to take a physical principle from nature and use that principle to make novel product. An example is the use of a nanostructure that reflects sunlight in such a way that the material shows a specific color. The same nanostructure, but with different materials can then be designed and made to provide color to the surface of the material. In this way, the need for painting is avoided.

A third way to apply biomimicry is to take a natural systems method and use the same method for an industrial system. An example is the natural system for closing the material cycles for carbon, oxygen, and water. By photosynthesis, carbohydrates are formed from carbon dioxide and water, which in turn are used by other natural species as an energy source by oxidizing them back to carbon dioxide and water. This third way has been developed into industrial ecology and its subsystem industrial symbiosis. This way is treated separately in this book in the section industrial ecology.

A fourth way to apply biomimicry is to be inspired by nature in the way it has solved problems. Perhaps this is the most fruitful method. It has the advantage that it is not rigidly held to the specific natural method. Such rigidness can lock in the designers to solving the problem in a fixed natural way. A classic example is the idea that a flying machine should be flying by moving wings such as birds do. This led to many failures. Only by abandoning this idea and separating the propulsion from the airlift, flying machines became a success.

The inspiration to improve the airplane wings for energy efficiency by looking at the design of bird wings, however, is still applied by wing designers.

For an elaborate description of biomimicry for product and process design and other, the reader is referred to references [20–22].

5.3.7 Design for renewable energy sources

5.3.7.1 Present bulk chemicals production from fossil fuel resources

Present industrial processes are still mainly driven by fossil sources. This is mainly performed by burning fossil energy resources such as natural gas in furnaces and using the heat in the form of steam or hot medium to provide the energy for reaction, separation, extrusion, and mass movement.

For the formation of hydrocarbon products and their derivatives, the reaction energy is provided as follows. First the hydrocarbons are transformed into olefins by the steam cracker process. This is a highly endothermic process and the required reaction energy is provided by burning fossil hydrocarbons in a furnace. By placing the reactor tubes in that furnace, the reaction heat is directly provided to the tubes.

The resulting olefins are then used in subsequent reactions to produce plastics and virtually all other chemicals. All these latter reactions are exothermic, so produce, the next product, heat. That heat stems from the energy stored in the olefins. The advantage of exothermic reactions is that the reactions run to complete conversion at moderate temperatures as the main contribution to obtain a negative free enthalpy is the negative reaction enthalpy.

5.3.7.2 Design for renewable energy-based product-process chain methods

When renewable energy sources are used to produce chemicals and other products, ways must be found to deliver energy for driving the reactions and for driving the other steps in the process.

Renewable energy is provided from the sun, directly or indirectly. Solar radiation is abundantly available but at a low energy density. It is so low that it must be concentrated first to be applied in industrial scale processes. A proven way is to apply photovoltaic cells to convert solar radiation into electricity which can be transported and concentrated.

Another way is using Concentrated Solar Power (CSP) technology. Solar radiation is concentrated using mirrors and it can then be used to generate very a high temperature, which can be transferred to a medium such as molten salt, which then can be used to drive a chemical reaction or converted subsequently to electricity. CSP is already in commercial operation in many locations [23].

Windmills are also used to harvest solar radiation and convert it into electricity. Yet another way is to grow biomass, which in turn can be used as a renewable energy source. It can also be directly used as feedstock for chemicals production.

In all cases renewable energy should be used efficiently and at low investment cost for the production of chemicals and other products. This requires new innovative solutions. An example of such a solution is the production of ethylene glycol from carbon dioxide, water, and solar radiation provided by [24]. As the primary feedstocks have little to no cost it is likely that this ethylene glycol production could be competitive compared to fossil-based production.

It is clear from this example that renewables-based products and process can be made cost-effective, but it requires a lot of considerations. Basic elements are solar radiation, electrolysis, and catalysis. The latter two may be combined. Murzin provides process design methods for products from renewables [24].

5.4 Design for specific targets

5.4.1 Introduction

This section introduces the reader into design for specific targets. Section 5.4.2 provides an elaborate design method for achieving energy efficient processes. The method is applied in the EngD program at Delft University of Technology. Section 5.4.3 helps the designer focus on the human aspects of the design result, be it a product or a process. Section 5.4.4 is about controlling the product quality all the time by using the Six Sigma method.

5.4.2 Design for energy efficiency

"Energy efficiency" is a generic quality that describes use of a minimum of external energy resources to obtain a desired service or an amount of product. For chemical products and processes the common energy resource is power, any form of fuel, or utilities that are derived from these two main resources. Their use is generally related to the amount of product produced or feed processed to get the "process specific energy" (in J/kg), as shown in Fig. 5.2. Aiming for energy efficiency, a low process specific energy minimizes both the environmental footprint (use of resources and CO_2 emission) and the operating cost to obtain the required external energy resources. For products, the energy efficiency of the intended service of the product may also be relevant, like fuel savings by low rolling resistance tires [25]. This section focuses on processes, and it gives a brief introduction to how to assess their energy efficiency and how to improve it.

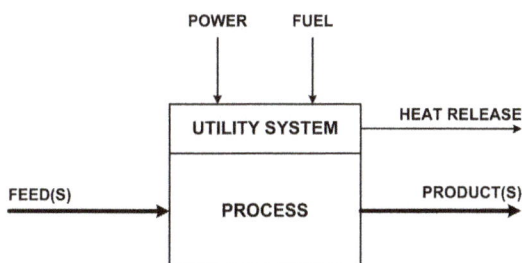

Fig. 5.2: Generic process input-output scheme.

Energy efficiency can be achieved in any form of energy that plays a role in the process of producing the desired product. Changes in composition and the physical state of material streams are achieved by chemical and physical processes. All of these cause changes in energy states and as we deal with practical processes, they have their specific irreversibilities. The most energy-efficient process is the reversible process. Thermodynamic approaches are available to determine the minimum energy resources required to perform the desired process in a reversible way. The ultimate low process specific energy found this way is the ultimate target to aim for in energy efficient design. This is the basis for "exergy analysis" and design approaches to minimize the exergy loss, also referred to as lost work [26, 27]. Exergy losses in a process can nicely be visualized in a Sankey diagram (Fig. 5.3), which shows the exergy supply to the process, the losses in each operation within the process, and the exergy sinks in products, waste streams, and losses to the environment [28].

The exergy analysis gives useful insight into the "irreversibilities" that are important in a process that is available. This may guide changes to the process to make it more reversible and improve energy efficiency. However, it relates energy efficiency to

**Enthalpy balance
extraction**

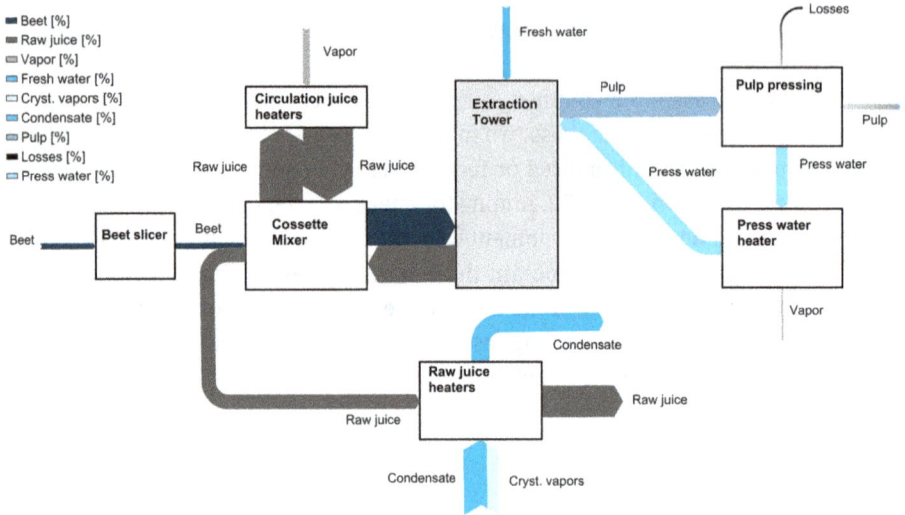

Fig. 5.3: Example Sankey diagram: sugar extraction (courtesy: Cosun R&D).

an ideal and unreachable target. For designers it is generally more relevant to get guidance about what is practically feasible. One source of practical guidance is the current state of the art: existing operating plants have a proven process specific energy. Values for known processes are available in the literature. They may also be available from governmental and consultancy sources that monitor the efficiency of specific processes and industrial sectors [29]. Typical process specific energy values can be used as alternative targets for new facilities, but these values should be treated as the *maxima that can be allowed* for a new process, as it will have to compete with the existing ones.

Traditional (petro)chemical bulk industries have been keen on energy efficiency especially since the 1970s. For these processes the main energy transformation is related to physical heating and cooling. This has resulted in a practical tool kit to analyze such processes, determine practical energy efficiency targets, and design the heat exchanger network, the system that performs the required physical heating and cooling. This tool kit is known as Pinch analysis [25] and is the most commonly used analysis and design method for energy efficiency in industrial practice. We will illustrate the approach using a simple example case.

Example 5.1 Example of reactor heating and cooling. Consider the scheme of Fig. 5.4, with the stream data given in Tab. 5.2.

Fig. 5.4: Simple reactor system with preheating and after cooling.

Tab. 5.2: Stream data of Fig. 5.3.

Stream	Source (°C)	Target (°C)	Mass flow (kg/s)	Heat capacity (kJ/kg/°C)
S1	50	100	5	2
S2	105	40	5	1.6

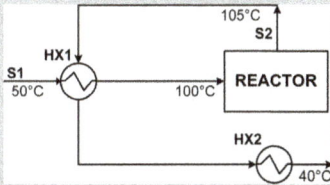

Fig. 5.5: Simple reactor system with feed-effluent heat exchange.

Cold feed S1 is heated up and fed to a reactor. The reactor product S2 is cooled down. We refer to Stream S1 as a "cold stream" as it needs to be heated up. We refer to Stream S2 as a "hot stream" as it needs to be cooled down. Exchanger HX1 has a duty of 5 kg/s × 2 kJ/kg/°C × (100 °C – 50 °C) = 500 kW, HX2 a duty of 5 kg/s × 1.6 kJ/kg/°C × (105 °C – 45 °C) = 520 kW. Both the heating and cooling duty may be supplied by an external source: steam to heat S1 and cooling water to cool S2. The process specific energy can now be defined as:

Process specific energy = (steam duty supplied)/product flow rate.
Steam duty supplied = 500 kW; product flow rate = 5 kg/s
Thus, process specific energy = 100 kJ/kg product

Obviously, it should be possible to make this process more energy efficient and lower the process specific energy. If we look at the overall heat balance, we find a heating requirement of 500 kW and a cooling requirement of 520 kW. The net result suggests a net cooling requirement of 20 kW. No heating would be required. Clearly, this approach is too simple and only based on the first law of thermodynamics (heat conservation), while it ignores the second law of thermodynamics.

This second law requires that for any feasible physical heat transfer the supply heat has to have equal or higher temperature than the demand heat. For a reversible heat transfer the supply (hot) and demand (cold) temperatures are the same, but in practice the exchanger area has to be limited in size, which brings up the requirement of a minimum temperature difference (well) above zero. This becomes clear from the generic heat transfer relation: $A = Q/U/DT$, in which A is the heat transfer area (m^2), Q the duty (kW), U is the overall heat transfer coefficient (kW/m^2/°C), and DT is the effective temperature difference (°C) that is equal to the logarithmic mean temperature difference for a pure countercurrent heat exchange [31].

The reactor effluent is a little hotter than the reactor feed and can be used to heat this feed. If we try to supply all the heat (500 kW) required in Stream S1, as illustrated in Fig. 5.5, it becomes clear that stream S2 is inadequate to do that. After cooling 500 kW, S2 is cooled to 42.5 °C, which is below the supply temperature of the cold feed S1. Apparently, there is a limit to the heat transfer from the hot product to the cold feed.

To find this limit we draw the hot supply and cold demand in an enthalpy – temperature plot, as illustrated in Fig. 5.6. In this plot we can graphically construct the feasible overlap where the hot (supply) curve is hotter (higher temperature) than the cold (demand) curve. When we allow the curves to touch, we allow the thermodynamic minimum approach of 0 °C and we get the ultimate heat transfer that is feasible. The resulting graph shows the following:

–   It is possible to transfer 440 kW from the hot product to the cold feed.
–   We will need 60 kW additional heating for the cold feed to reach the target reactor inlet temperature.
–   We will need 80 kW additional cooling for the hot product to reach the target product temperature.

This is the best thermodynamic solution for the given system but as the temperature approach is 0 °C it will require an infinite transfer area. For any practical design it may be used as a reference. It will be necessary anyway to have three exchangers as shown in the scheme depicted in Fig. 5.7.

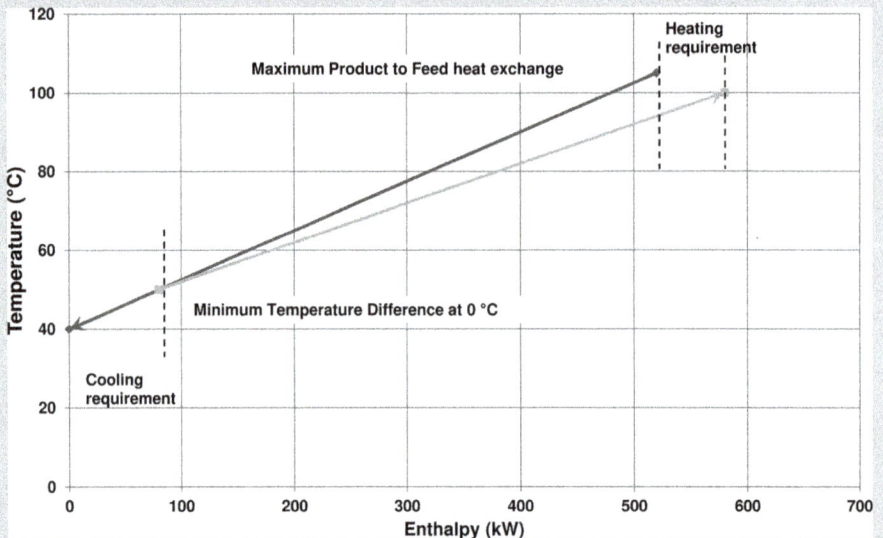

Fig. 5.6: Enthalpy–temperature plot example 5.1 with minimum temperature difference of 0 °C.

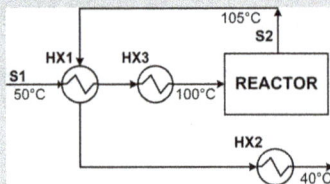

Fig. 5.7: Heat-integrated reactor system.

For a practical design, the temperature approach should be selected to allow the application of real heat exchangers. Choose, for example, a minimum temperature difference of 10 °C. Note that a realistic value depends on the exchanger type that can be applied; for the more common types the minimum difference is 10–20 °C, but more.

Advanced exchangers allow a minimum difference as low as 1 °C. With the selected value of 10 °C we can do a similar graphical construction; see Fig. 5.8 to determine the heat that can be transferred in the enthalpy–temperature plot. The resulting hot-to-cold process stream transfer is 360 kW, the heating requirement is 140 kW, and the cooling requirement is 160 kW.

Obviously, we can do this exercise for any other minimum temperature difference. If we increase the temperature difference, we will reduce the process-to-process heat transfer and increase the need for additional heating and cooling. This will make the process less energy-efficient. Additionally, we will need less transfer area for process-to-process heat exchange (S2 to S1) as the duty goes down and the temperature differences go up. Meanwhile, the transfer area for the additional heating and cooling (by utilities) will increase as these duties go up. The net effect is case dependent.

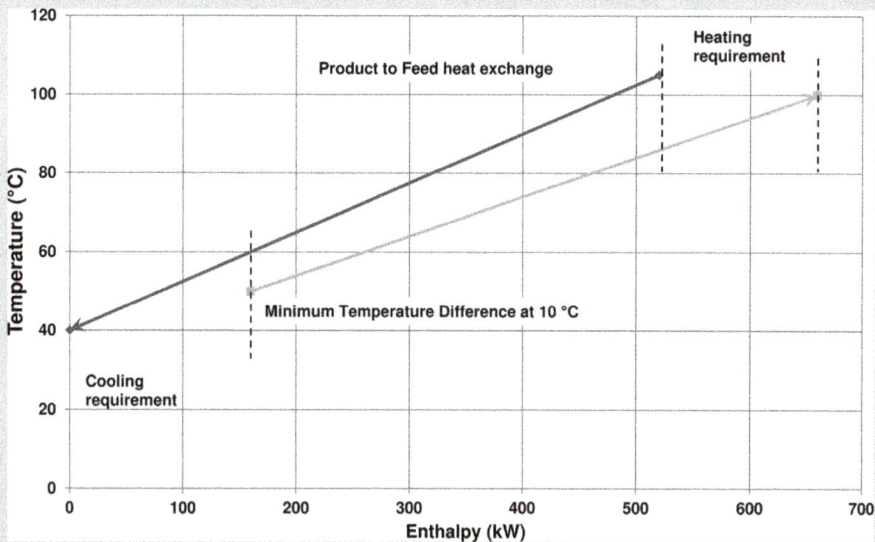

Fig. 5.8: Enthalpy–temperature plot with minimum temperature difference of 10 °C example 5.1.

Example 5.1 shows that a heat curve plot can be used to determine realistic targets for process-to-process heat exchange and the need for external heating and cooling resources. Pinch analysis is an extension of this simple analysis for processes with any number of hot and cold streams. To make this Pinch analysis all cold process streams are lumped in a cold composite stream. Plotted in a temperature–enthalpy plot this gives the "cold composite curve," Fig. 5.9, which shows the heat that is required in the process at each temperature. In the same way all hot streams are lumped to the "hot composite curve." Graphical construction, similar to the heat curves of the example, as illustrated by Fig. 5.10, allows the determination of the maximum overlap, depending on the minimum temperature difference that we allow between the hot and cold composite curves. The point at which the composite curves are closest together to the selected minimum temperature difference is called the "pinch," which is considered the key area in designing an energy efficient heat exchanger network.

Pinch analysis is the thermal analysis of a process to determine realistic targets for process-to-process heat transfer, heating, and cooling by external energy resources similar to the analysis for one hot and one cold stream in the example. Commercial software like Supertarget [32], Aspen Energy Analyzer [33], and various process simulation packages support such analysis and provide easy interfaces to generate these curves from a process model. The composite curves are a powerful visualization to show what heat recovery (process-to-process heat exchange) is possible and what quality of external energy resources or related utilities are necessary. It also gives an indication of how difficult it will be to get the desired heat transfer between the hot and cold process streams to and from the utilities.

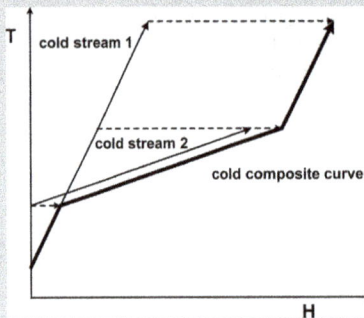

Fig. 5.9: Cold composite curve.

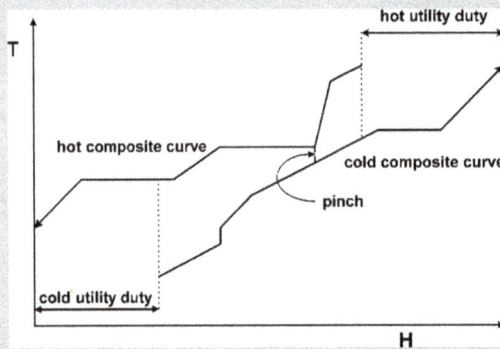

Fig. 5.10: Typical composite curves showing pinch, process-to-process heat exchanger, heat demand, and release to external sources.

Figure 5.11 illustrates that if curves are far apart and the pinch is a narrow region (left) the heat transfer will be easier than when the curves are close together and have a large parallel region (right). Based on this insight it is possible to design an efficient utility system that converts the primary external energy resources to the utilities required in the process. Also, it is possible to derive from the composite curves process modifications that will ease or enhance the heat transfer and finally allow for a more energy-efficient design.

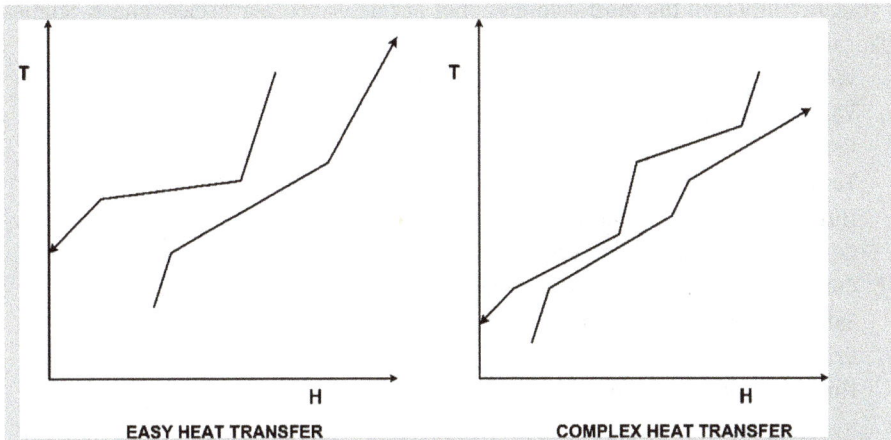

EASY HEAT TRANSFER   COMPLEX HEAT TRANSFER

Fig. 5.11: Composite curves for easy and complex heat transfer.

Once the process is fixed and the utilities selected, the composite curves offer guidance in designing the heat exchanger network to heat balance the process following the Pinch Design Method. Most pinch-related literature focuses on this last design stage, but in practice this network design is only a small part of the process design work. Pinch analysis is useful in the entire process design work, and it is recommended to apply it early when the first process concepts are evaluated to identify the scope in improving the energy efficiency of the process, especially when physical heat transfer is important in this process. Excellent explanations of Pinch analysis and the Pinch Design Method can be found in Smith [30] and Kemp [34].

5.4.3 Design for human factors

Processes and products are sometimes two sides of the same coin. A process facility is in essence a large machine that is operated and interacted with by personnel and can thus be regarded as a sociotechnical system. Several large scale incidents with fatalities have occurred in the industry where human factor aspects such as poor human-machine interface, lack of understanding of human error, lack of situational awareness, etc. were the cause. Human factors are all those things that enhance or improve human performance in the workplace. Human Factors Engineering (HFE) focuses on the application of human factors knowledge to the design and construction of sociotechnical systems. The objective is to ensure systems are designed in a way that optimizes the human contribution to production and minimizes potential for design-induced risks to health, personal, or process safety or environmental performance [35].

Good HFE will increase Health Safety and Environmental (HSE) performance and reduce the likelihood of error, which in turn helps improve the operational cost of the facility. There are a variety of HFE studies that can be performed depending on the project phase. The international Association of Oil and Gas Producers (IOGP) [35] has produced

a guidance in which the most common ones are described per project phase. Studies that can be performed during the basic engineering phase are for example:
- Working Environment Health Risk Assessment (WEHRA)
- Valve Criticality Analysis (VCA)
- Vendor package screening
- Task Requirements Analysis (TRA)
- Human Machine Interface (HRI) requirements analysis
- Control room requirements analysis
- Control system and alarm management analysis
- Safety critical task inventory
- Critical task analysis
- Human error ALARP demonstration

An HFE study called the "Working Environment Health Risk Assessment" (WEHRA) is discussed here as an example and is typically performed during basic engineering and detailed engineering studies.

The goal of the WEHRA is to identify work tasks that may be associated with potential risks to health. The purpose is to uncover these risks, investigate if they can be avoided with proper design, and suggest measures to reduce the risk to an acceptable level. The first step is the mapping and creation of an activity chart, where the nature of the activity, area of work, duration, and frequency of the work tasks are described (see Tab. 5.3).

Tab. 5.3: Example WEHRA activity chart for a single item.

No.	Area	Equipment	Activity	Duration	Frequency	No. of persons
1	Cellar Deck	Choke	Replacement of actuator	8 h	Two times per lifetime for each choke	One on board the installation

To identify the events/tasks that may lead to any potential health hazard, a ranking table is used. In this table, all activities involving personnel are listed and evaluated per working environment factor used. The scoring and evaluation are done in a meeting, where several disciplines can provide their input. The working environment aspects that are covered are:
- Noise
- Vibration (whole body, hand-arm)
- Ergonomics
- Indoor climate
- Outdoor operations/climate
- Lighting
- Radiation

- Chemical substances and preparations
- Biological agents
- Psychosocial and organizational conditions

A qualitative risk ranking is performed for each working environment factor (see Tab. 5.4). Evaluations are noted down and actions formulated for follow-up in the design.

For further reading, the reader is referred to the textbook by Edmonds [36].

Tab. 5.4: Example row of a ranking outcome, which is created for each activity in the WEHRA activity chart.

Noise	Vibration	Ergonomics/ access	Climate	Lighting	Radiation	Chemical	Biological	Psych./org.
2	1	5	2	1	1	1	1	1

5.4.4 Design for Six Sigma

The term design for Six Sigma is used in two ways. The first is just about making the design such that the Six Sigma (sigma stands for standard deviation) value of the product output is within the agreed specifications with the customers. This means that the chance that at any time the product is outside the specification limits is extremely small. This method is applied in the detailed design stage by using all statistical information on input parameters and choosing the design parameters such that this input variation is so limited that the product variation defined by Six Sigma is within the product specification.

The second way is a method to improve the quality of the generic business processes related to design and innovation used inside a company. It is derived from the quality improvement method Six Sigma. The latter is focused on improving existing business processes using statistical data and information.

For further reading, the reader is referred to textbooks by Cudney and Agustiady [37] and Yang and El-Haik [38].

References

[1]   Ulrich KT, Eppinger SD. Product Design and Development, 6th edn, New York: McGraw-Hill, 2016.
[2]   Akao Y. Quality Function Deployment: Integrating Customer Requirements into Product Design, Steiner Books, 2004.
[3]   "ISO 16355-1:2015". ISO. Retrieved 10 October 2023.

[4]   Pugh S. Total Design – Integrated Methods for Successful Product Engineering, United Kingdom: Addison-Wesley Publishers Ltd, 1991.

[5]   Guler K, Petrisor DM. A Pugh Matrix based product development model for increased small design team efficiency, Cogent Engineering, 8(1), 1 Jan 2021, 1923383.

[6]   Hedrich P, Brötz N, Pelz PF. Resilient product development-a new approach for controlling uncertainty, Applied Mechanics and Materials, 885, 20 Dec 2018, 88–101.

[7]   Jänsch J, Birkhofer H. The development of the guideline VDI 2221-the change of direction. In DS 36: Proceedings DESIGN 2006, the 9th International Design Conference, Dubrovnik, Croatia, 2006.

[8]   Birkhofer H, Jänsch J, Kloberdanz H. An extensive and detailed view of the application of design methods and methodology in industry. In Samuel A, Lewis W (Hg). ICED 2005 Proceedings, Australia, Melbourne, 2005.

[9]   Pelz PF, Groche P, Pfetsch ME, Schaeffner M. Mastering Uncertainty in Mechanical Engineering, Springer Nature, 2021.

[10]   Graessler I, Hentze J. The new V-Model of VDI 2206 and its validation, at-Automatisierungstechnik, 68(5), 27 May 2020, 312–324.

[11]   Singh MD, Joshi JG. Mechatronics, PHI Learning Pvt. Ltd., 7 Oct 2006.

[12]   Braungart M. Eco-effectiveness cradle-to-cradle design method, plenary lecture, Watervisie Congres te Rotterdam op 18 februari 2016.

[13]   Braungart M, McDonough W. Cradle to Cradle – Remaking the Way We Make Things, New York: North Point Press, 2002.

[14]   MacArthur E. Foundation, circular economy scheme. resourced from; from: https://www.ellenmacar thurfoundation.org/.

[15]   Cramer J. Circular cities webinar. resourced from: http://www.nzwc.ca/events/circular-cities-amsterdam/Documents/Webinar-JacquelineCramer-Mar14-17.pdf.

[16]   Ayres RU, Ayres LW. A Handbook of Industrial Ecology, Cheltenham USA: Edward Elgar Publ. Ltd, 2002.

[17]   Chertow MR. "Uncovering" industrial symbiosis, Journal of Industrial Ecology, 11(1), 2007, 11–30.

[18]   Wu Q. Industrial ecosystem principles in industrial symbiosis: By-product synergy. Chapter 14. In Harmsen J, Powell JB (eds). Sustainable Development in the Process Industries: Cases and Impact, Hoboken NJ: John Wiley & Sons, 2010, pp 249–264.

[19]   Magan A, Olivetti E. By-product synergy networks: Driving innovation through waste reduction and carbon mitigation, Chapter 6. In Harmsen J, Powell JB (eds). Sustainable Development in the Process Industries: Cases and Impact, Hoboken NJ: John Wiley & Sons, 2010, pp 81–108.

[20]   Benyus J. Biomimicry: Innovation Inspired by Nature, New York: William Morrow Paperbacks, 2002.

[21]   Baumeister D. Biomimicry Resource Handbook: A Seed Bank of Best Practices, Createspace, New York: Independent Publishing Platform, 2014.

[22]   Martin-Palma RJ. Engineered Biomimicry, New York: Elsevier, 2011.

[23]   List of solar power stations. sourced 24th of October 2023, https://en.wikipedia.org/wiki/List_of_ solar_thermal_power_stations.

[24]   Murzin DY (ed). Chemical Engineering for Renewables Conversion, Advances in Chemical Engineering, Vol. 42, 2013, 2–371.

[25]   Pike E. Opportunities to improve tire energy efficiency, The International Council on Clean Transportation, White paper 13, July 2011.

[26]   Dincer I, Rosen MA. Exergy: Energy, Environment and Sustainable Development, 2nd edn, Amsterdam: Elsevier, 2012.

[27]   Smith JM, Ness van J, Abbott H, Swihart M. Introduction to Chemical Engineering Thermodynamics, 8th edn, McGrawHill, 2018, Chapter 16: Thermodynamic Analysis of Processes.

[28]   Sankey-diagrams, http://www.sankey-diagrams.com.

[29] GEA. Global Energy Assessment – Toward a Sustainable Future, Laxenburg, Austria: Cambridge University Press, Cambridge, UK and New York, NY, USA, and the International Institute for Applied Systems Analysis, 2012.

[30] Smith R. Chemical Process Design and Integration, Chichester: John Wiley & Sons, 2005.

[31] Thulukkanam K. Heat Exchanger Design Handbook, 2nd edn, CRC Press, Boca Roca, 2013.

[32] Supertarget, http://www.kbcat.com/energy-utilities-software/supertarget.

[33] Aspen energy analyzer, http://home.aspentech.com/en/products/pages/aspen-energy-analyzer.

[34] Kemp IC. Pinch Analysis and Process Integration – A User Guide on Process Integration for the Efficient Use of Energy, 2nd edn, Oxford: Butterworth-Heinemann, 2007.

[35] IOGP. Human factors engineering in projects, Report, 454, 2011, Resourced from https://humanfactors101.files.wordpress.com/2016/02/human-factors-engineering-in-projects.pdf.

[36] Edmonds J. Human Factors in the Chemical and Process Industries: Making it Work in Practice, 1st edn, New York: Elsevier, 2016.

[37] Cudney EA, Agustiady TK. Design for Six Sigma: A Practical Approach through Innovation, Cleveland Ohio: CRC Press, 2016.

[38] Yang K, El-Haik BS. Design for Six Sigma: A Roadmap for Product Development, 2nd edn, New York: McGraw-Hill, 2009.

6 Scoping the design

6.1 Introduction to scoping the design

Scoping the design involves defining the design goal, its constraints, it scope, and its context. These set of actions are carried out in discussion with the client of the designer as described in general in Chapter 4. That chapter is well suited for experienced designers. This chapter, with its specific guidelines is particularly suited for students learning to design. The focus of this chapter is on the concept and feasibility innovation stages but advice for the other stages will also be provided.

6.2 Defining design goal and name

6.2.1 Purpose design goals and names

A design goal is a clear short statement on what should be achieved by the design. In a business environment this will be obtained after several rounds of discussion with the client. In design education it will, in a simple form, be provided by the teacher. The students should then redefine it, so that each design team member agrees to the design goal and also the teacher. The design goal statement is crucial in communicating what the design is about with anyone. A vague design goal will lead to a mediocre design. A clear design goal is a good start to a superior quality design. A good name will further enhance the communication inside the team and with others outside the team.

After the design goal is defined the design should be given a name. The discussion in the design team to generate and decide on the design name will also help create good team spirit.

6.2.1.1 Design goal setting

The design goal should be a very clear short statement that the internal client (business case owner) and all participants of the innovation group understand and agree upon. It is the shortest summary of what the design is about. For product concept design the goal can be formulated in terms of what functions the product has to have. An example is: Design a chewing gum that does not foul the streets. For process design it will state how much of what product has to be produced.

For developing a breakthrough innovation it is very important to define a challenging design goal statement. It should state the aim of the project in very concrete terms. It takes time to formulate this problem statement, but you only can work efficiently if you know what your goals are. If it is not concrete, sooner or later conflicts

https://doi.org/10.1515/9783110782127-006

will arise between people working in the concept stage and they cannot decide which is a better solution [1].

If the goal statement is not different enough from the goal statement that gave rise to the current commercial process than the end result of the whole innovation process is that you just rediscover what was good about the original process. All innovation work has then been in vain [1].

The goal statement should therefore contain a strong quantitative reason for change. The best way of achieving this is by explicitly referring to the best commercial process or product and stating in the goal and also in the constraints how much better the novel solution should be. This means that a good reference case needs to be defined. Section 6.4 discusses this subject in more detail. The main purpose of this challenging design goal is to stimulate design team members to come up with new solutions.

The design goal stated in the concept stage may be modified in the beginning of subsequent stages. For product designs this modification will often be needed, due to more insight from the interaction between the market desires and the product composition. For process designs, often the goal stays the same or is only slightly modified.

6.3 Defining the design scope (system levels, boundaries, and context)

A major choice when scoping the design is the choice of system level in which the design will take place. All system levels relevant to design and a conscious choice of the system level for which the design should be made are described in Chapter 2. This choice will involve a discussion with the client. Unconsciously a system level may already have been selected by the client. The designer can then show the other system levels to the client and suggest advantages for including other system levels into the design scope. In a designers' course the teacher may first choose a system level in which the students have to make a design and then chose a higher system level and again ask the students to make a design. In this way students become aware of the importance of choosing the system level.

Even within a system level, the boundaries for design still have to be chosen. The boundaries define which elements are outside the design objective and are treated as context (a given or a constraint). Also, these boundaries will be defined in discussion with the client. A simple example of a process design is whether to include storage of feedstock and product in the design scope or not.

After the system level for which the design will be made has been chosen, it is useful to define the other system levels that form the context of the design. If, for example, it is decided to take the chemical reactor and the catalyst level as design scope of the process then the other higher system levels will not structurally be changed by the new design. However, the size of mass and energy flow from and to the surrounding system levels may change due to the new reactor and/or catalyst design, if the amount of by-product relative to the main product is reduced.

6.4 Defining constraints (specifications)

6.4.1 Design constraints

A design constraint in general consists of type, a parameter, and its upper and lower limits. These constraints may also be called design specifications. A good starting point for defining constraints is to use the generic criteria check list provided in Chapter 4 and decide for each criterion what type of constraint and what quantified limits should be included. For a concept design by students the reduced constraint list SHEETS of Chapter 4 can be used.

In addition to these there will be constraints imposed by the company strategy. If the company has an explicitly written down innovation strategy, then defining the strategy-related constraints is easy. This may however be not the case. The written strategy may not be available, or the strategy may not be sufficiently defined to cover the specific innovation design and project. Once the design is made and presented a conflict with the implicit strategy may be revealed. Then the strategy may be further defined, and the design constraints adapted to this new situation.

> **Example**
> Here is an example from the author's experience. A team had to design a sustainable process for ethylene production. The design solution by the team was to take biomass as a feedstock. Ethylene was produced via fermentation and dehydration. When the design result was presented to top management, the reply was that an existing output stream from crude oil should be used as feedstock and that the sole use of biomass as feedstock was outside the company strategy. That latter statement was not written down till that time. Shortly after, it was included in the company strategy.

This example shows that also a strategy can be underdefined. A stage-gate approach in which concept design is presented to management in the discovery stage will quickly reveal this under-definition, resulting in a more defined strategy and thereby avoids large research and development efforts on projects that do not fit the business strategy.

In the discovery stage the constraints will be limited to that it should fit the company strategy and perhaps that potentially value for the company should be created. For the concept stage the minimum set of constraints that should be used is SHEETS:
- Safety
- Health
- Environment
- Economics
- Technically feasible
- Socially accepted

In addition, it is becoming more and more common to include a constraint on the contribution to sustainable development. A full set of constraints inside a company will be longer. The full set provided in Chapter 4 may be used as a checklist to make sure that no constraint type is forgotten.

It is good to consider however that at the start of the concept design the problem will still be underdefined and ill-defined. In some cases the problem is first defined by the client of the designer and during several meetings, due to the knowledge the designer, the problem is better defined. Some preliminary solutions may be presented to the client to test whether the problem is properly defined. If so, the problem is specified further into a goal and constraints (specifications). This subtle game between designer and client has been described by an experienced designer as: "Our job is to give the client, on time and on cost, not what he wants, but what he never DREAMED he wanted – and when he gets it he recognizes it as something he wanted all the time." [1]. Read also in Chapter 4 more about this subtle game. For students, learning that design problem definition is not just accepting what the teacher provides as the design problem to work on, but to redefine it, is very important. In education dedicated effort has to be made on this aspect.

For the feasibility stage a complete set of constraints should be defined and discussed with internal and external clients, as now the product and the process for commercial scale production has to be defined. The same holds for the development and detailed design stages. The table on all aspects relevant for a design as presented in Chapter 4 can be used to generate a complete set of criteria and can also be used as a checklist to determine whether an item was forgotten.

6.4.2 Product specifications

Product specifications are of tremendous importance in both product design and process design. In product design the specifications will be developed through all innovation stages by interaction with marketing and customers. It is crucial that the developed specifications reflect the customer needs, differentiate the product from competing products, and are technically as well as economically realizable. In the initial project stages the specifications represent the hopes and aspirations of the team, often without being constrained by what can be technologically achieved. During the course of the project the specifications are further refined to include actual technological constraints and expected production costs by making difficult trade-offs between desirable product characteristics.

For consumer products in particular it is important that the product to be developed is appealing to all senses. Here is the list of senses to ensure that for each sense product specifications are made:

- Sight
- Feel
- Hearing
- Smell
- Taste

6.4.3 Process specifications

Process specifications will also have to be made. These will start with product output specifications, which are mostly obtained from the business part (marketing and sales) of the organization. If the product is radically new then not all characteristics of the new product are captured in quantified specifications. Some product specifications are then likely to be defined/provided later on, when the product prototype has been tested by the client, resulting in additional specifications. The amount of product to be made will also be part of the output specification.

For radically new products with functional concept specifications, the process designer is advised to make multiple process designs based on assumptions on the product composition and discuss the process designs and their potential consequences for product performance properties with the product developer.

Input process specifications also start from the business part of the company that provides which type of inputs are allowed and desired and also which part of the supply chain is now taken inside the design scope. These pieces of information will be used to define the input specifications. In addition, the process concept designer will further define and refine the input specifications using the relations with the process output specifications based on yield figures. As a starting point a 100% theoretical molar yield of product on feedstocks may be used or empirically obtained molar yield data from scouting experiments. The designer should be aware that often the precise yield dimension (molar or mass based) is not reported in the experimental results. A check on the precise yield definition will be needed. How the output and input specifications have been obtained will also be part of the design report. Chapter 13 contains more details of what and how outputs and inputs of a process should be reported.

6.4.3.1 Identifying reference cases to improve design constraints

Identifying a reference case and defining constraints for the new design with respect to this reference case is a very powerful method to make sure that the final design solution is far better than the reference case. The reference case is the best existing commercially available product and/or process, upon which the design solution should deliver a significant improvement.

For a radically new product design defining a meaningful reference case can be complicated. Often the novel product has no single commercial reference product.

The reference cases may be obtained by observing the functionality of the novel product and then finding existing products fulfilling the same functionality as group, so that a combination of existing products is selected as the reference case.

For a novel process design the reference case selection is relatively easy. It should be the best process in commercial operation for the same product. If that process is in the company where the innovative design takes place then all information will be easily available. If the process is operated by a competitor, then information is less easy to obtain. In some cases the information can be gathered from environmental reports on that process that had to be made public to the local society.

For a novel product or process concept design it is important to set a few hard-to-meet constraints related to the reference case. An example is setting the constraint that the novel design should be better than the reference case by a factor 4. The reason for this requirement is the same as for defining the product goal, namely only setting challenging constraints for the initial new concept design will ensure that the final design in the detailed stage will still be superior to the reference case.

A final remark on constraints: It is not very important whether the statement on the requirement to improve significantly relative to the reference case is put under the design goal or in the constraint section, or in both. The only guideline is: the design goal should be short. The number of constraints should be many.

6.5 Generating basic design data

For a design, input information to facilitate the design will be needed. This type of input information is called basic design data. These data need to be generated. This is done by searching in literature and patent sources, by determining best estimates, by calculations, and by carrying out experiments. Not all this information will be generated prior to the design start. Some will be generated during the design, when it appears to be needed to complete the design. For each subsequent innovation stage more of these basic design data with increasing level of detail will be generated. The ultimate challenge here is to generate just the minimum amount of needed data required to make reliable decisions in each stage, as generating more will result in unnecessary project expenses for data that may never be used when a project will not be continued. This chapter provides guidelines on generating these data. The focus is on the first innovation stages up to and including development.

Basic design data includes not only data, but also information, knowledge, understanding, and wisdom as defined by Ackhoff [2]. Here is a short elaboration on the elements covered by basic design data:

–   **"Single discipline"** design data is data that can be easily understood by people skilled in that discipline, such as physical property data, safety data, and economic data. Single discipline data in general consist of a parameter (symbol), an amount (figure), a dimension, and may contain a reference. Examples are: the density of water is 1,000 kg/m^3 (at 277 K); a further distinction is between scale-independent data, such as physical properties, and scale-dependent data, such as investment cost, which depend on the scale (production capacity) of the design.

–   **"Information"** means relations between data. An example is reaction kinetics, providing information on the relation between reaction rates, concentrations, temperature, and catalyst type.

–   **"Knowledge"** means applying the data and information to the design. For example, selecting a best residence time distribution concept, such as plug flow, using reaction kinetics information for reactor concept design.

–   **"Understanding"** means appreciating why the design choices have been made. It is cognitive and analytical. It is about explicit reasoning why the whole or a major design is made as it is. It involves for instance the design choices made on a particular separation and reaction combination using safety and economic criteria.

–   **"Wisdom"** means evaluated understanding. It calls upon data, information, knowledge and understanding, and in particular, on moral and ethical codes. Wisdom is in general very little described in basic design data, but when it is done, it is of enormous importance.

Example

An example from the author's experience on the role of wisdom is a design for shale mining and oil production in Morocco in the 80s by the company I worked for. One of the design questions was, what are the environmental constraints? It appeared that the Moroccan government had a general set of environmental emission constraints for business activities. Experienced process designers in the company felt unsecure about this set and initiated an investigation into the local environment for the mining and the conversion process. The investigators reported that a sanctuary for migrating birds was near the spot of the mining and that this could be disturbed by the mining and processing activities. The company then decided to make sure that this would not happen and also decided to apply European environmental standards for the design – the reasoning being that causing environmental problems in Morocco could damage the company's global image of a being a responsible business.

It is very important that generated data are reported, discussed, stored, and made retrievable to all project members. Usually such a data base can best be established in the concept stage. Chapter 7 deals with the subject of reporting, storage, and retrieval in detail. Table 6.1 provides an overview of potential sources to generate basic data relevant for chemical product and process design. In the following sub-chapters, guidelines for basic design data generation in each innovation stage are provided.

6.5.1 Ideation stage

Main objective in the ideation stage is to identify which data are available and need to be obtained when deciding to pursue an idea further in the concept phase. Data generation in the ideation stage is therefore limited to those that can be easily obtained from literature sources or simply and quickly derived using simple equations and correlations or best guesses made by an experienced researcher or process development engineer. Physical properties, safety, health, and environmental data will be obtained from literature sources. If not available, educated guesses will have to be made. Consulting experienced employees of the company is often the quickest way of obtaining the required data and information. Economic data such as earning power can be derived by subtracting feedstock cost from sales prices; for more information, see Chapter 10 on economics.

6.5.2 Concept stage

The main challenge in the concept phase is to just generate the required data to be able to decide on the most promising design concept(s). There is a huge risk in wanting to generate lots of nice-to-know data, which are in the end discarded because a different concept is chosen or the project is abandoned as a whole. In the concept stage quick decisions have to be taken and the budget is limited, so the emphasis should be on rapidly finding these data, even at the expense of accuracy and reliability.

Some physical chemistry data are always needed in the concept stage. Crucial lacking physical and physical-chemical data may be generated by specific experiments. These experiments can be done in-house but specific experiments such as physical basic data can also be obtained from outside specialized laboratories. Alternatively modern flow sheet computer packages that have a library of physical property models can be used to quickly generate physical properties. Care should be taken in using these models. It needs a specialist to choose the best model for a particular property. It may also involve additional experiments to validate the calculation results. More detailed physical chemistry data measurement and modelling studies are usually performed in the feasibility stage.

Generation of chemical kinetics data in the concept stage for process concept design is in general restricted to:
- the chemical stoichiometric equations of the main reactions involved;
- order of magnitude reaction times required for specified temperature and concentration windows;
- experimental results of proof of principle experiments showing products made and yields of product on feedstocks for a particular catalyst, some temperatures, and reaction times.

Economic data in the concept stage is in general limited to purchase prices of feed-stocks and sales prices for products. For new products these prices will not be available. Pricing new products in the concept stage is an important competence for innovative product companies. The reader is referred to the classic article by Dean [3] and to a McKinsey article [4] to get some insight into the art of new product pricing. Detailed methods for economic evaluations of novel processes and products are found in Chapter 10. Here we simply state the following guidelines:

1. Estimate the new unit product value for the customer and do not base it on manufacturing cost.
2. Determine the feedstock cost per unit of product.
3. Estimate the earning power EP_0.

Safety, Health, and Environmental (SHE) data are also very important firstly for a rapid analysis of preliminary concept designs and for evaluating concept designs when finalized. The product or a process that is inherently unsafe will not pass the concept stage-gate. A product or process with no SHE information will also not pass the concept stage-gate, or if does then it will not pass the feasibility stage-gate.

Tab. 6.1: Sources for basic design data.

Data type	Reference books	Web-based sources
Physical chemistry data	– *The Properties of Gases and Liquids*, McGraw-Hill, 2004 – *Materials Science and Engineering Handbook*, 4th edition, CRC, 2004 – *CRC Handbook of Chemistry and Physics*, 97th edition, 2016 – *Dechema Chemistry Data Series*, Dechema (2004)	– Properties of organic compounds – Chemexpert Chemical Directory – NIST Chemistry Webbook – Beilstein online database – Knowledge center (Bio) Chemical Engineering

Technology and scientific information	– *Ullman's Encyclopedia of Industrial Chemistry*
	– *Kirk Othmer's Encyclopedia of Chemical Technology*
	– *Encyclopedia of Chemical Processing and Design*
	– *Riegel's Handbook of Industrial Chemistry*
		– www.sciencedirect.com
		– www.scopus.com
		– SciFinder
		– Wikipedia
		– www.google.com
		– Patents
		  – http://nl.espacenet.com/
		  – www.patsnap.com
		– www.nexant.com
		– www.ihs.com
		– www.intratec.us
		– www.cheresources.com

Tab. 6.1 (continued)

Data type	Reference books	Web-based sources
SHE data	– *BP Process Safety Series* – *Handbook of Chemical Health and Safety* – *Handbook of Chemical Compound Data for Process Safety* – *Chemical Safety Handbook* – *Property Estimation Methods for Chemicals* – *Handbook of Environmental Health* – *Handbook of Reactive Chemical Hazards*	– KCCE: databases – ChemWatch (product safety) – INCHEM (chemical safety and environmental health) – International chemical safety card (by CAS number) – MSDS – http://chemiekaarten.sdu.nl/chkonline/ – http://hazard.com/msds – http://www.ilpi.com/msds – EU Directives and Law (for industry) – http://eur-lex.europa.eu/ – http://ec.europa.eu/enterprise/ – www.msdssearch.com
Economic data (market, pricing)		– http://www.icis.com/chemicals – www.nexant.com – www.ihs.com – www.orbichem.com – www.intratec.us – www.business.com/directory/chemicals – www.alibaba.com
Equipment cost	– *Product and Process Design Principles*, 3rd edition, John Wiley & Sons (2010) Chapter 22 – *Plant Design and Economics for Chemical Engineers*, 5th edition, McGraw-Hill (2004) Chapters 12–15 – *Dutch Association of Cost Engineers*	– www.matche.com/equipcost/ – www.mhhe.com/engcs/chemical/peters/ data/ce.html – www.equipnet.com – www.ipps.com – www.alibaba.com

6.5.3 Data generation feasibility stage

Data generation in the feasibility stage is focused on detailing the required information for the selected design to such reliability levels that it can be used to establish a feasible commercial scale product and/or process design and when needed also a pilot and/or demo plant design. Alternative technical options for each process step will be contemplated. Selection criteria will be generated, as well as best options selected. Equipment will be sized and stream compositions calculated. All these data need to be documented, stored, and made retrievable. For a plant design inside existing processing site it is important to have all local plant and site-related boundary conditions documented. For reliable experimental physical property generation and modelling, large enterprises will have their own specialist group. Smaller companies will turn to specialized research institutes to obtain key pieces of information. Similarly much more detailed and reliable chemical kinetics data need to obtained, as the best reactor concept with its connected separation concept plus recycle can be designed only with sound kinetics. Here is a checklist of required chemical kinetic data:

- (Bio)chemical reactions
- Possible reactants/pathways
- Stoichiometry
- Conditions (pressure and temperature)
- Yields, selectivity versus conversion
- Product distribution
- Side reactions
- Reaction rates
- Equilibrium
- Phases (gas, liquid, multi)
- Exothermic/endothermic
- Residence time, space velocity

Also, market and economic information will be detailed further to establish the business case of the project and assess its economic viability. Important aspects to be included are:

- Local society and cultural dimensions of market
- Market share, competitive position
- Desired production rate
- Desired product purity
- Product sale price versus purity specifications
- Feedstock materials compositions and their prices
- Location
- Climate
- Depreciation agreement or policy
- Feedstock prices
- Product sales prices
- Investment cost estimate
- Utilities availability, specifications, and price
- Process cost structure
- Raw materials availability and composition

A minimum set of safety, health, and environmental data for the feasibility phase comprises:

- Safe operating window for the process parts
- Toxicity of all components applied for product and process

- Life cycle assessment of environmental impact
- Waste disposal options
- Waste streams/emissions
- Environmental legislation

Chapter 11 provides comprehensive information on safety and health data and Chapter 12 provides sustainability data needed for evaluation.

6.5.4 Development stage

During the development stage a large amount of data for many design object aspects has to be generated on top of the already generated data in previous stages. These data concern:
- The commercial scale product and process and their related product prototype and pilot and/or demo plant
- Data generated by prototype and pilot plant testing to validate the feasibility of the novel product and process
- The marketability of the novel product
- Intellectual property protection

Table 6.2 provides a checklist of items for which basic design data have to be generated in the development stage.

Tab. 6.2: Basic design data items of the development stage.

Item	Obtained from
Safety performance: product and process: see Chapter 11	Literature and experiments
Health performance product and process: see Chapter 11	Literature and experiments
Environmental performance LCA data: see Chapter 12	Literature
Sustainable development performance: see Chapter 12	Literature
Economy performance: see Chapter 10	Calculations
Product technology data	
Process technology data	
Technology performance process	Pilot plant test results
Technology performance product	Prototype product test results
Product market data	Market surveys
Product sales data	Market and client surveys
Intellectual property	Design rights, trademark, patents, copyright

Process performance data

Phenomena that cannot be predicted from modeling are: corrosion, fouling, foaming, catalyst decay, solvent decay, and others. These potential problems should be part of the reasoning for having the pilot plant; see also Chapter 3 on reasons for having pilot plant. The pilot plant and the testing program should be such that observations and measurements on corrosion, fouling, foaming, and catalyst decay can be made.

ⓘ 6.6 Exercises

Exercise 6.1: Design goal, scope, and constraints
(1) State whether the following statements are true or false:
   (a) A weak problem statement could lead to the rediscovery of conventional processes.
   (b) Only a concrete problem statement can lead to breakthrough innovation.
   (c) Company strategies are always written down prior to concept design.
   (d) SHEETS are the minimum set of constraints that need to be used for defining the scope of design.
(2) Choose the correct order of activities in a design scoping exercise is:
   (a) Choose system level, choose boundaries, choose context.
   (b) Choose boundaries, choose system level, choose context.
   (c) Choose context, choose boundaries, choose system level.
(3) Assign whether the following are related to product or process specifications.
   (a) Sweetness of Coca Cola
   (b) Redness of powder coating
   (c) Texture of check stick for dogs
   (d) Greenhouse gas emissions
   (e) Sustainability

Exercise 6.2: Design data
(1) For data types listed below, assign in which stage (Concept, Feasibility, Development) information/knowledge/data are definitely needed:
   (a) Experimental proof of principle of reactions (yields, temperature, catalyst)
   (b) Side reactions
   (c) Reactor residence time
   (d) Scale-up
   (e) IP
   (f) Ballpark estimates of reactions in operating window of pressure and temperature

(2) Match the following references to data type:
   (a) Ullman's Encyclopedia
   (b) CRC Handbook of Chemistry and Physics
   (c) Handbook of Chemical Health and Safety
   (d) Alibaba website
   (e) Dutch Association of Cost Engineers
   (f) Orbichem website
(3) State whether the following statements are true or false:
   (a) When local regulations/legislations regarding environment are general, it is better to always stick to the European environmental policies for processes.
   (b) In the development stage it is common to obtain data through experiments in case they are not available from literature.
(4) Which among the following are part of the SHE data?
   (a) Earning power
   (b) Product requirements
   (c) Toxicity of components
   (d) Environmental laws
   (e) Catalyst deactivation rate
   (f) GWP
   (g) MAC values

References and further reading

[1]   Dorst K. Understanding Design, 175 Reflections on Being an Designer, Amsterdam: BIS Publishers, Amsterdam. 2006.
[2]   Ackoff RL. From data to wisdom, Journal of Applied Systems Analysis, 16, 1989, 3–9.
[3]   Dean J. Pricing policies for new products. Harvard Business Review, 54(6), 1976, 141–153.
[4]   Marn MV, et.al. Pricing new products, McKinsey Quarterly August 2003, Resourced from: https://www.mckinsey.com/business-functions/marketing-and-sales/our-insights/pricing-new-products. Last accessed: 30 November 2023.

7 Executing designs

7.1 Introduction to executing designs

This chapter describes methods and heuristics for executing novel product and novel process designs for innovation classes 4 and lower, as defined in Chapter 3. Its focus is on designs for discovery and concept stages and is particularly useful for MSc students and young professionals. For later stages, Chapter 10 on design optimization should be consulted. Figure 7.1 shows all major steps in executing any design in any stage. All steps (except scoping, which is discussed in Chapter 6) are discussed in this chapter.

The synthesis step is about generating design ideas and options. All kinds of methods are available to generate these options for product and process design. The designer should feel free to generate any idea. In Section 7.2, this subject is treated in detail.

The analysis step is about a rapid initial assessment of the generated ideas. This subject is treated in detail in Section 7.3.

The design steps synthesis and analysis have strong interactions. An initial synthesis result is analyzed by how well it meets or fits the design scope. The difference between the analysis and evaluation steps is that in the analysis step, the synthesized design options are often modeled and qualitative and/or quantitative data generated (e.g., mass and heat balances) and simulation outcomes are compared with the design performance requirements. This subject is treated in Section 7.4.

Evaluating the design means that the whole design is first checked on meeting all essential criteria. Second, the design is checked and adjusted to meet the desired criteria in a balanced way. This is described in Section 7.5.

Finally, the design is reported in writing, so that it can be used for the stage-gate assessments and as a starting point for the next stage. This is described in Section 7.6.

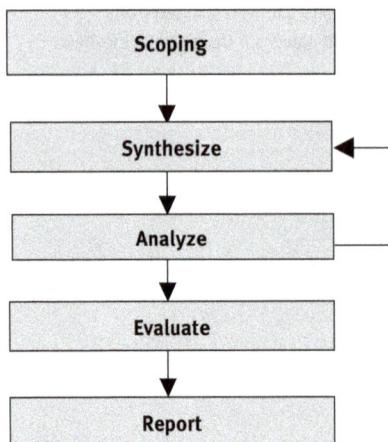

Fig. 7.1: Design steps.

https://doi.org/10.1515/9783110782127-007

How to analyze and optimize a selected design by making models and simulating a design performance is covered in detail in Chapters 8 and 9. Specific heuristics on how to evaluate a design for the important criteria of safety, health, environment, economics, technical feasibility, and social acceptance are found in Chapters 10–13. How to report on heuristics is found in Chapter 14.

7.2 Synthesizing preliminary design solutions

Generating preliminary solutions by design is called (design) synthesis. This synthesizing can be done in many ways. For combined products and processes synthesis in general, heuristics information is provided in Section 7.2.1. Heuristics for synthesizing general products and specific structured products are provided in Section 7.2.2. Heuristics for synthesizing processes are provided in Section 7.2.3.

7.2.1 Synthesizing design solutions using heuristics

Synthesizing design options in the ideation and concept stage is about generating preliminary solutions. There are many ways to generate preliminary solutions, which are called heuristics. These can be divided into more general heuristics that contribute to generation of the input for preliminary design solutions and more specific heuristics that can be used to guide the team in making choices between alternatives.

7.2.1.1 General heuristics
Here are some practical, general heuristics to generate (synthesize) options for preliminary design solutions:

A: Look at the design scope generated by the methods of Chapter 6. Read the design goal. Then scribble down any idea that comes to mind that may partly meet the design goal. This scribbling may be a simple sketch.

Ask yourself: what does the solution have to do to meet the design goal? Write down what the solution must do. Verkerk et al. [1] calls this the qualifying function. Do not bother about determining how the solution could work. If you think you need more than one function to meet the design goal then write down those functions.

B: Look for information on existing solutions (literature, patents, interviews of people) that may partly fulfill the qualifying function. Write down these partial solutions. Deconstruct existing solutions into constituting functions; so make an abstract description of functions.

C: Fill in some columns on the left-hand side of the Delft Design Matrix of Chapter 4 with keywords.

D: Put the design aside and do something completely different. Anytime an idea comes to mind, note it down directly in your smartphone, voice message, or a piece of paper at hand.

E: Organize brainstorming, or brain writing sessions (or any other creativity method as proposed by Tassoul [2]) to generate solutions.

F: Use any of the following words: substitute, combine, adapt, modify, put to other use, eliminate, or rearrange part-solutions to improve the solution [3].

G: Generate families of (part-)solutions with each family having the same "ancestor." An ancestor is the common element of one family of solutions [3].

H: Apply the "TRIZ" creative problem-solving methodology [4]. This methodology is particularly fruitful for new product designs. It provides 40 principles, all of which can be used to generate new product design options. It is good in generating partial solutions first. Later these serve to generate complete product designs.

7.2.1.2 Specific heuristics

Specific heuristics are based on the knowledge (rules of thumb) of experienced designers that was developed from their findings in similar projects from their past. Since 1950, these rules of thumb have been collected and combined into more generally applicable heuristics. For examples of heuristics that can be applied in process synthesis the reader is referred to Section 7.2.3. Many more heuristics with varying level of detail can be found in handbooks on design or rules of thumbs.

However, before using heuristics it is important to take note of what a heuristic is and where it came from. The definition of a heuristic is:

A heuristic is a rule of thumb which is usually true.

The main takeaway from this definition is that the designer should always ask her-/himself what makes the heuristic true for the design problem under consideration. Based on this evaluation a justified decision can be made whether or not to apply the heuristic. Often you will also find contradicting heuristics. When applicable, their use is perfect for the generation of design alternatives, so always consider all applicable heuristics and evaluate their impact on the possible designs. Ignoring part of them will create the high risk of a suboptimal design result.

When applying specific heuristics it is important to recognize that most of them have been established during the development of the fossil-based industries. During those ages energy costs were usually much lower than capital-related costs. Given the current developments with respect to climate change and renewable energy sources, today's designers will encounter many situations where the straightforward application of established heuristics do not yield the best design solutions. A clear example is that due to the higher prices of renewable energy much more investments are possible for energy saving/recovery compared to 10 years ago. We cannot stress enough how important it is to apply existing heuristics with great care and use them to learn why choices in design solutions today differ from those made in the past.

7.2.2 Synthesizing product design

This section is about product synthesizing design in the ideation and concept stages using simple heuristics. For novel product design of innovation class 4 all heuristics of Section 7.2.1.1 can be applied. When the new design is a modification of an existing product a lot can be learned from deconstructing the existing product into functions.

Example 7.1 Chewing gum product design

Here is an example from a group of students that for the first time in their life had to make a product design. The design goal was to design chewing gum that would not foul the streets. They looked for information through literature searches, patent searches, and by asking a chewing gum manufacturing company to provide publicly available information about chewing gum. They deconstructed the conventional chewing gum into the gum that has an elastic plastic chewing function and taste additions. They concluded that the plastic gum of polypropylene is the part that fouls the street. It stays on the street for 28 years on average if it is not removed.

They then looked for a polymer material that would depolymerize into water-soluble components and found a polyvinyl alcohol with the desired hydrolyzation rate. They also made the chewing gum attractive by putting the taste components in micro globules. These depolymerized in the mouth and hence the taste components were slowly released into the mouth [5].

7.2.2.1 Synthesizing design: physically structured products

Structured products are different from simple homogenous products. Homogeneous products are uniform down to the molecular scale. Structured products are homogeneous down to a certain size often down to a few microns and are then heterogeneous. An example is mayonnaise. It is an oil-in-water emulsion. The dispersed droplets have the size of 2–10 μm. (Discrete products such as cars and devices are heterogeneous on every scale, and are not treated in this chapter.)

Simple homogeneous products can be gases, liquids, and solids, such as polymers and ceramics. Particulate matter such as crystalline powders also qualifies as homogeneous products. Physically structured products are, for instance, emulsions, foams, and gels (see Chapter 8, Tab. 8.5 for more examples). They have one or more thermodynamic phases, with at least one of the phases having distributional features at the microscales. Some chemical species in the product may have a distributed molecular structure (e.g., polymeric chain length distribution). In addition, the product may be made up of some coexisting thermodynamic phases with a geometrical distribution on a microscale, such as particle size- and/or shape distributions. Structured products have properties that are related to both thermodynamic product phases: the chemical composition of these phases as well as their internal spatial distributional structure (e.g., droplet size distributions in emulsions). Designing physically structured chemical products and a more precise characterization of the inner structure are presented in the following section.

7.2.2.2 From functional product specification to product structure

The complexity of the internal product structure arises from the presence of multiple constituents and from the distributive properties of its constituents. Examples are:
– supramolecular structures such as polymers with molecular chain length distribution
– a thermodynamic phase present as particles or droplets of varying size
– dispersion of multiple phases

The joint effect of all distributive and lumped properties of the internal structure determines the practical performance features of the product. These performance features (and their product costs) determine the competitiveness of the product on the market in meeting customers' expectations. An example of practical performance requirements for a structured product such as a crystalline material is given by Bermingham [6]:
– For downstream handling: ability to filter, to wash, to dry, and to let flow, freedom from dust, mechanical strength;
– For customer applications: no caking in storage, dissolution rate, mechanical strength, porosity, freedom from dust, aesthetic appearance;
– For food, some practical performance requirements [7, 8] are, in order of importance: safety > nutritional value > taste > smell > mouthfeel.

Cussler and Muggeridge [9] and Wesselingh [10] provide more examples.

When developing and designing a new product and its associated manufacturing process in industry, an approach from market analysis to process design and back to customer expectations, as indicated in Fig. 7.2 can be of help.

The second box: product functions and performance requirements are about transforming customer wishes and expectations into functions (what has to be achieved). To support such transformations, a systematic procedure has been developed for product design in general. This procedure is called "House of Quality" [11], also known as "Quality Function Deployment." This method is used here to transform (deploy) in steps to the bottom box of Fig. 7.1.

In the second box, a transformation is to be made from product functions to the selected components that will comprise the product. These components can be of a physical, chemical, biological, mechanical, or electronic nature. During product design

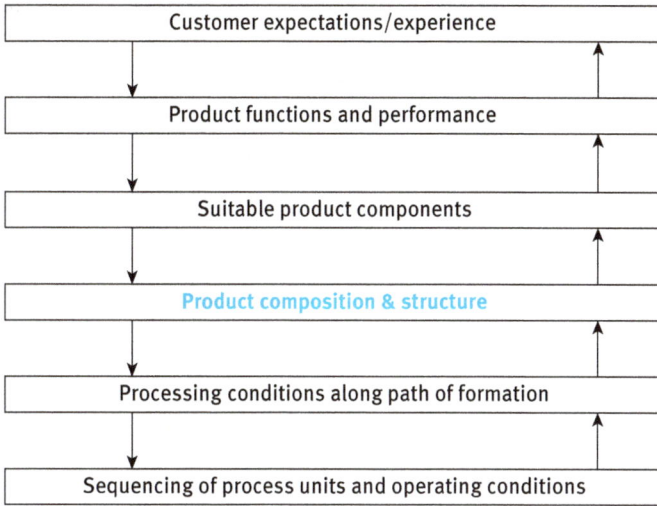

Fig. 7.2: Product and process design information downflow and physical causal effects upflow.

and development, the product functions are materialized into a physical structure for the product.

The bottom-up arrows in Fig. 7.2 show the physical cause and effects. The selection of process units with associated operating conditions and equipment design and their sequencing into a process flowsheet determine the path of processing conditions inside the process units when going from feed to product. The processing conditions involve the distributions of thermodynamic phases, pressure, temperature, concentrations, residence times, and so on. The processing conditions along the path from feed to product influence the formation of the internal structure of the product. This internal structure involves chemical composition per thermodynamic phase, the spatial dispersion of the phases, and any internal distributive properties of the thermodynamic phases of which the product is made up (e.g., droplet size distribution).

The interactions between the "down" flow of design information and the "up" flow of physical cause and effect indicate that having proper information about the physical and chemical structure of a chemical product is essential to be able to synthesize the processing plant. The internal product structure is truly the conceptual linking pin between product design and process design. Knowing such a structure enhances the effectiveness of the design process as the physical features of the product serve as a functional specification for the process design.

Three central physical features are provided by Cussler and Moggridge [9, p. 130] for a chemical product:

1. Structural attributes, involving physical and mechanical properties based on internal product structure

2. Changes in thermodynamic equilibrium, induced by external variations such as temperature, and acidity (pH)
3. Key rate processes that can occur (reactions, mass, and heat transfer, fluid flow)

These features capture "internal product structure" by means of four main items (pages 129–130):
1. Chemical composition
2. Physical geometry
3. Chemical reactivity
4. Product thermodynamics

When putting a product to its application, it is necessary to trigger the product activity by means of external controls, such as:
- Solvents
- Temperature change
- Chemical reactions
- Physical changes, like pressure, electrical field, and surface tension

For example, the acidic environment of a stomach triggers the decomposition of the coating of a drug and then its chemically active substance is released.

The reader is invited to also use the Delft Design Matrix of Chapter 4 to further synthesize the product and its process, as it is noticed that in all product design books, the outline of the internal product structure is short and informal. These outlines are not sufficient for conceptual product and process design purposes. It is especially not so when process simulations are used, requiring a quantitative representation of the product structure. In Section 8.4.3, a mathematical modeling approach for multilevel structured products to analyze the product concepts is given.

7.2.3 Radically new process synthesis design

7.2.3.1 Introduction to process synthesis concept design
To meet sustainable development goals often radically new processes are needed, based on renewables or based on recycled materials as inputs, with no waste stream outputs, requiring far less energy, which are inherently safe, and low cost. Here a generic process concept design method is presented that has been proven to result in superior process designs, which furthermore have been implemented at commercial scale.

7.2.3.2 Process synthesis function design
The design method described here is based on Siirola's process synthesis method [12]. However, the first step is obtained from Douglas process concept design [13], and the

author has added his experience in presenting function design and in the use of streams.

All steps are summarized in Tab. 7.1.

Tab. 7.1: Functional process concept synthesis design.

Step	Description
1	Draw process box, define output streams, and input streams compositions
2	Define reaction function block as island inside process block
3	Define separation function blocks
4	Select streams for solvent, extraction, stripping, separation functions
5	Define heat exchange functions
6	Define form function to meet product form requirements
7	Integrate functions

Step 1: defining input and output process design streams is getting the product stream composition (specifications) and the input stream composition. In concept design these compositions will in general not be known completely. But desired product purity for some components will be known. The same holds for the input composition. The precise process capacity will not be known. A reasonable guess of the desired process capacity will be good enough. By simple mass balances the input can be adjusted to fit it to the design output composition and mass flow sizing.

It is useful to draw a process box to represent the whole process with arrows to and from this box representing the input and the output flows. This box helps the design and research team enormously in focusing on what are the defined input and outputs.

Each function within the process box is also represented by a box with input and output streams as arrows. Function boxes are thus connected by these stream arrows.

Step 2: If the molecular composition of the product is different from the input, then a chemical reaction will be needed to eliminate this difference. Hence, a function chemical reaction is needed in the process. This reaction function can be placed as a block inside the input-output box as an island for the time being.

If for the reaction certain chemicals are needed to be mixed a function mixing should be stated as a block mixing in front of the reaction block.

Step 3: If the reaction function output composition is different from the desired process product output, then a separation function can be defined to eliminate that difference. The other output stream of de separation is just drawn at first instance.

It may be recycled to the mixing block or to the reactor block, or to a second process output stream if it meets the desired composition of a second process output stream. Or a second separation step can be defined to meet the second process output stream.

In general, steps 2 and 3 are the most important steps as these two steps determine the product yield on feed. If for instance the product reacts further to undesired by-product then a low single-pass conversion followed by product separation and re-cycle of unconverted feed will reduce this by-product formation enormously.

Here are some heuristics (HR) obtained from Levenspiel [14], Seader et al. [11], and Douglas [13] about reaction and separation selection functional concepts:

HR1: For a single irreversible reaction, any order in feed component except zero; choose a plug flow concept reactor.

The background to this heuristic is that the reactor volume required is lowest for a plug flow residence time distribution reactor.

HR2: For a single autocatalytic reaction such as in fermentation choose a back-mixed concept reactor function.

The background to this heuristic is that in a back-mixed concept the product concentration is high throughout the reactor, so that the autocatalytic effect is maximized.

HR3: For a single exothermic reaction choose back-mixed reactor followed by plug flow reaction function.

The background to this heuristic is that back-mixing facilitates heating up the feed inside the reactor. Plug flow facilitates deep conversion in a small volume.

HR4: If an unwanted reaction is of high order in feed component B, then distribute feed B along plug flow reactor.

The background to this heuristic is that in this way the concentration of B is kept low and thereby the unwanted reaction is kept low.

HR5: If a consecutive reaction of the product to an undesired waste component occurs, choose a plug flow reactor with small conversion followed by separation and re-cycle unconverted component back to reactor.

HR6: If the product inhibits the reaction; chose a plug flow reactor concept with small conversion followed by separation and recycle unconverted component back to reactor.

HR7: Remove hazardous, corrosive and troublesome components first.

HR8: Favor separations that directly match the products and/or divide the feed as equally as possible.

HR9: Remove components in order of decreasing percentage of the feed.

HR10: Do the easy separations first and process zeotropic (low relative volatility) mixtures at the end.

Step 4: This is about the choice of media. If a solvent is needed for instance for keeping the product dissolved, then any process stream of the process can be considered for this function.

This choice of media (solvent, absorption liquid, etc.) is often not considered in process synthesis. Often a medium is chosen for lab-scale experiments from available solvents, be it for dissolving a solid solvent or for crystallization medium, an absorption medium. For stripping, often nitrogen is chosen as stripping gas. But these media have all kinds of adverse effects on cost, safety, health, and environment. Additional separation to recover the media will be needed, as well as additional storage. Diffusive emissions to the atmosphere will increase due to the additional equipment with their flanges and pumps.

It is far better to consider using available process streams such as feed streams, product streams, and process streams inside the process for a solvent, absorption liquid, or crystallization medium. In bulk chemicals production where processing costs are important, only available media inside the process are used, so it can be done. Here are some media heuristics (MH):

MH1: Need a solvent: Consider all process streams including feedstock and product streams to be used as solvent.

The background to this heuristic is that if a solvent is introduced additional cost of storage and additional cost on Safety, Health, and Environmental measures will be needed.

MH2: Want to strip a component from a stream: Consider using distillation or an available process gas stream for stripping.
The background to this heuristic is that if a stripping agent is introduced then that stripping agent will need to be cleaned for reuse, so additional cost will be involved.

MH3: Want to wash a component: Consider using distillation or extracting the component with an available process stream.
The background to this heuristic is that if an additional washing medium is applied then that medium needs to be cleaned for reuse, so additional cost will be involved.

Step 5: Now temperature ranges, pressures ranges, and phases are selected to obtain feasible reactions and separations. To that end separation functions are more defined; for instance, distillation or extraction is selected, and simple flow sheeting is performed to define the function in some detail.

Step 6: Eliminate the difference between the required product form and the product output stream defined by step 1. An example is a desired plastic in the form of particles, while the product form from steps 1–5 is a (hot) paste. The functions required are then a forming step and a cooling step.

Step 7: Now options to combine and integrated functions should be explored. This can be combined with the use available streams as a solvent or extractant, or a stripping agent, so use step 4 again.

Figure 7.3 shows a generic functional process design, for a continuous process, where also mass transport is needed, so this is also defined as a function.

It will be clear from the description of steps 2–7 that this design method is not about a simple linear step-by-step approach but needs several iterations to obtain a good process concept design or a few process concept design alternatives.

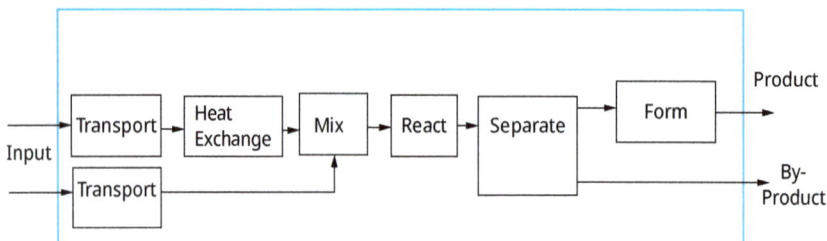

Fig.7.3: Process functions design example.

Example 7.2 Function block flow sheet for structured brainstorm sessions

The author used function block flow diagrams facilitating a structured brain writing session. The company involved wanted the best process separation transforming a complex feed with molecules of type, A, B, C, and D into 4 products. Each product should contain A, B, C, or D with a maximum specified trace amount of the other feed components. The company also wanted to know all feasible process concept designs, so that in the end a well-defendable patent could be written.

I draw all conceivable process block flow diagrams based on binary separations, one of which is shown in Fig. 7.4. With simple logic, 15 different block flow diagrams could be defined with total 25 different binary separation blocks.

7.2.3.3 Designing and selecting unit operations for functions

For each function block defined in the previous section a unit operation can be selected in the concept stage. However, in the feasibility stage or in the development stage, different unit operations for the functions may be selected. For this latter selection, experienced process engineers will be involved. In the concept stage, just a preliminary selection often by young chemical engineers will be made to facilitate investment cost estimation and to communicate the technical feasibility of the process concept.

For unit operation selection Tab. 7.2 can be of help. A comprehensive source of unit operations is Perry's Chemical Engineers' Handbook [14]. However, newly established unit operations based on process-intensified principles, as found in the book by Stankiewicz should certainly be considered [16]. Here some practical advice and heuristics are provided for unit operation selection.

For the mass movement function, a pump can be chosen for liquid transport and for gas a compressor can be chosen. However, other options should also be considered. For liquids, downflow by gravity is an option. If the liquid stems, for instance, from the top part of a distillation, then the liquid can flow to other places by gravity. Gas flow can be induced by a pressure difference created by evaporation.

Tab. 7.2: Functions and technical and unit operation options.

Function	Unit operation options
Mass movement	Gravity, pump, compressor
Mixing	Static mixer, mechanical mixer
Reaction	Reactor
Separation	Distillation, extraction, absorption
Enthalpy change	Evaporation, condensing, heat exchanger
Forming	Extruder, Prill tower

For selection of a unit operation for separation, the following heuristic (HU) is available from PDC [17], Seader et al. [11], and Douglas [13]. Start from the top of the list and consider whether it is applicable. If it is not applicable go down to the next option, and so on.

HU1: When a phase separation is feasible because the feed has two phases, choose decantation.
HU2: If two phases can be generated by a temperature shift: choose a flash, condensation, or distillation.
HU3: If a high product purity is needed and not a high yield: choose
  – membrane separation,
  – liquid-liquid extraction,
  – extractive distillation,
  – adsorption.

HU4: When a solid product is desired, use crystallization.
HU5: Favor condensation to remove high boilers from non-condensables.

For the function-integrated blocks established process intensified unit operations may be suitable. The book by Stankiewicz contains several of these [16]. A classic example is reactive distillation, combining reaction and distillation.

If established unit operations for the function integration are not available, then a special design can be made. This is best done by a team of chemical engineers and mechanical engineers by drawing options and making models predicting the hydrodynamic behavior and mass and heat transfer and reaction behavior. Experimental testing may also be involved.

For reactive distillation, the following heuristics (HRD) are available [18]. Reactive distillation is feasible when:

HRD1: The reaction time is less than 30 min and the distillation temperature profile is compatible with the reaction temperature. By changing the distillation pressure, the distillation temperature may be changed to obtain the desired temperature profile for the reaction. For reactive distillation hydrogenations applied in refineries was used.

HRD2: There is a consecutive reaction forming an unwanted by-product.

HRD3: The reaction is an equilibrium reaction.

HRD4: The distillation has an azeotrope.

Example 7.3 Biofuel process with function integration and use of streams

A process design containing function integration and the use of streams is the biofuels BTL process [18]. Its process block flow diagram is shown in Fig. 7.4.

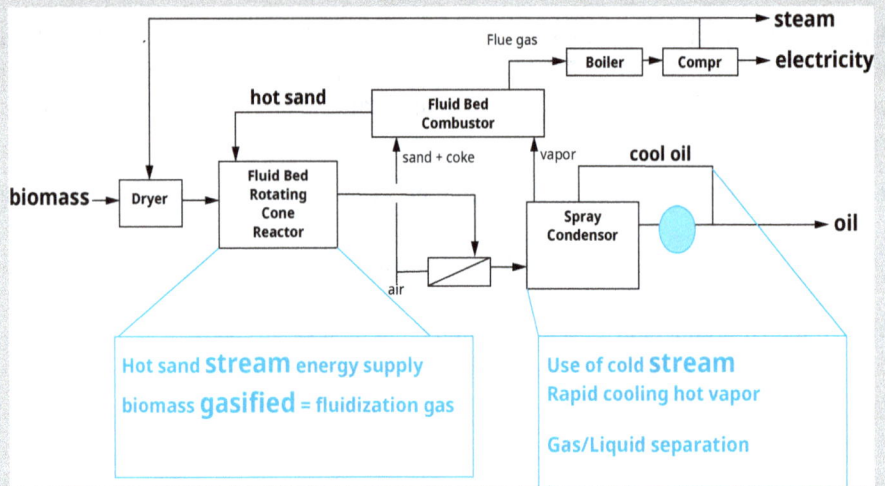

Fig. 7.4: Biofuel BTL process: Use of streams.
Harmsen J, Verkerk M, Process Intensification, deGruyter, 2019.

The biofuel process combines heat transfer, reaction, and mass transport functions in a rotating cone fluid bed reactor. Biomass is fed to the reactor, where it is brought in contact with hot sand whirled around by the rotating cone. The biomass gasifies rapidly and thereby creates sand to fluidize. The sand and biogas are separated. The biogas is rapidly cooled in a spray condenser. The cooling liquid is the cooled biofuel. The non-condensable gas is separated in the spray condenser. It is combusted in the fluid bed combustor. This rapid gasification and quenching results in a high liquid biofuel yield. The function integration results in low investment cost.

This process has been developed by pilot planting and cooperation with the Engineering Procurement and Contracting company and has been implemented several times commercially [19].

Example 7.4 Function block flow sheet for selecting unit operations by structured brainstorm sessions

The company involved wanted the best process separation concept for transforming a complex feed with molecules of type, A, B, C, and D, into four products. Each product should contain A, B, C, or D with a maximum specified trace amount of the other feed components. The company also wanted to know all feasible process concept designs, so that in the end a well-defendable patent could be written.

To achieve these two goals, I drew all conceivable process block flow diagrams based on binary separations, of which one is shown in Fig. 7.5. With simple logic 15 different block flow diagrams could be defined with a total 25 different binary separation blocks, described in Tab. 7.3.

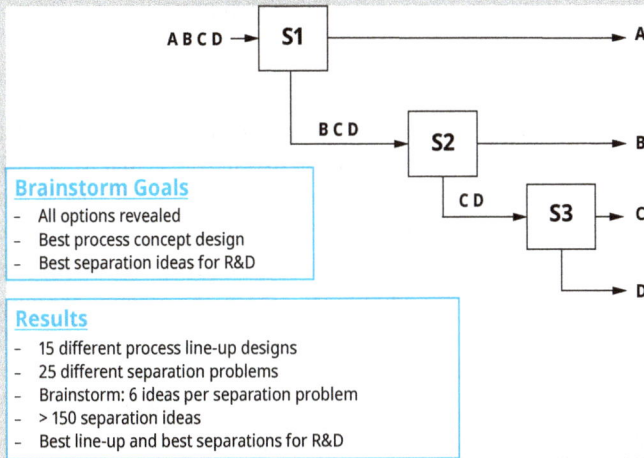

Brainstorm Goals
- All options revealed
- Best process concept design
- Best separation ideas for R&D

Results
- 15 different process line-up designs
- 25 different separation problems
- Brainstorm: 6 ideas per separation problem
- > 150 separation ideas
- Best line-up and best separations for R&D

Fig. 7.5: Use of functions to generate ideas.
Harmsen, et.al., Product and Process Design- Driving Innovation, deGruyter, 2018.

Tab. 7.3: All alternative flow sheets and all binary separation functions for four product streams from one input stream.

Flow sheet label	First separation and label	Second separation and label	Third separation and label
FS1	A/BCD S1	B/CD S2	C/D S3
FS2	A/BCD S1	BC/D S4	B/C S5
FS3	A/BCD S1	BD/C S6	B/D S7
FS4	AB/CD S8	A/B S9	C/D S3
FS5	ABC/D S10	AB/C S11	A/B S9
FS6	ABC/D S10	AC/B S12	A/C S13
FS7	ABC/D S10	BC/A S23	B/C S5
FS8	B/ACD S14	AC/D S15	A/C S13
FS9	B/ACD S14	A/CD S16	C/D S3
FS10	B/ACD S14	AD/C S24	A/D S20
FS11	C/ABD S17	AB/D S18	A/B S9
FS12	C/ABD S17	AD/B S19	A/D S20
FS13	C/ABD S17	BD/A S25	B/D S7
FS14	AC/BD S21	A/C S13	B/D S7
FS15	AD/BC S22	A/D S20	B/C S5

Then in a brain writing session, six experienced chemists and process developers generated a list of unit operations for each separation block. This resulted in each function being separated into five to seven unit operation options. Thus, in total, over 150 unit operation options were generated.

In a separate session the process diagram and unit operations were ranked using qualitative criteria and the top few were selected for a research program.

7.2.3.4 Choice of batch processing versus continuous processing

The choice of batchwise processing versus continuous processing is important as it determines most of the subsequent design choices. The heuristic rule is that as regards economics, continuous processing is chosen when the production rates >5 kt/year and that batch processing is chosen when the production rate <0.5 kt/year is batch [13].

The basic reason batch is chosen for small scale production rates is that most or all process steps are carried out in the same equipment. Hence, the amount of equipment is reduced and thereby the investment cost. For large-scale processing the size of the equipment for batch processing becomes so large that many parallel pieces of equipment have to be installed. As charging and discharging time means that batch processing takes time, meaning that the process equipment stands idle for considerable time, continuous processing is more economical. The advantage of continuous over batch for large-scale capacities is, therefore, clear and holds in general.

For single process steps such as a reaction step, however, the picture is often far more complicated. To illustrate this, the following case is discussed. Batch reaction processes are often carried out in mechanically stirred reactors in which cooling is carried out by the reactor wall and/or by evaporating a solvent. The solvent vapor is condensed to liquid in an overhead condenser and then returned to the reaction liquid. For such a system, a continuous reactor with high heat exchange capacity is economically more favorable. Schwalbe made a comparison for an industrial case where an existing batch mechanically stirred tank reactor for a production of 500 kg/year of a fine chemical was compared with a new continuously operated microreactor. The latter had a payback time of 0.4 years and overall cost reduction of 2.3 M€/year, so a cost reduction of 4,600 €/kg [20] was possible.

A second reason a continuously operated reactor is chosen is related to safety. During batch processing, the reactor is closed. If a runaway reaction occurs the pressure buildup can be high and rapid, causing an explosion. Batch processes have indeed shown far more explosions than continuous processes. A continuously operated reactor is open via the output stream, so pressure buildup is far less severe.

For more complex processes other choices can be made. A choice can be made to operate a certain process step batchwise and the other process steps continuously. A hold-up vessel after the batch step is then needed to ensure continuous feeding to the continuous section.

The fed-batch variant in which part of the feed is batchwise and the other part is fed (semi) continuously is considered same as fully batch. The choice heuristics still hold.

If batch is chosen, then the fed-batch option can be chosen subsequently. This is because, for instance, from a reaction selectivity analysis, distributed feed of one component provides a higher selectivity.

For complex processes the choice between continuous processing versus batch processing is best taken in the feasibility stage and not in the concept stage. In the feasibility stage, experienced process engineers can work out batch and continuous processing in some detail and then make a choice using economics and safety criteria. For more detailed descriptions of batch design and modeling, the reader is referred to references [21–23].

7.3 Analyze concept designs

7.3.1 Introduction to analysis

In concept design interim design solutions need to be quickly analyzed on meeting the design goals and criteria. This analysis then will lead to design modifications, so from the analysis step the designer goes back to the synthesis step. Guidelines for this analysis described in Section 7.3.2.

In concept design often alternative solutions are also generated, which need to be analyzed and compared to each other to make preliminary selections. To that end Section 7.3.3 contains methods.

7.3.2 Analyze preliminary designs

7.3.2.1 Qualitative analysis of preliminary design solutions

In the discovery and concept stages large numbers of ideas should be rapidly analyzed, while little information on each idea will be available. This rapid analysis is best carried out with a group of experienced people. First, they should generate a set of criteria to facilitate rapid analysis. Here are three steps to generate the criteria and perform the rapid analysis.

Step 1: Ask the experts to go back in their memory to similar situations in the discovery stage and then to remember which ideas were successfully implemented, and then what features of the ideas were the key success factors. From these success factors some criteria can then be directly defined.

These criteria are then specific to the analysis problem at hand and specific to the company where the innovation takes place. The criteria often include capabilities of the company to develop and implement certain novel products and processes and not others.

Step 2: Show the experts the list of generic criteria for analysis of Tab. 4.8 in Chapter 4. Also show them the SHEETS criteria list of Chapter 4. In addition, other criteria such as strategic fit and intellectual property protection should be added, from the business perspective.

The total set of criteria is then made from the results of steps 1 and 2.

Step 3: The experts individually attribute a value from 1 to 5 for each idea for each criterion. For each idea, the total score is obtained by multiplication of the scores for each criterion. By multiplying the scores, a balanced total score will be obtained, as the idea that scores well on each criterion will obtain a higher final score than ideas that score high on a few criteria and low on others. Hence, using multiplication reveals ideas that score well over the range of criteria and thereby these ideas are more robust in meeting the criteria and are likely to have a higher chance of success in the end.

It is important that the experts score individually without corresponding with each other. In that way the spread in results will also provide information on the reliability of the results. If experts discuss scores, they will talk each other into more averaged values.

This method of developing criteria by experts and in particular, step 1 of drilling into the source of past experiences, is based on empirical findings on experts by Kahneman, which he calls: "the outside view." Even experts do not naturally use all their experiences of the past, unless this experience is specifically called upon by these series of questions to generate the outside view. [24, p. 245–247].

The results of this rapid analysis can then be used to go back and synthesize additional solutions or generate modifications to the ideas, so that they score higher on criteria that so far had a low score.

7.3.2.2 Quantitative analysis of interim concept solutions

In the concept stage ideas are designed in some detail. Here is an important question regarding these interim concept solutions: Is the concept sufficiently better than the reference case for all criteria? The answer to this question will determine the success of the innovation project. If the new design is only a little better, say shows a 20% lower cost, then later on in the project, due to needed changes the difference between the new concept and the reference case will become smaller and in the end the project will fail, because the small benefit of the new concept will be insufficient to bear the risk of introducing it at commercial scale.

Use of a reference case for analysis

Analyzing a new concept is helped enormously by having a defined reference case to compare with. The reference case should have the same function as the innovation product or process or the combination thereof. The reference case should furthermore be a commercially established product or process, and it should be the best or belong to the best-in-class group. If for that best-in-class not all information is available, then the case should be constructed with some plausible assumptions.

The new design should be far superior to the reference case for at least one criterion and not worse in any of the other criteria. A new design which is far better for some criteria, but worse for other criteria will eventually be abandoned, as weighing different criteria to obtain a conclusion is always debated and, in the end, top management will decide that the project does not have sufficient support. So, a new design is to be equal or better for all criteria and far better in at least one criterion. If not, an alternative design should be chosen to further work on. Selecting between alternatives is discussed in Section 7.3.3.

A few simple analysis methods for each criterion type are provided below.

Safety and health

A rapid analysis of safety and health issues of the novel design can be carried out by a simple comparison of the novel design with the reference case. The novel design should be safer than the reference case. Often this can be achieved by eliminating the most unsafe parts of the reference case. For products, safety is mostly concerned with toxicity. Using the toxicity index for each ingredient in the product will be sufficient to draw a conclusion. Chapter 11 provides the toxicity index method.

Environmental analysis

For a rapid analysis of intermediate results and for an assessment of concept design results for the concept stage, a semiquantitative method based on large improvements on a single impact type, the rapid LCA method of Jonker can be used [25].

The method consists of a few steps.

Step 1: Define the functional unit of the design as the same as for the reference case.
Step 2: Select the most critical environmental impact type.
Step 3: Compare the environmental impact of the design with the reference case.
Step 4: Conclude whether the design is sufficiently better or not.
Step 5: Improve the interim design until its meets the environmental criterion. Chapter 12 provides more information on evaluating environmental performance.

Sustainable development analysis

A rapid analysis on contributions to sustainable development can be done by asking whether the design provides a novel solution for the needs of people, whether it would be socially accepted, if it would be environmentally friendly, and if it would be economically attractive. If some answers are negative, then the design should be improved. If some answers cannot be given, then more work on the design should be done. Chapter 12 provides more information on sustainable development.

Economics: rapid analysis in the concept stage

A rapid economic analysis in the concept stage can be simply based on adding the cost of all feeds for the product and comparing that with the anticipated market price. If the total feeds cost price is higher than the market price the design should improve on this aspect. Chapter 10 provides several other methods for analyzing interim design solutions.

For all other criteria no formal rapid analysis methods are available. Asking the opinion of an experienced expert for an interim solution regarding marketing or other criteria is a rapid method.

Role of modeling and analysis

By making a model and then simulating the design behavior, an analysis can be made of its technical feasibility. The model also generates other output parameter values, useful for many other analysis items. Chapters 8 and 9 provide detailed information on how to model, simulate, and analyze product and process designs.

Next step after analysis

The analysis result could be that the design should be further developed to facilitate the analysis. The analysis result could be that the interim design does not meet certain criteria. In that case the interim solution should be improved for the aspects of concern. Hence, the designer goes back to the synthesis step to improve the design. The analysis could be that the design is good enough for the next design step, a full evaluation. This is discussed in Section 7.4.

Role of modeling in analysis

Product and process modeling play an important role in determining whether the concept meets certain criteria, as with modeling, detailed information on product compositions, mass, and energy flows are provided and the performance effects of changing design parameters can be easily determined. Chapter 4 discusses the roles of modeling and Chapters 8 and 9 provide methods for modeling, simulation, and optimization of products and processes.

7.3.3 Best selection from alternative concept solutions

In the discovery stage a large number of interim solutions will be generated from which the most promising options should be selected with little information at hand. For efficiency reasons, these ideas should first be ranked. Only the top ideas are then further worked out. This holds to a lesser extent also for the concept stage.

It is recommended to develop ranking criteria for the discovery stage that are generated by experienced R&D people from inside the company. In particular, the criterion for technical feasibility should be defined such that it is meaningful in the particular company's context, because bringing ideas to successful implementation strongly depends on the innovation capabilities of that company. The intuition of those innovation experts can be considered as reasonably dependable because it has been built up in a "regular environment," namely the R&D area of their company. This insight stems from Kahneman [24].

However, it is also important to have at least one expert from outside the company in the evaluation panel. Preferably it is someone with experience in innovation in more than one company. This "outside view" as defined by Kahneman brings additional knowledge to the table based on a broader experience [24, p. 245].

An important element of activating the intuition and memory of experts is to prompt their memory. This is done as follows. The leader of the evaluation session asks the experts to go back in their memory on which projects were successful and what characteristics in the early discovery or concept stage appeared to be key success factors [24, pp. 245–246]. The author has executed this memory prompting several times by asking the experts to go back in time in their memory and think of projects that in the end were successful. Every time the experts fell silent for minutes and then produced stories and appropriate criteria that even surprised them. They did not know that they knew them.

First, criteria for the ranking should be generated. The generic table for problem definition and evaluation can be used as a starting point. It is not very practical to have many criteria. Six in total is often the maximum in order to keep the ranking manageable.

Safety, health, and environment may be combined for one criterion. Social acceptance, market opportunity, strategic fit, and economic attractiveness may also be combined in a criterion. Technical feasibility to develop the idea to commercial realization is also likely to be a criterion.

They should also carry out the description of the scaling of the criteria. A scale of 1–5 is in general sufficient for each criterion. A meaningful short statement for each score value helps to speed up the scoring for each criterion later. According to Kahneman, the use of weight factors will not improve the overall quality of the judgment, because the human mind unconsciously also makes a judgment of the outcome of the whole and will adjust the score using weight factors. In fact, putting the criteria list together and then directly asking the experts to give the total score will result in similar conclusions to the result that adds up the scoring of individual criteria scoring [24].

The valuation of each idea or concept for each criterion is in general best carried out by scoring the experts individually for each idea. In this way the errors of the experts are independent, so that the averaged final score value is more dependable than when the scores are obtained by agreement. In this way, one avoids the halo effect of people tending to form a consensus [24].

However, for complex process alternatives and for complex product ideas this individual scoring may be inferior to a plenary session in which first a discussion is started on how the idea would perform. In this discussion each member's imagination is stimulated to envisage how the idea would perform and then plenary scoring is, in the opinion of the author, of more value.

The overall score is best obtained by the multiplication of the score values per criterion. In this way, alternatives that score a reasonable value for each criterion get a higher overall score than alternatives that score highly on one or two criteria and low on all others. This means that the top alternatives are more robust toward future uncertainties since they score well on all criteria and not on merely one or two. This is a better method than adding up the individual scores as advocated by Kepner and Tregoe [26]. The latter method compensates for low scores with high scores, and alternatives often obtain the same total sum score, although their score distribution is different.

The use of weight factors for criteria is not recommended for this intuitive scoring of alternatives, as the human mind unconsciously "corrects" the score for the weight factors [24].

More sophisticated methods such as analytical hierarchy process (AHP) and data envelopment analysis for ranking alternatives with multiple criteria require far more effort to obtain a ranking result. Moreover, the ranking result obtained is harder to communicate to the management, who must provide the budget to further develop the top alternative.

7.3.3.1 Selecting best process concept option

In the discovery and sometimes also in the concept stage, many process options are (and should be) generated. The design is often only a simple block flow diagram. Inputs and outputs are known from scouting experiments. Process concept alternatives can be obtained from literature searches.

For efficiency reasons these concepts should be ranked and only the best selected for further development. Ranking with a panel is, however, not always immediately available. For the individual designer, the following simple method has been developed to still select the best option. The simple selection method is carried out as follows. A large number of process alternatives are obtained from literature searches. Each alternative solution is then first tested on whether for any of the safety, health, environment, and social acceptance aspects there is a major problem foreseen. If so, the alternative is filed and, for the time being, not further investigated. For the remaining alternative process alternatives, block flow diagrams are made and the yield of product on feedstock is researched in literature. Then an economic ranking can be made based on this design information, as provided in the next paragraphs.

7.3.3.2 Economic ranking of process concepts

For evaluating process routes with different yields of product on feedstock and different number of process steps quickly and economically this book is useful. The feedstock cost per unit product follows directly from the product yield on feedstock:

Feedstock cost/ton product = Feedstock price ($/ton)/yield

The investment cost can be estimated by the number of process steps involved and the size of a recycle stream, described in Section 10.4.

If the reactions form an equilibrium between product and feedstock, then also an additional separation will be needed and a recycle stream from the separator back to the reactor. The equilibrium constant K can be estimated as follows [27]:

$$K = e^{-\frac{\Delta G}{RT}} \tag{7.1}$$

The value of the free enthalpy value, ΔG, can be looked up, and K can then be determined. If K is much larger than 100, then single pass deep conversion is feasible. If K is around 1 then single-pass conversion is not feasible and an additional separation step has to be added to the total separation steps.

7.3.3.3 Overall selection between alternative process designs

The selection method is shown in Fig. 7.5. First a preselection is made on Safety, health, and environmental criteria. If for instance a highly toxic agent is used in an alternative process option, then that option may be discarded for that reason.

Then for the remaining alternatives the product yield on feedstock is noted down. If there are feedstock type differences between the alternatives, then the feedstock cost per ton of product is determined from the feedstock price and the product yield on that feedstock. The total number of process steps can be determined from the stoichiometry equation of the reaction involved. If an equilibrium reaction is involved a separation and a recycle stream are involved.

The best process route has the lowest feedstock cost per ton of product and the lowest number of process steps. If the best selection cannot be made, because the lowest feedstock and the lowest number of process steps belong to different alternatives then the lowest feedstock cost option should be chosen. In all other cases, the top two or three options should be further investigated in detail by more detailed process design and cost analysis, using information from Chapter 10.

An example of a process concept design with the essential information needed for the ranking method is shown in Fig. 7.6 [28]; the data are from Rasrenda [29].

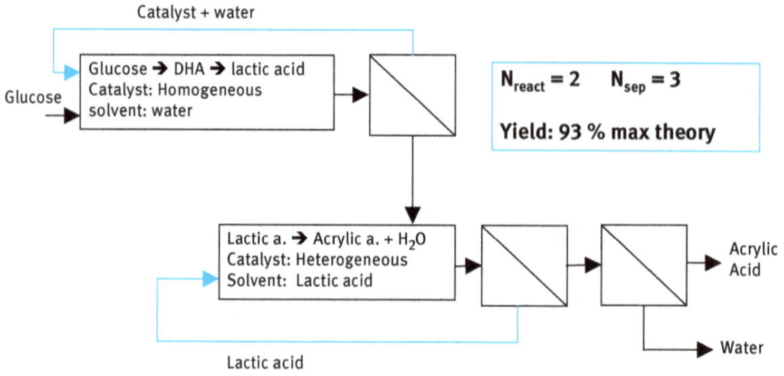

Fig. 7.6: Block flow diagram to Acrylic Acid: Route C.

7.4 Evaluate designs

7.4.1 Introduction

Evaluating designs in a commercial organization will in general be done before the gate from one innovation stage to the other by the designers. The whole design including evaluation will then be used to decide to pass the gate or to stop the project, or to revert the project to the stage for further work. Part of the evaluation will also be a selection between alternative designs and a decision on which alternative to recommend for the next stage.

Criteria for the evaluation in general will concern: Safety, health, and environmental, as well as economic, technical feasibility, and contributions to sustainable development goals. Furthermore, in the concept stage some business aspects, such as strategy fit and foreseen benefits, foreseen cost, market share, and risks will be considered.

Chapter 3 provides for each stage (gate) details on what information should be provided and in particular what experimental proof is needed for each stage-gate to be able to judge the technical feasibility of the innovation. Specific evaluation methods for later stages for economics, health, and safety, environmental, social, and sustainable development are provided in Chapters 10–13.

Section 7.4 is about evaluating the design. Section 7.4.2 describes balancing the design for al criteria. Section 7.4.3 is about making the design robust to future uncertainties, to competition, and to intellectual property rights. Section 7.4.4 describes a recommended selection of one design alternative to the stage-gate panel, so that an informed decision about passing the stage-gate to the next stage can be made. Section 7.4.5 is about reporting the design.

7.4.2 Balancing the design

Balancing design, also called trade-off design, is about making small design changes such that they meet all criteria equally well. Hence, if a design just meets one criterion, while the design is far away from the other criteria boundaries then the design needs to be modified, such that it is away from the boundary of that particular criterion, while it may be a bit closer to the other criteria boundaries. The reason for this balancing design is that further down the innovation road the design just meeting the one criterion may be vulnerable to changes in that criterion. Making changes in the design will then be more costly. If for instance this criterion change happens in the development stage, the pilot plant or the product prototype needs to be changed, with a lot of cost involved.

This balancing design action happens in the feasibility stage, where for the first time a commercial scale design is made that meets all criteria, defined as a feasible design. This feasibility is strongly related to minimizing the risk of failure; hence the focus on minimizing the overall risk.

The subsequent sections deal with criteria types, optimization over all criteria, and increasing robustness toward future uncertainties by using scenarios.

The design criteria set in the scoping step of the design are often further defined as the criteria that must be met, otherwise the design will not be accepted, and desired criteria for which the design should be optimized.

When a design nears its completion, a lot of knowledge will have been generated revealing the relations between design decisions and design performance. Then it is time to balance the design such that it becomes robust to uncertainty in the value of certain parameters such as feedstock prices. This robustness can be obtained to a certain extent by choosing design parameters such that the design output or performance is away from the most critical constraint boundaries and that the optimum is a smooth optimum and not peaky. Polar graphs are useful in communicating the balanced design. They show both the design and its criteria in one graph. The effects of uncertainty ranges in design parameters on the design performance and uncertainty ranges of the criteria can be shown in a single figure.

7.4.3 Make design robust to future uncertainties

7.4.3.1 Increasing robustness towards future uncertainties using scenario sets

As the design will be implemented in the future, and then it will be in existence for many years, it is important to take not only the present status of the evaluation criteria and the present context into account, but also future development in criteria such as more stringent or less stringent environmental criteria and new competitive alternatives, by which the product or the process can rapidly become out of fashion or cannot compete on price. This then may lead to a premature end of the product and process.

However, by envisaging potential future developments the design may be modified so that the design is robust to these future developments, the implemented design may last longer.

A systematic way of making designs robust to future uncertainties is by using a scenario set containing several opposing future scenarios. A schematic example of such a scenario set is provided in Fig. 7.7.

The time horizon of the scenario set should be at least 20 years in the future, as new processes, and also new products last at least 20 years. Placing the scenario set in the distant future has the additional advantage that it creates a feeling of freedom to think wildly about the future. Each scenario should be daringly extreme, so that it really challenges the design on success under each scenario.

The individual scenarios A1, A2, B1, and B2 have to be described as stories of conceivable world futures using the extremes of the horizontal and vertical axes. The top vertical axis extreme, sustainable development means a future world where sustainable development goals are set and met and conditions for all people to live, and work are fair. The bottom vertical axis extreme, unsustainable and uneven growth means a future world where conditions for society, environment, and economics are disruptive. The horizontal axis extreme, global cooperation means that countries and companies agree to harmonious cooperation making regulations and standards uniform worldwide. The horizontal axis extreme, isolated problem solving, means that global cooperation between countries and companies does not take place.

The individual elements of each scenario now have to be filled in by the designers, preferably with others with different backgrounds. A brief description of the parameters in the center of the scenario set, which should help generate the four scenarios follows.

The element society concerns at least its growth or decline, its education, and its health, also disruptive events such as war can be considered. The element economics concerns at least its (lack of) economic development, market type (free, competitive, or government controlled). The element technology is about breakthrough novel technologies (or not). Also agriculture techniques for food production should be taken into account here.

All element descriptions within each scenario should be connected to each other into a conceivable story of the future. Each scenario should be given a name. This helps to make each scenario internally consistent.

When the scenario set is complete the design should be placed inside each scenario. Also, the reference case should be placed in each scenario. Then the design should be assessed on how it will perform under each scenario. Comparing it with the reference case in each scenario will help determine whether it will be successful or not.

If the design is not successful in one or more scenarios it should be changed in such a way that it will be successful in each scenario. When that is finally the case then the conclusion can be that the improved design is likely to be robust to future uncertainties.

The scenario template is for a global context. However, it can be adapted to a geographical region if the design is not for a global market but for a local market. The

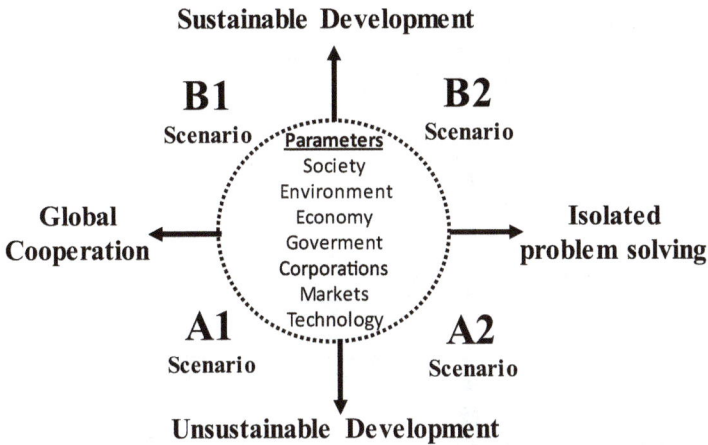

Fig. 7.7: Scenario set template.

example below shows such an adaptation. It also shows that a scenario set can be quickly made even by students and also design testing and improvements are often easily made.

The reference case can also be used in the scenario set of Section 7.4.3 to see how the reference case performs in future, compared to the new design. This often gives great insight in the robustness of the new design compared to the reference case.

Pitfalls in using a scenario set

Pitfall 1: Attributing a likelihood to each scenario

The likelihood is not only hard to define scientifically, but it also reduces the power of using the scenario set. As soon as a scenario is considered to be less likely than others, and the design for that scenario fails, the stress to change the design will be lower and may be even neglected. Thus, the value of the whole exercise to get a robust design for all scenarios in the set is lost.

Pitfall 2: Preferring a certain scenario

Often designers and others express a preference for a certain scenario. For politicians this is rational as they try to change the context by making laws and regulations. However, for designers all scenarios should have the same weight in testing the design on its robustness.

Example 7.5 Scenario set made and its use for obtaining a robust design

This example shown in Figure 7.8, is obtained from EngD students Sharangdhan Bhave, Ved Dubhashi, Marianna Kaloutsi, and Rocio Villagran Vargas, at TU Delft for their paracetamol process design in 2022 [30].

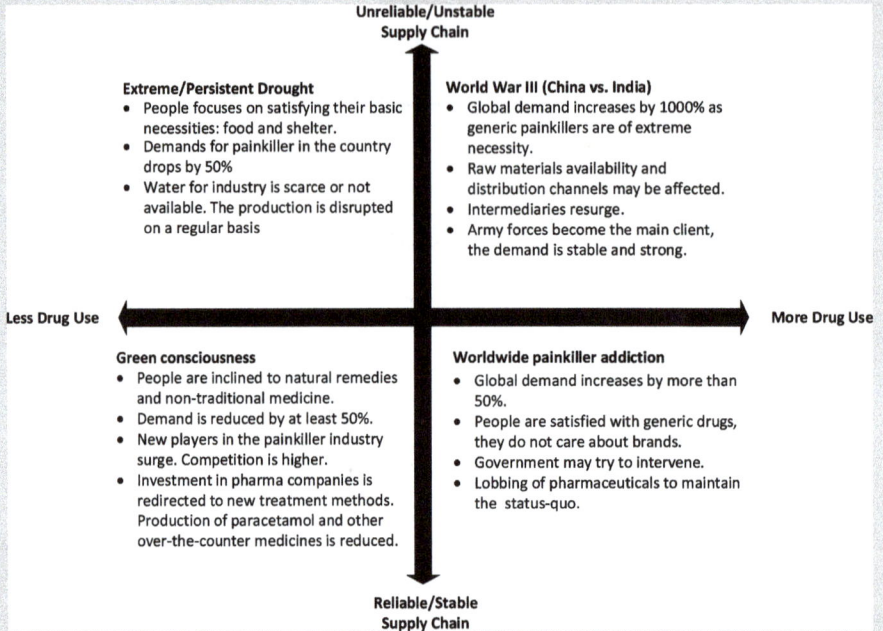

Unreliable/Unstable Supply Chain

Extreme/Persistent Drought
- People focuses on satisfying their basic necessities: food and shelter.
- Demands for painkiller in the country drops by 50%
- Water for industry is scarce or not available. The production is disrupted on a regular basis

World War III (China vs. India)
- Global demand increases by 1000% as generic painkillers are of extreme necessity.
- Raw materials availability and distribution channels may be affected.
- Intermediaries resurge.
- Army forces become the main client, the demand is stable and strong.

Less Drug Use ←→ **More Drug Use**

Green consciousness
- People are inclined to natural remedies and non-traditional medicine.
- Demand is reduced by at least 50%.
- New players in the painkiller industry surge. Competition is higher.
- Investment in pharma companies is redirected to new treatment methods. Production of paracetamol and other over-the-counter medicines is reduced.

Worldwide painkiller addiction
- Global demand increases by more than 50%.
- People are satisfied with generic drugs, they do not care about brands.
- Government may try to intervene.
- Lobbing of pharmaceuticals to maintain the status-quo.

Reliable/Stable Supply Chain

Fig. 7.8: Scenario set for Central and East Asia to test new paracetamol process design.

In order to improve the design for the whole scenario set depicted in Fig. 7.8, and in particular for scenarios green consciousness and extreme/persistent drought, a set of design changes have been made.

Design improvements to obtain robustness to future uncertainties

The design improvements are the following:

- The present design makes use of steam to dry the product and continue with the tableting process. In case of disruption of water supply, which can be caused by drought, the process cannot deliver the desired production rate. To tackle this, it was decided to replace steam with hot air provided by a solar concentrator. The air is pretreated to meet the industry standards. Also, potatoes need 25–50 mm of water per week. This cannot be guaranteed in case of an extended drought. Wheat, on the other hand, needs 20 mm of water per day with the option of planting wheat engineered to survive long periods of drought. Therefore, starch will be obtained from this crop.·

- In case the demand increases, the process should have a higher maximum production capacity. The original design was a batch design, especially the mixing of ingredients that happens in multiple steps. We change that part of the design to continuous mode. For that, a continuous power blender for the initial mixing is designed. This increased its capacity.

- The marketing of the paracetamol product should be flexible. In case of a reduction in human consumption, either because of an inclination for natural medicines or a restriction from the government, the company should be able to maintain their sales. Paracetamol is also given to dogs as painkillers, and it is licensed as Pardale V. The tableting process will have the flexibility to pack the product as Paracetamol or Pardale V according to the market demand.

– It is important to note that the company should make the consumer aware of the potential risks of consuming large quantities of paracetamol. For that, the packaging will have a legend related to responsible consumption of painkillers.

With these three design modifications, the process is likely to be successful in all the scenarios. In extreme/persistent drought, the process will be able to run smoothly even if there is a shortage of water for industry. In green consciousness, the product is rebranded, and the production can be maintained. Also, the process is still successful in World War III and Worldwide Painkiller Addiction scenarios. It is concluded that the modifications strengthen the process and make it resilient to future variations.

7.4.4 Robustness to competition by comparing with reference case

The use of a reference case is described for the analysis step in Section 7.3.2.2. The reference case should also be used for the evaluation step. However, now the reference case should be described extensively, so that a thorough evaluation of the new design with the competing reference case can be made.

Finding all relevant information about the reference case may require considerable effort. For reference products it may take a full chemical analysis of the competitors' product to evaluate it with the newly designed product with its performance tests.

For a new process design evaluation, the best commercial scale process should be found. This may involve reading public environmental impact reports of the competitors' process. In some cases, it may even involve making a design of the competitors' process from limited public information available. However, this effort will pay off in knowing the present performance of the best competing product and process, so that improvements by the novel product and process can be determined for all evaluation criteria.

7.4.5 Intellectual property (IP) creation and protection

The new design should be protected so that other companies cannot use it for free. Intellectual property is therefore to be created. Designs can be protected in various ways. It can be done by patents, copyright, or design rights. Chapter 3, Section 3.6.5 provides more information. Most important is to have IP protection before the design is made public in any way.

On the other hand, the design should also not infringe patents by other companies or institutes, so a patent search should be applied, and a patent expert should be consulted to check for this item.

7.4.6 Selecting alternative for next stage

Often in concept design alternative designs have been evaluated. The research team should then make a recommended selection between the alternatives, to guide the stage-gate panel in deciding to pass the selected design to the next stage for further development. This selection is best carried out by using the same criteria for the alternatives.

7.5 Report designs

Any design consists of words, symbols, drawings, and pictures to report the design result to others, so reporting orally and in writing is extremely important. This truth is sometimes forgotten especially in the discovery stage. A brilliant idea may come up during a discussion in the corridor. If it is not written down and reported to others it will be lost. Chapter 13 elaborates extensively on the importance of reporting designs and provides heuristics for doing so for each part of a design in each stage.

7.6 Exercises

7.1 Section 7.2.3.2 provides 10 heuristics on process synthesis function design. For the first 4 the background is explained. Explain the reasoning/background of the other 6 heuristics (HR5 – HR10)

7.2 Explain the reasoning/background for the unit operation heuristics (HU1 – HU5) in Section 7.2.3.3.

7.2 Explain the reasoning/background for the reactive distillation heuristics (HRD1 – HRD4) in Section 7.2.3.3.

References and further reading

[1] Verkerk MJ, et al. Philosophy of Technology, London: Routledge, 2007.
[2] Tassoul M. Creative Facilitation, Delft: VSSD, 2009.
[3] Korevaar G. Sustainable Chemical Processes and Products: New Design Methodology and Design Tools, Delft: Eburon, 2004.
[4] Altshuller G. TRIZ Keys to Technical Innovation, Worcester USA: Technical Innovation Center, 2005.
[5] Harmsen GJ, Chewing gum that does not foul the streets, oral presentation, and abstract Proceedings ECCE 2007, Copenhagen, 2007, Conf CD ISBN 978–87-91435-57–9.
[6] Bermingham SK. A Design Procedure and Predictive Models for Solution Crystallization Processes. Development and Application, PhD Thesis, Delft: Delft University Press, 2003.
[7] Bruins. Phase equilibria for food product and process design, Fluid Phase Equilibria, 158–160, 1999, 657–671.

[8]   Wesdorp L. Good food requires engineering?! Simultaneous product and process development, Keynote lecture on ESCAPE-12 Symposium, The Hague, on CD-ROM supplement. In: Grievink J, van Schijndel J (eds). Computer-Aided Chemical Engineering, Vol. 10 Amsterdam: Elsevier, 2002.

[9]   Cussler EL, Moggridge GD. Chemical Product Design, Cambridge: Cambridge University Press, 2001.

[10]   Wesselingh JA, et.al. Design and Development of Biological, Chemical, Food and Pharmaceutical Products, Chichester: John Wiley & Sons, 2007.

[11]   Seider WD, et al. Product and Process Design Principles, Hoboken NJ: John Wiley & Sons, 2009.

[12]   Siirola JJ. Strategic process synthesis: Advances in the hierarchical approach, Computers & Chemical Engineering, 20, Jan 1 1996, S1637–43.

[13]   Douglas J. Conceptual Design of Chemical Processes, New York: McGraw-Hill International editions, 1988.

[14]   Levenspiel O. Chemical Reaction Engineering, 3rd edn, New York: J. Wiley& Sons, 1999.

[15]   Green DW, Perry RH (eds.). Perry's Chemical Engineers' Handbook, 8th edn, New York: McGraw-Hill, 2008.

[16]   Stankiewicz AJ, Moulijn JA (eds). Re-engineering the Chemical Processing Plant, New York: Dekker, 2000.

[17]   Schembecker G. Process Synthesis Course, Breda: Process Design Center, 2000.

[18]   Harmsen GJ. Reactive distillation: The frontrunner of industrial process intensification: A full review of commercial applications, research, scale-up, design and operation, Chemical Engineering & Processing: Process Intensification, 46(9), 2007, 774–780.

[19]   Harmsen J, Verkerk M. Process Intensification: Breakthrough in Design, Industrial Innovation Practices, and Education, Walter de Gruyter GmbH & Co KG, Jul 20 2020.

[20]   Schwalbe T. Chemical synthesis in microreactors, Chimia, 56, 2002, 636–646.

[21]   Diwekar U. Batch Processing Modeling and Design, New York: Taylor & Francis Inc, Abingdon, 2014.

[22]   Sharratt PN. Handbook-of-batch-process-design, New York: Springer, 1997.

[23]   Korovessi E, Linninger AA. Batch Processes, Boca Raton: CRC Press, 2005.

[24]   Kahneman D. Thinking, Fast and Slow, London: Penguin Books, 2012.

[25]   Jonker G, Harmsen J. Engineering for Sustainability A Practical Guide for Sustainable Design, Amsterdam: Elsevier, 2012.

[26]   Kepner CH, Tregoe BB. The Rational Manager, Princeton NJ: Kepner-Tregoe Inc, 1981.

[27]   Moore WJ. Physical Chemistry, 4th edn, Englewood Cliffs N.J: Longmans Green and Co, 1962.

[28]   Harmsen J. Best Renewable Routes to Adipic Acid, Acrylic Acid and Propionic Acid, Oral and Abstract, AIChE Spring Meeting & 12th Global Congress on Process Safety, Houston, USA, 2016, 10–14.

[29]   Rasrenda CB. Platform Chemicals from Biomass, PhD Thesis, Groningen University, 2012.

[30]   Bhave S, Dubhashi V, Kaloutsi M, Vargas RV. Paracetamol Process Design, EngD Course Sustainable Product, Process, and Systems Design, TU Delft, 2022.

8 Product modeling and optimization

In this chapter, the role of mathematical models in the design of products is described as also the roles of other types of models – verbal, schematic, and physical. The information in this chapter is in line with the design levels and design steps of Delft design map (DDM), as indicated in Tab. 8.1.

Tab. 8.1: Delft design map (DDM) parts related to modeling and optimization of this chapter.

Stage-Gate™:			Concept				Feasibility			Development			
	Framing		Supply chain imbedding				Process technology			Process engineering			Final
Design levels ⇒ Design steps/ activities ⇓	PF		CW-PC	PC-PF	I-O	SP	TN	UN	PI	ED	OI	FO	(Final)
1 Scope													
2 Knowledge													
3 Synthesize													
4 Analyze													
5 Evaluate and select													
6 Report													

8.1 Verbal, schematic, mathematical, and physical models

In Section 4.6, the role of (mathematical) modeling in design is described. This section also emphasizes the use of the nonmathematical types of models for both product and process design activities. Before going into these different model types, the creation of models by designers is the heart of the design activities, as all these models together make up/describe the design alternatives and the final design (be it from different perspectives).

This has been characterized by Van Overveld [1] (lecturer, General design methodology at TU Eindhoven), who in his very inspirational way has taught this in the TU Delft and TU/Eindhoven post-Master EngD designer programs (see Section 14.2.4). Van Overveld characterized "designing" and "a design" as:

- "Designing is: taking decisions to make stakeholders happy, according to an intentional plan, leading to a new 'artefact', an improvement to an existing artefact or a new or improved part of an artefact."
- "'A design' is a representation of the (new/improved) artefact, in the form of one or more models that are mutually consistent. Each model focuses on one aspect of the artefact (geometry, physics, chemistry, operation, . . .) and communicates

https://doi.org/10.1515/9783110782127-008

the artefact with the stakeholders, documents the decisions, helps the realization and aids in the use of the artefact."

Different model types are also created and used in the (bio)chemical product and process design field. These models describe, in their own way, parts of the design results and decisions. The most used model types are also classified in general systems theory by Skyttner [2]. Next to mathematical models, iconic/physical models, schematic models, and verbal models are also used. Engineers get well exposed to mathematical models and schematic models like graphs and charts, but receive less education and training in creating schematic and physical/iconic models. Reporting on design context, scope, knowledge, analysis, evaluation, and selection decisions is not recognized as a model as well. All these models serve different purposes and are used during and after the design process. For communication purposes, the physical and schematic models play a paramount role.

Viewing the design as a collection of (mutually) consistent models also helps in defining the scope and deliverables for the design team (in specifying what types of models and to what level of detail these models will be delivered).

Examples of these four model types will be given in Sections 8.2 and 8.3 and in Tab. 8.5 and 8.6, and references are given to examples of these models used in this book.
- Physical (or iconic) models. These models:
    - Look like the real system they represent (car, building, manufacturing plant, product prototypes, mock-ups, etc.)
    - Can be small scale or full scale
- Schematic models. These models are pictorial representations of conceptual relationships of the designs of:
    - Molecular structures drawings, (bio)chemical reaction pathways
    - Product structure drawings (spatial distribution of phases with each size, shape, form, etc.)
    - Product structure drawings in different stages of manufacturing or use
    - Graphs showing relationships (suggested or measured) between the product and process variables (reaction rate versus concentrations, heat transfer rate versus temperature differences, payback time versus capital investment, etc.)
    - Tables (containing symbols and data), pure component properties, stream specifications, process stream summary, etc.
    - Diagrams (house of quality (HoQ), process block diagram, process flow diagram, etc.)
- Verbal models. These models contain words (organized in lines of text or tables) that represent systems or situations that (may) exist in reality. These words provide:
    - Design problem context
    - Knowledge relevant to solving the problem, analyzing, and evaluating performance
    - Assistance in construction of (parts) of the design (equipment CAD drawing)

- Assistance in operating, maintaining, and repairing (manuals)
- Justification to design decisions and recommendations
  Verbal models are organized in lines, bullet lists, (text) tables, paragraphs, chapters, etc.

And finally:
- **Mathematical models**
  - Are a reduced representation of reality
  - Most abstract and precise models in a structured language
  - Such models are systematic and precise indeed, but they will often be restricted with respect to their domain of valid application, that is, the underlying modeling assumptions must be applicable to a particular domain and, ideally, an experimental validation of the model or of its critical sub-models must have taken place.

These "models" of an artefact (verbal, iconic, schematic, or mathematic) are representations of (parts of) the original artefact. A model has a correspondence to the actual artefact and is created/synthesized to solve design issues/problems of the real artefact. The design issues can be analyzed/evaluated by manipulating the model (because manipulating the actual artefact is not possible, as it is not there yet). So, the models should be a good representation of (part of) the real artefact such that the results of the manipulation of the model can be mapped back to the artefact. The complete set of models represents the artefact's design.

8.2 Process design: schematic and mathematical models useful for product design

In (bio)chemical product and process design, all four types of models are used, but only some of them have become standard tools. Chemical or process engineers – historically educated and trained in process design and mathematical process modeling and simulation – have been exposed to the use of a limited class of schematic models.

Chemical engineers value mathematical models very much as a tool in process design and they are extensively being used in process and equipment design; they have been so since a long time. The mathematical modeling of manufacturing processes (composed of unit operations like chemical reactions, separations, including thermodynamic, kinetics, and transfer phenomena models) has a long history and is extensively used in process simulators and mathematical optimization software in all innovation phases. This will be extensively discussed in Chapter 9. In the initial innovation phases, mathematical models start simple (e.g., mass balances based on reaction stoichiometry and conversion/yield assumptions) and develop in complexity for those processes that show potential to succeed and warrant more detailed (and extensive modeling). Some of the mathematical process models often linked to engineering diagram or drawings are process block schemes (PBS), process flow diagram/scheme (PFD/PFS), and piping and instrumentation diagram (P&ID). These drawings, often produced to internal and/or external standards, may form the basis for mathematical modeling or a vehicle to capture the mathematical models' quantitative results.

Examples of the PBS and PFD of the TAME (tert-amyl methyl ether) production process are presented in Figs. 8.1 and 8.2. 3D CAD process and equipment drawings are also widely used by process design and engineering companies.

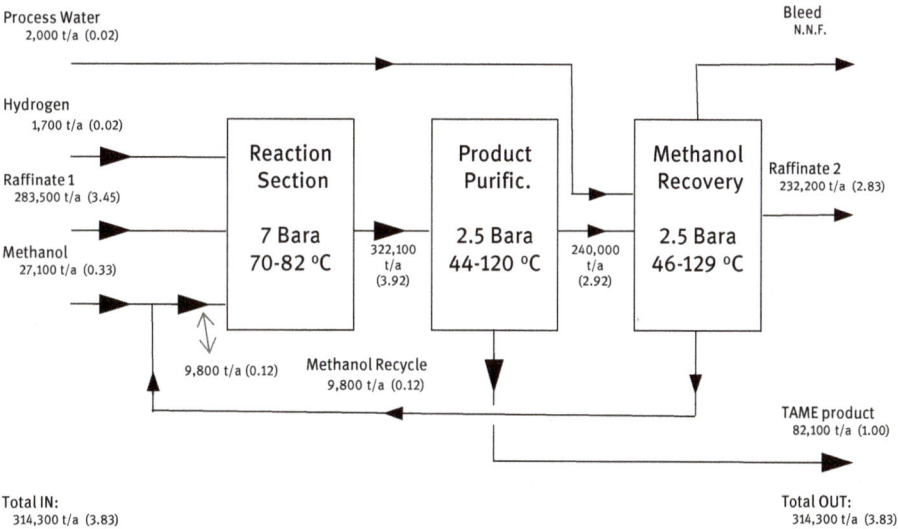

Please note:
- Values between brackets ( ) are t/t values.
- Blocks and character fonts could have still smaller size.
- Values normally are round off, however without loosing consistency on balance.
- All streams ENTER at the LEFT hand side of the block scheme.
- ALL streams LEAVE at the RIGHT hand side of the block scheme.
- N.N.F. or NNF = Normally No Flow.
- Indicate temperature and pressure inside blocks and for all streams entering (LEFT) and leaving (RIGHT) the block scheme (not done in the above scheme).
- Indicate stream numbers for each stream (cross-referencing with Process Stream Summary).
- Never put arrows at the end of the connecting lines; for better readability of the scheme put the arrows at the middle of the line.

Fig. 8.1: Process block scheme (PBS) example – TAME (tert-amyl methyl ether) production process [3].

These engineering diagrams and drawings also play a key communication tool within the design team as also to other stakeholders outside the team. They form key input documents for preliminary economic analysis (gross margin calculations, based on product production and raw materials usage, see Chapter 10), for equipment lists and equipment data sheets (PFD), and for HAZOP studies (P&ID) (see Chapter 11), to name a few. More detailed 3D CAD equipment and plant drawings are used for layout optimization and equipment and plant construction purposes. These standardized drawings

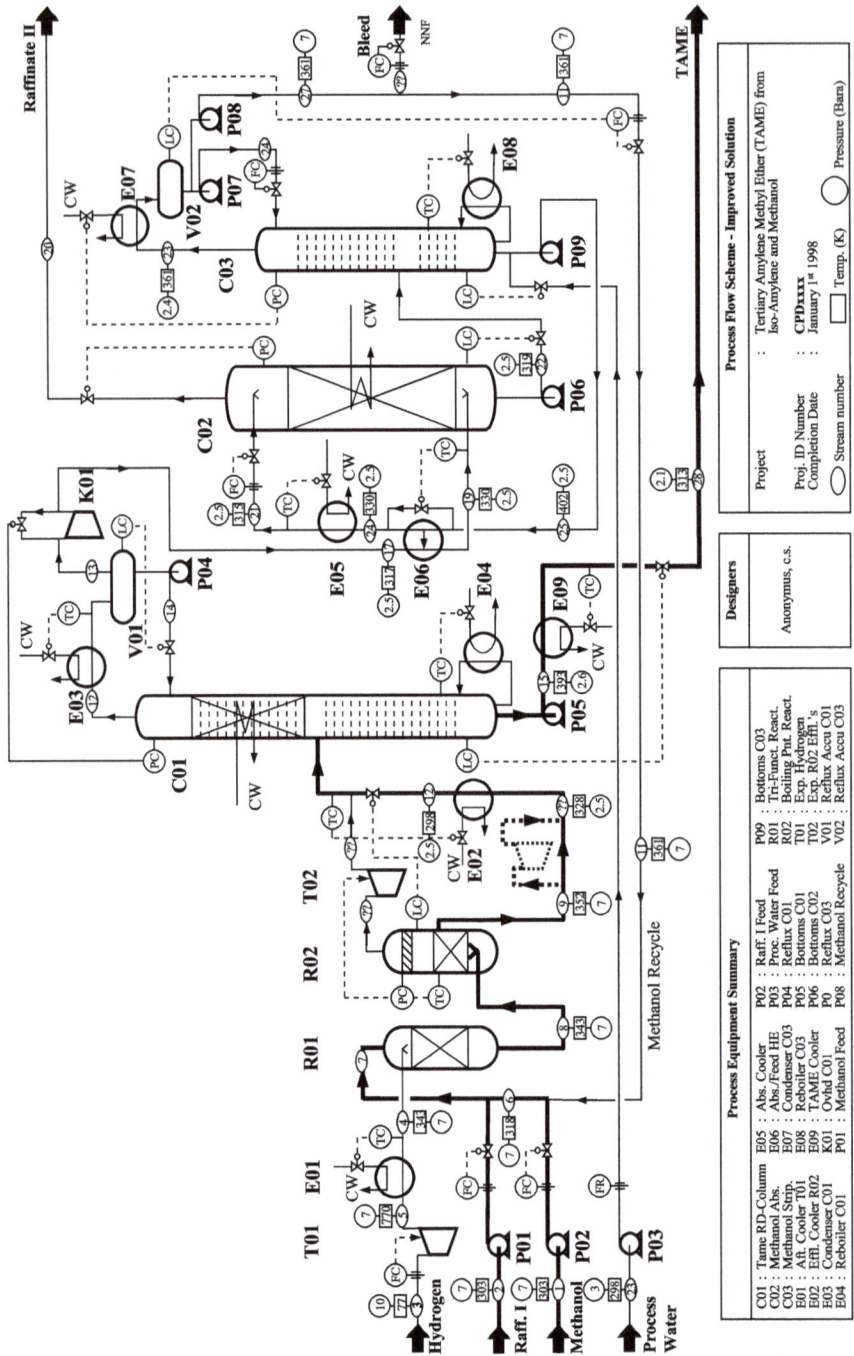

Fig. 8.2: Process flow diagram/scheme (PFD/PFS) example – TAME (tert-amyl methyl ether) production process [4].

are widely used as a communication tool between (often) different companies (brand owner, engineering and procurement companies, equipment manufacturing and plant installation companies).

The standardization of engineering drawings and their use has not (yet) cascaded (back) to the design of chemical products. Chemical products that are process-like devices (fuel cells, lab-on-chip, and sensor-on-a-chip) are also easily represented by the above tools, as the device function and performance delivery resemble a chemical process.

For other products (like biofuels, engine lubricants, coatings, laundry detergents, flexible packaging, etc.) products drawings are much less standardized (as they are often only used in a limited range of stakeholders) and are therefore not very widely applied.

For product design teams, drawings form a very important – but often vastly underestimated – tool for new product design and development. It is in mainly design knowledge capture and the communication thereof that the roles of drawings/sketches are paramount. Drawings, as the first mental models representing the interactions of the product (and its components) with its surroundings during its entire life cycle, are key. They can be used not only for illustrating key product (component) interactions, but can also be basis for mathematical modeling, communication, analysis and synthesis vehicles, for generating the optimal product design. Therefore, the skill of making drawings/sketches in product design is viewed as a key competence every designer should learn to master.

8.3 Product design schematic models

8.3.1 House of quality model for consumer function and property function

In reviewing the different product and process design phases (as presented in Section 4.3), the use of product-related drawings/sketches and their links to mathematical models will be shown here.

The initial design levels after the project framing level are: "consumer wants – product concept" and "product concept – (product) property function." In the "consumer wants – product concept" level, the translation of consumer needs and wants (voice of customer) into quantifiable product attributes takes place. These needs, wants and attributes can be vastly different for any (chemical) product, as illustrated in Example 8.1. The quality function deployment (QFD) method using the HoQ design tool has been applied very successfully for this.

QFD is a structured, yet flexible methodology to define customer needs or requirements and translate them into a development process to yield products that meet those needs. It will help you to:

- Understand customer needs: As customers often do not even know what they want or need, QFD aims to understand their needs better than they do themselves.
- Predict the product's value perception of customers: Understanding how customers judge your product's value is key throughout the entire development process.
- Obtain stakeholder buy-in: All organizational departments need information about the customer needs to enable them to establish processes to develop, market, and deliver to those needs.
- Develop customer-needs-based goals: It is impossible to measure how well the product fulfills customer needs without establishing performance goals such as performance metrics, concepts, design characteristics, process parameters, and production controls.
- Document requirements: Without documentation, customer needs can be misinterpreted from department to department.
- Provide structure: As product development can easily get stuck in inefficiency, the logic and structure of QFD can support development to take a deliberate approach.
- Prioritize resources: By pinpointing the most critical areas for the bottom line and for the customer, resources can be allocated properly to those areas.

Example 8.1: Ice cream that customer wants

Almeida-Rivera et al. [5] present a nice overview of the relationship between consumer wants/perception (great taste, scoopable directly from freezer, healthy) to the property function of ice cream: (see Fig. 8.3).

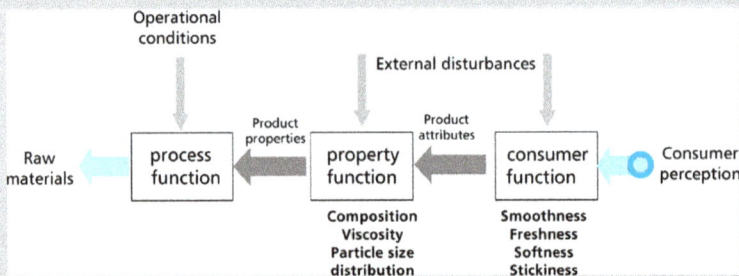

Fig. 8.3: Consumer and property functions in the consumer wants and product attribute design spaces [5].

In this diagram, at first, these consumer likings of an ice cream are translated by a "consumer (or quality) function" to quantifiable product attributes (e.g., smoothness, freshness, softness, and stickiness). The consumer liking/perceptions of a product are mostly obtained through questionnaires and interviews. Careful preparation of these customers' interviews requires good understanding of the product in use. Especially for product innovations, it is a good practice to visualize the various customer/product interactions. This starts, for example, with the purchase of the ice cream in the supermarket, the transport and storage prior to consumption, the consumption activities, and finally, waste disposal. The product attributes are scored against an "arbitrary" scale, which represents the relative performance of the attribute against consumer expectations.

In the second step, these quantifiable attributes are then linked by a "property function" to measurable product properties, based on the composition and product microstructure. For the ice cream case study, product properties include water crystal size and volume fraction, fat crystal size and volume fraction, and air cell sizes and volume fraction. A schematic representation of an ice cream's microstructure is depicted in Fig. 8.4.

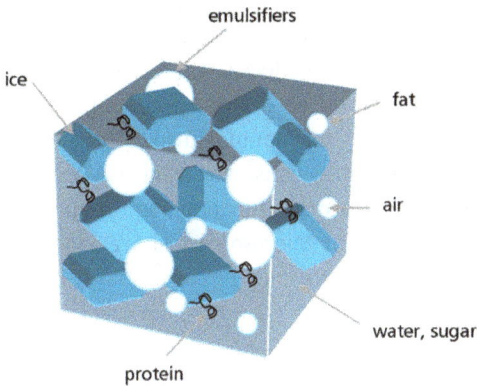

emulsifiers

ice

fat

air

water, sugar

protein

Fig. 8.4: Schematic representation of the microstructure of ice cream [5].

The flow of design decisions runs from the consumer product qualities to the process function and the selection of raw materials. This flow is reverse of the cause-effect chain of physical events when making a product. While the logic of the flow of design decisions is in itself impeccable, yet a practical "information gap" problem arises. This problem is due to the ever-increasing degrees of freedom when going with the design decision flow, that is, in every step of determining the functions (consumer = ≥ property = ≥ process), there are relatively few specifications of what should be the outcomes of a function, while there is an excess of inputs to a function. Consequently, there is an abundance of degrees of freedom. The question arises, how one can handle/fix this degrees of freedom excess? This information gap problem can be solved in a meaningful way when a lot of side information is added per function. That is, additional restrictive requirements on the function inputs must be imposed along with an optimization of the performance of a function. Finding meaningful restrictions and setting suitable optimization goals are essential parts of the design process.

The QFD method can be used to translate consumer wants (consumer voice) into product attributes requirements. The backbone of QFD is the House-of-Quality (HoQ), which relates customer-defined requirements to the product's technical features needed to meet them [6]. The HoQ process involves:

1. Identifying customer requirements
2. Identifying technical requirements
3. Comparing customer requirements to design requirements
4. Evaluating competitive efforts
5. Identifying technical features

The consumer's requirements are the input to the first HoQ-1 (Fig. 8.5), which translates the voice of the customer into product attributes, as seen by the manufacturer. These product attributes become the input for the second HoQ-2, where they are connected to product property requirements. In the third HoQ-3, these product property requirements are mapped onto process requirements for manufacturing. These three subsequent steps (HoQ-1, HoQ-2, and HoQ-3) – translating consumer wants via product attributes and (physical) product properties to the process function requirements – are schematically shown in Fig. 8.6. HoQ templates for Microsoft Excel are available at the QFD ONLINE website [7].

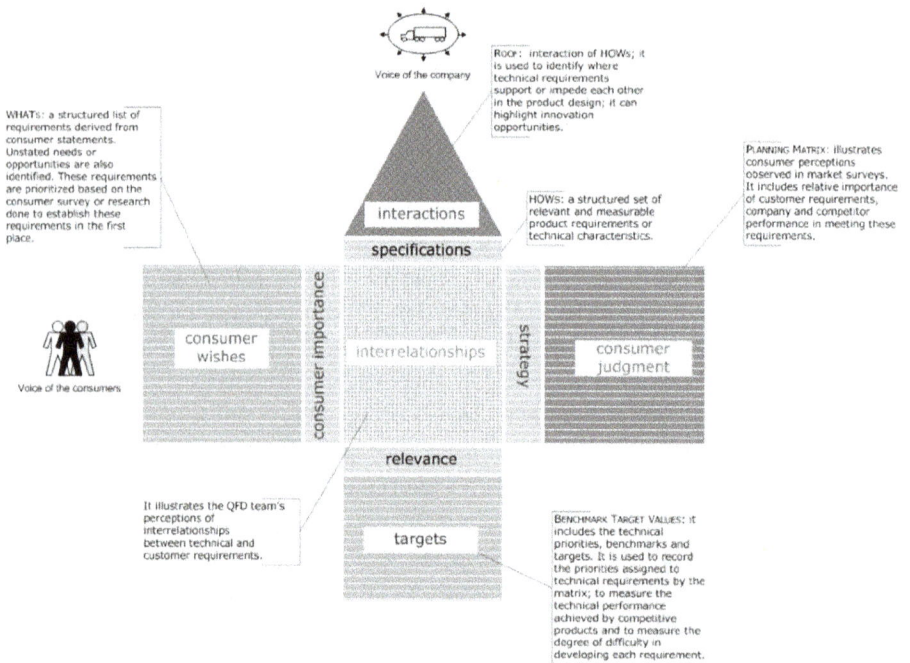

Fig. 8.5: Schematic representation of the use of house of quality (HoQ) 1 in the quality function deployment (QFD) methodology – translating consumer wants (voice of consumers) into product attribute requirements [5].

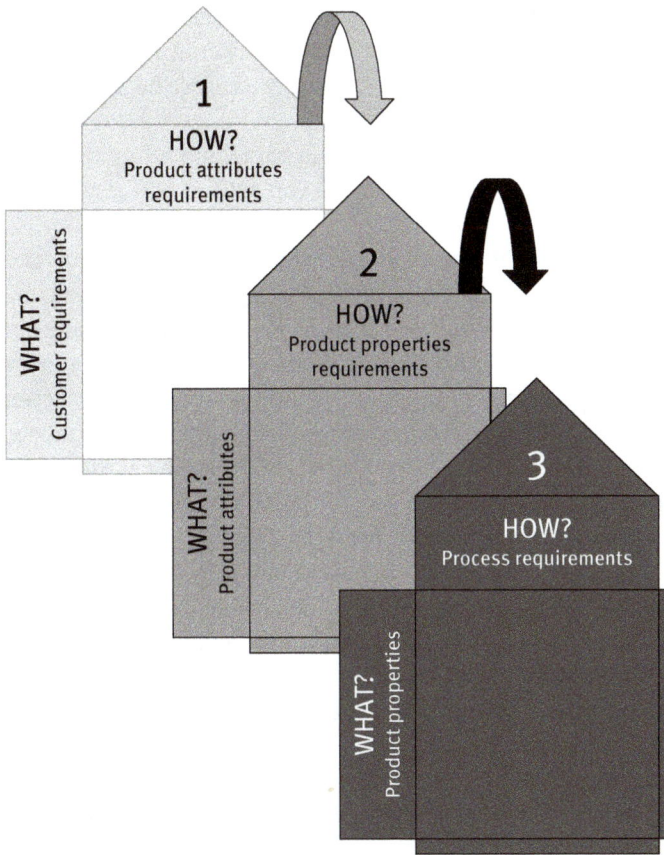

Fig. 8.6: Schematic representation of the use of houses of quality (HoQ) 1, 2, and 3 for translating consumer wants into product attributes, product properties, and process requirements.

8.3.2 Understanding the product application/use process: visualizing system splits

In support of the HoQ QFD method, it greatly helps to visualize and to draw the product in the environment(s) of applying/using the product. The feel of the mouth, creaminess, and iciness (sensory) product attributes should be met and not change during the life cycle phases, prior to consumption. More importantly, by quantifying the different conditions across the manufacture and in-use processes (temperature and relative humidity during storage, transport, display, shopping, home storage and consumption, including time of exposure, etc.), potential risks for product quality can be identified and modeled. If (prolonged) exposures to higher temperatures, prior to ice -cream consumption, is identified, product/environmental interaction drawings and mathematical models can be used to quantify, substantiate, and mitigate this risk. Transport phenomena models involving the ice cream material, the ice cream wrap, carton packaging box, and heat transport models can be used. Relevant physical transport and material properties (heat conductivity (moist) air, ice cream, packaging material, density and specific heat, heat of melting, etc.) come into the equations.

By applying the approaches in this paragraph and in paragraph 8.4.1, the understanding and quantification of the key product application process mechanism can greatly enhance product (re)-design.

Wesselingh et al. [8] emphasize the essentials of product/systems interaction and promote drawing system splits to start the modeling approach. They advocate making different splits of the product/system interactions to get an overview of the product in use and during the different phases of its life time. They use the example of a laundry washing machine and a laundry detergent product. Different systems can be chosen to get different viewpoints of this system and make simple mathematical models, such as an input/output mass and energy balance analysis of a washing machine (IN: soiled dry textile, clean cold water, laundry detergent, electricity power; OUT: wet clean textile, dirty water (containing soil and detergent components, heat losses).

Other ways to split this system is to subdivide the laundry process into process steps (add laundry, add detergent into dispenser, add water, heat up, wash, rinse, centrifuge, and take out laundry. The system can be quantified by taking simple heat and mass balance equations and using systems variables like mass and temperature of textile, water, detergent, required washing temperature, power supply, heat-up rate, washing time, rinsing time, and centrifugation.

A closer look can be taken on the textile composition and structure. How is the textile structure? Is it solid fibers, spun into yarns, or woven into textile? What are the volume fractions of voids in the yarns and between the yarns in the woven textile? Fiber density and diameter, the yarn density and the total surface area of the fibers that need to be wetted and covered with surfactant molecules for efficient removal of soil particles or stains need to be quantified.

The interaction of the detergent product in water and its contact with the textile in the washing machine can also be visualized. The textile tumbles repeatedly in the water, water (with dissolved surface active components) (partly) refreshes the liquid between the fibers and yarns. By the bending and stretching of the textile, refreshing of dirty liquid by fresh liquid takes place, and by the surface-active activity of the detergent ingredients, soil is removed from the surface of the fibers and is kept suspended in water. Antiredepositing ingredients avoid resettling of the soil on the fibers. The liquid water refreshing rate can be computed/estimated, based on a simple transport phenomena model for convective flow of water in/out the textile between the yarns (by bending) and a much slower transport of water/dirt/chemicals into the yarns (between the fibers) by diffusion. The rate-determining step for the overall process is the diffusion in the yarns to the fibers (as can be derived from regime analysis and characteristic times estimations (covered in Section 8.4.1).

Also, the soil, stains, the detergent product, and structure itself can be sketched and modeled. They can be split into different stain types (soluble and insoluble particles, fat/oil, anionic surfactants), when not scavenged. Identification of the molecular identity, quantities, and solution behavior in water are all important. The sketches/drawings can capture and visualize the key phenomena to be considered in the use and life cycle phases of the product.

8.4 Mathematical models for consumer and property functions

In this section, different approaches to create mathematical models for the consumer function and product property function are presented.

8.4.1 Characteristic times and regime analysis

A very convenient way of quantifying the different processes/mechanisms taking place in applying the product and to finally deliver the performance is to compute the so-called characteristic times (or time constants) for these mechanisms. In the 1960s, this practice of developing time constants for rates of changes in chemical engineering systems developed at the disciplinary interface of physical transport phenomena and process dynamics. In the 1980s, this methodology of determining characteristic times within a process was applied to the scale-up of bioreactors by using the "scale-up by scale-down" approach. This method was used to calculate the various characteristic times for mechanisms (mixing, mass and heat transfer, oxygen depletion, product formation, and biomass growth) in stirred aerated fermentors [9, 10] and to establish the rate-limiting mechanism(s). The rate-limiting mechanism determines the regime such as the oxygen- or substrate-limiting regime in (aerobic) fermentations. The process was researched and optimized in a scaled-down fermentor, in which the rate-limiting mechanism(s) on this laboratory-scale were the same as in the large-scale fermentor to be designed.

This approach has been further developed and taught over the years by Kossen, Luyben, Heijnen, and Van der Lans of TU Delft's Biotechnology department, and is still taught to MSc and EngD students at the Delft University of Technology's biotechnology and chemical engineering fields.

This regime analysis method using characteristic times is also very versatile for understanding and improving product application/use processes, and the design of the microstructure and composition of the product.

Characteristic times or time constants can be defined as the ratio of a capacity over a corresponding rate, where capacity and rate are to be used in a rather general way. Some examples of time constants that can be derived from differential mass and energy balances are presented in Tab. 8.2.

Characteristic times for the laundry detergent example in the previous section can be calculated. For the steps involved in the washing process, characteristic times can be estimated for: heating up water, heating up textile, disintegration and dissolution of the detergent powder into the water, diffusion time of chemicals/water into yarns, time for convective transport of water/chemicals into and out of the woven yarns, enzymatic conversion of fat, proteins, and bleach reactions. Based on the results and equations, the rate-limiting step(s) can be identified. By experimentation – in a properly designed scaled-down experimental setup (operating at the regime where the same step is rate-limiting), the "rate" constants can be determined. The experimental results and regime insight can be used to modify the product design (composition and/or structure), and the in-use process of the product. It may prevent making noneffective changes (and cost) in products and processes that are not limiting.

It helps greatly to use schematic "drawing to scale," where possible, so that the designers appreciate the different length scales of the product and systems it interacts with. Order of magnitude estimations for the different system mechanisms' time constants/characteristic times are sufficient. By determining the slowest of the rate-determining process(es), one can focus on these to improve the product and product application process, and assume that the order of magnitude of faster mechanisms take place "instantaneously." This improves understanding of the system and also simplifies the mathematical modeling required. In reference [11], some further examples of product/system splits and modeling are presented:

- Detergent requirements (p. 38)
- Flow properties of toothpaste (pp. 64–66)
- Running an oil tanker (pp. 77–80)
- Injection needle (pp. 95–97)
- Designing a capsule for controlled release (pp. 119–120)
- Design of Rockwool (pp. 170–172)

A very useful source for transport phenomena data and models historically dispersed over many literature sources was gathered and published by Janssen [12]. It provides a good starting point for preliminary estimation of the rate key (transport) processes in the application of products, depending on the structure and size.

A comparable approach to quickly estimating the material and component properties for component substitution in existing products is also presented by Cussler and Moggridge [13]. They present examples of component substitutions, based on thermodynamics (like solubilities) and kinetics (chemical conversion rates and mass and heat transfer rates).

Tab. 8.2: Characteristic times for transport phenomena and (chemical) kinetics.

Time constant (s)	Nominator	Denominator
Travel time: $\tau = L/v$	L (m): travel length	v (m/s): travel velocity
Residence time: $\tau = V/F$	V (m³): volume	F (m³/s): volume flow rate
Diffusion time: $t_d = L^2/D$	L (m): diffusion length	D (m²/s): diffusion coefficient
Conversion time: General: – $t_c = (CV)/(r_cV)$, $t_c = C/r_c$ Zero order: $t_c = C/k$ First order: $t_c = 1/k$	V (m³): volume C (kmol/m³): – Reactant concentration C (kmol/m³) 1 (–)	V (m³): volume r_c (kmol/m³/s): – Conversion rate per volume k (kmol/m³/s): – Zero-order rate constant k (1/s): – First-order rate constant
Mass transfer time From gas bubble into liquid: – $t_{mt} = 1/(k_l.a)$ Depletion of gas bubble: – $t_{mt} = m/k_l a$	1 (–) m (–): ratio C^*_g/C^*_l C^*_g: reactant concentration in gas corresponding to the reactant saturation concentration in liquid (C^*_l)	$k_l a$ (s⁻¹) or [(m/s). (m²/m³)]: – k_l: liquid side mass transfer coefficient (m/s) – a: specific surface area (m²/m³)
Heat transfer (equilibration) time: – $t_{ht} = Vpc_p/UA$	Vpc_p [J/K] – [(m³).(kg/m³).(J/(kg/K)])	UA [W/K] [(W/(m².K)).(m²)] $(1/U = 1/a_I + (pc_p)/\lambda + 1/a_e)$
Momentum transfer time (shear stress relaxation): – $t_{shear} = (L^2\rho)/\eta$	L (m): shear layer thickness ρ (kg/m³): density	η: (kg/(ms)) viscosity

8.4.1.1 Estimation skills – Fermi problems

Estimation skills like these are key for a quick appreciation of the important and less important phenomena to be considered in the design of the product and its application process/use.

Another very nice way for designers to train and apply estimation skills (or "back-of-the-envelope" calculations) is to solve the so-called Fermi problems, named after the physician Enrico Fermi. Fermi estimates deliver the order of magnitude estimates. Information with this accuracy is in many cases often sufficient to make a decision – for example, to neglect a phenomenon or to study it in more detail.

8.4.1.2 "Systemic" time constants – eigenvalues

Estimation of time constants of the separate rate phenomena can give much local insight into a particular phenomenon. Unfortunately, having a full set of such time constants as such is hardly indicative of the behavior of the overall system (product/process) in which various phenomena interact. Suppose there are N resources in a process (e.g., species mass and energy) that can interact with each other and each interaction represents a physical phenomenon with a rate effect (reaction, phase transfer, temperature change, etc.), then $\frac{1}{2} \times N \times (N - 1)$ time constants can be obtained, related to the rate phenomena. The alternative is to order the rate parameters [from which the time constants were derived] in a (square) matrix.

> **Example 8.2**: The Jacobian matrix
> The Jacobian matrix is an ordered set of linear derivatives of an equation with respect to all variables for all equations of a set of conservation equations for the resources in a process (unit). Such a matrix has $N \times N$ entries and each entry can be turned into an equivalent time constant. However, mathematical analysis shows that the overall behavior of the process/product system is determined by the eigenvalues of such a matrix. One obtains only N eigenvalues. The joint effects of $\frac{1}{2} \times N \times (N - 1)$ time constants of the system phenomena are condensed into N generalized or "systemic" time constants.

The benefit of computing and analyzing such eigenvalues is that it extends the knowledge that can be inferred from the full set of rate-related time constants. For instance, eigenvalues will reveal:

- Any linear dependency between resources, meaning that a resource varies as a linear combination of some other resources (occurrence of zero-value eigenvalue(s));
- Stiffness: some resources respond very swiftly to changes, while others lag (very high ratio of the largest eigenvalue over the smallest nonzero one);
- Oscillatory behavior (occurring for an even number of complex-valued eigenvalues with nonzero imaginary parts)
- Instability of the process/product system
  (some eigenvalues have positive real parts causing exponential growth)

These patterns of behavior are hard to fathom from just seeing a set of real-valued time constants.

8.4.2 Data-driven nonlinear product modeling: artificial neural networks

Due to the absence of fundamental relationships between the consumer perception of the product's sensory attributes (e.g., smoothness, softness, and stickiness) and the product's physical attributes (composition, ice crystal size distribution, air cell sizes, etc.), the "product property function" can be formulated by means of data mining techniques.

In 2008, Bongers [14] presented a method for relating the consumer perception of an ice cream to the product physical attributes. This approach was based on large sets of experimental data (and artificial neural networks (ANN)). The "product property function" involved a reduced set of physical product properties that could predict the consumer perception of the ice cream product. Based on this work, Unilever has been able to introduce an innovative lower caloric ice cream with superior consumer perception by reducing fat content, controlling ice crystal, and air bubble sizes by improved processing.

A similar approach was adopted by Almeida (2007) in the rationalization of the ingredients in the manufacture of a mayonnaise-like emulsified product. By means of data mining techniques (partial least square and neural network techniques), only relevant product attributes were classified in sensorial and analytical attributes. These attributes were then used to create alternatives for the product's microstructure and ingredients composition.

The neural networks technique was also used by Dubbelboer et al. [15] in the formulation of the property function for high-internal-phase emulsions.

Another class of chemical product design tools – to study and model product property functions – are DOE (design of experiments) software tools MODDE 11 (Umetrics), JMP® Statistical Software and Design-Expert®. Next to the design of experiments capability, these tools also perform results analysis (model-building and regression) and are increasingly used to build product property models.

8.4.3 Scientific models for product state and behavior

8.4.3.1 Constituents
Before embarking on a description of the possible constituents of a product, it is helpful to introduce a common distinction between two different types of products. Following this distinction, the class of products to be considered for design will further defined. Two product types can be discerned in relation to their extensive form:
-   *"Matter"*-like, being continuously extensive (like any fluid)
-   *"Device"*-like, discrete entity with predetermined dimensions (like solid objects)

Fluid phase products are continuously extensive. Single molecular gases or liquids qualify, as also blended single phase fluids and formulations, and multi-phase systems like emulsions, foams and gels (see Tab. 8.5). This is independent of how the product is made (batch-wise or continuous manufacturing) or how the product is packaged. With respect to solid-phase products, both forms can occur. If the size of particulates of the solid phase is very small and there are numerous ones that can move more or less freely, the product becomes fluid-like again, such as a crystalline powder. It is not an individual particulate that matters in the use of the product. The particulates are used collectively as a kind of "bulk" material. Therefore, the product is classified as extensive.

However, if the solid-phase product has to be handled on the basis of individual pieces rather than as a "bulk" and the solid-phase product has sufficiently large (macroscopic) dimensions, say, well above one millimeter, the product will be classified as discrete. A slab of steel and an electronic chip are examples of discrete products, even when the size is adjustable in the manufacturing process. Also, the multiple solid-state layers on a wafer will be considered as discrete components in the device-like product.

In this paragraph, the characterization of a structured product is written primarily with the continuously extensive product in mind. At the end of the paragraph, some supplementary remarks will be made concerning the device-like discrete products. Here, the following assumptions are made about the product:

- Is continuously extensive
- Is made up of several chemical and/or biological species
- Has one (pseudo-)continuous thermodynamic phase
- Is dispersed when more than one (fluid) phase is present
- Has an interface between each set of two phases, which acts as a distinguishable entity having properties other than those in the "bulk" of the adjacent phases
- Has diffusion of species and thermal energy in and between the phases
- May have chemical reactions between the species

The build-up of constituents in a multiphase structured product is visualized in Fig. 8.7, where the relation between the phase composition, multiphase structure of the mixture, and performance is shown.

8.4.3.2 Product structure

Regarding the geometric aspects of the "bulk" structure, it is observed that in many products, the bulk is made up of a dispersion of one or more thermodynamic phases with multi-level distribution (see Fig. 8.8).

Examples of such dispersed systems are emulsions (liquid phase in another liquid phase), foams (large volume fraction of gas bubbles in a dispersing liquid), bubbly-liquids (dilute suspension of gas bubbles in liquid), gels (semirigid dispersion of solids in a fluid). Dispersion may repeat itself at increasingly smaller scales. Usually, one level of dispersion is considered in chemical equipment modeling, but for structured products like food, cosmetics and pharmaceuticals, the use of multiple levels may be relevant. For repetitive dispersions, the spatial complexity of the system will further increase. A discrete volume entity can act as a continuous phase in which even smaller discrete volume entities are spatially distributed, for example, a droplet of a dispersed phase (β) is contained in a continuous phase (α), which is in turn contained in a droplet of the dispersed phase (β).

However, a droplet of the dispersed phase (β) may contain even smaller droplets of the phase (α) and of another phase (γ). This dispersion of phases might repeat itself at successive levels in increasing detail. The extreme case of a many-level spatial distribution of dispersions (or even a continuous distribution) is not considered. The repetitive occurrence of dispersed phases has implications for the calculation of the overall phase ratios in the product. This is explained by means of a small example.

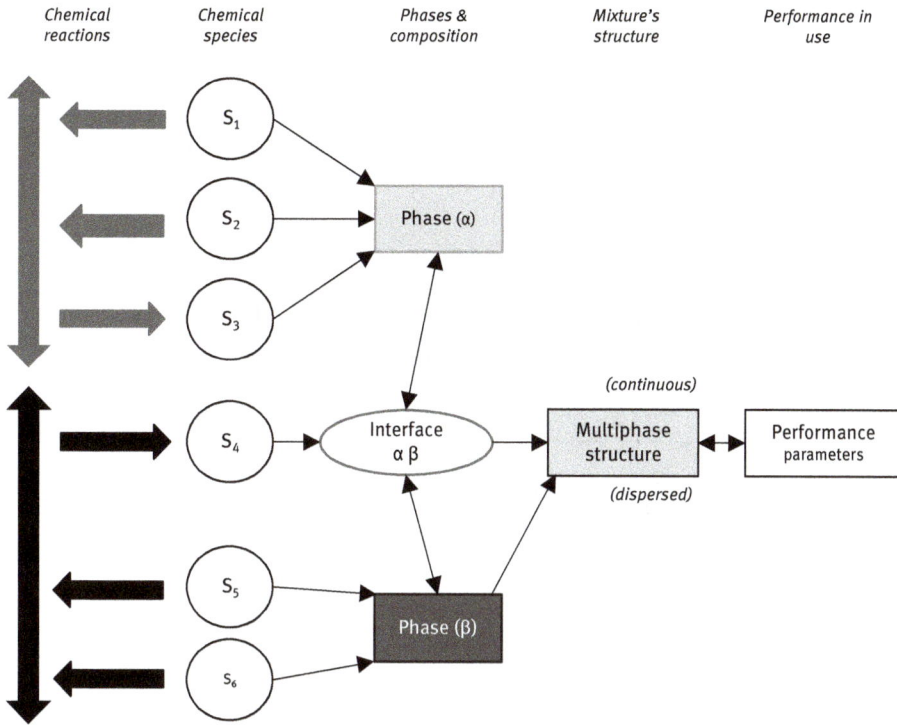

Fig. 8.7: Product composition, structure, and performance for fluid phases.

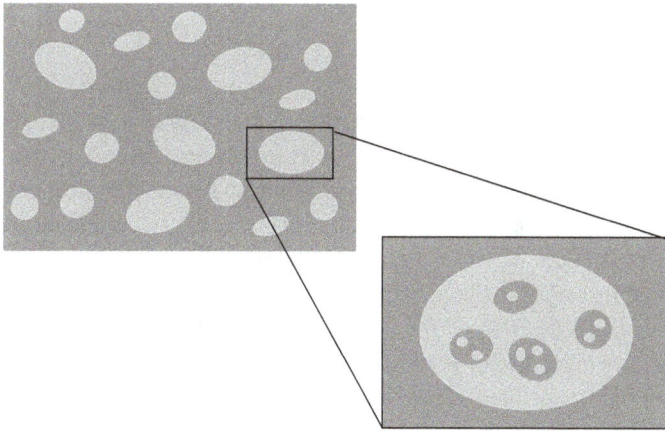

Fig. 8.8: Multilevel/repetitive phase dispersions (oil/water/oil/water).

Example 8.3: Computing the overall phase ratios
The first column in Tab. 8.3 indicates the continuous phase. The matrix row pertaining to a continuous phase shows which mass fractions of the other (two) phases are present (as dispersed phases) in that continuous phase, for example, when a volume element of phase β is the continuous phase, it will typically contain 0.2 mass fraction of phase α and 0.3 mass fraction of phase (γ). This information allows computing the overall phase ratios in a simple way.

Tab. 8.3: A matrix representation of the phase ratios.

Continuous phase	Distributed phase, α	Distributed phase, β	Distributed phase, γ
Phase α	Continuous	$\varepsilon^{(\alpha)}{}_\beta = 0.3$	$\varepsilon^{(\alpha)}{}_\gamma = 0.1$
Phase β	$\varepsilon^{(\beta)}{}_\alpha = 0.2$	Continuous	$\varepsilon^{(\beta)}{}_\gamma = 0.3$
Phase γ	$\varepsilon^{(\gamma)}{}_\alpha = 0.1$	$\varepsilon^{(\gamma)}{}_\beta = 0.4$	Continuous

Consider a two-level dispersion between three phases, where at the first level of dispersion, phase α is the continuous phase, while the other two phases (β) and (γ) are dispersed. At the second level of dispersion, droplets of phase (β) contain dispersions of phases (α) and (γ), and droplets of phase (γ) contain dispersions of phases (α) and (β).
Computation of relative amounts yields:

$$\varepsilon^{\text{overall}}{}_\alpha = [1 - \varepsilon^{(\alpha)}{}_\beta - \varepsilon^{(\alpha)}{}_\gamma] + \varepsilon^{(\alpha)}{}_\beta * \varepsilon^{(\beta)}{}_\alpha + \gamma\varepsilon^{(\alpha)}{}_\gamma * \varepsilon^{(\gamma)}{}_\alpha = 0.67 \tag{8.1}$$

$$\varepsilon^{\text{overall}}{}_\beta = \varepsilon^{(\alpha)}{}_\beta * [1 - \varepsilon^{(\beta)}{}_\alpha - \varepsilon^{(\beta)}{}_\gamma] + \varepsilon^{(\alpha)}{}_\gamma * \varepsilon^{(\gamma)}{}_\beta = 0.19 \tag{8.2}$$

$$\varepsilon^{\text{overall}}{}_\gamma = \varepsilon^{(\alpha)}{}_\beta * \varepsilon^{(\beta)}{}_\gamma + \varepsilon^{(\alpha)}{}_\gamma * [1 - \varepsilon^{(\gamma)}{}_\alpha - \varepsilon^{(\gamma)}{}_\beta] = 0.14 \tag{8.3}$$

Note that the overall phase ratios do sum up to unity.

8.4.3.3 Interfaces between dispersed phases

Each volume body of a phase will have bulk-like features (e.g., in the bulk of the dispersed entities, like droplets) as well as interfacial features for the contact surfaces between phases. When describing interfaces between two thermodynamic phases, one can adopt two viewpoints. In the "macroscale" view, the surface is considered to be an idealized two-dimensional, smooth object of negligible thickness. In the "microscale" view, it is seen as a curved three-dimensional body with a finite thickness (see [16]).

A special interface is the exterior surface of a product. The exterior shape of a fluid-like product with low or moderate viscosity takes the form of its containment vessel. There are usually no exterior surface functions expected of such products. However, a fixated exterior surface will occur for solid phase products or fluid products with a very high viscosity. Then, the form of the surface and surface functions to be performed are made part of the design.

8.4.4 Product structure matrix

One can think of a multitude of combinations of thermodynamics phases that are possible for creating multiphase products. Table. 8.4 provides a schematic overview of the various options.

Tab. 8.4: Product structure matrix for multiphase chemical products (adapted version, original from Prof. A.A. Broekhuis, RU Groningen, Netherlands).

Dispersed phase ⇒ Continuous phase ⇓	Gas	Liquid	Solid
Gas	Mixtures	Aerosols/mist	Dust
Liquid (organics, water, metals, polymers)	Solutions Bubbles Foams	Solutions Emulsions Gels	Solutions Colloids/sols Suspensions Pastes
Solid (organics, metals, ceramics, polymers	Adsorbents Foams	Adsorbents Blends Compounds	Blends, alloys, powders, granules, composites **Device-like (discrete):** Solid films, coating, multiple, stacked layers

A representative example of a complexity of a multiphase dispersed system is ice cream. Ice cream is both an emulsion and foam at the same time. The continuous phase is water with a high concentration of sugars. The liquid water contains a population of small ice crystals. The ice crystals form the first dispersed phase. The second dispersed phase is air, which is foam-like, dispersed in the form of small bubbles in the continuous liquid phase. The third dispersed phase is fat, distributed as small droplets in the water phase, so-called fat globules. The stability of this emulsion is controlled by means of emulsifiers. Actually, emulsifiers are being used to partially destabilize the fat emulsion and so cause partial coalescence. These partially coalesced fat droplets tend to collect at the waterside of the interface with the air bubbles and so help to stabilize the foam of air in water. The coalesced fat can form a spatial 3-D network.

A vast amount of scientific knowledge and detailed scientific models have been developed for describing the behavior of (chemical) products and their components.

A very nice overview of the chemical product design-related theories is presented by Bröckel et al. [17], members of the European Federation of Chemical Engineers (EFCE), in the section group "Product Design and Engineering." The following product design-related theories and basics are presented:

- Interaction forces between particles (by Schubert)
    - From forces between atoms and molecules, to adhesion forces between particles in both gaseous and liquid environments
- Fundamentals of crystallization (by Ulrich, Jones)
- Emulsification techniques for the formulation of emulsions and suspensions (by Schuchmann)
- Characterization of disperse systems (by Polke, Schäfer)
- Modeling of chemical systems to predict product properties (by Gani, Abildskov)

Wesselingh et al. [18] presents an elegant approach to use simple but very illustrative and useful models to quantify various product physical structures (in nm to mm range). These models can be used for a first order of magnitude quantification of product structures in terms of volume fractions, surface area of dispersed phases, number of surfactant molecules needed to form monolayers on the interface, etc. Also, adsorption mechanisms and rheology are explained and quantified.

Bröckel et al. [19] present a number of illustrative examples of product design, like:
- Bio-diesel (by Meier)
- Fats, oils, and waxes (by Reiz and Kleinebudde)
- Starch and starch-based products (by Leeb and Schuchmann)
- Gelatin (for food, pharmaceutical, and technical applications by Babel)
- Sugars (by Häusler)
- Synthetic precipitated silicas (by Schmoll)
- Heterogeneous catalysts (by Kunz et al.)
- Peroral products (by Mäder)
- Carotenoid products (by Leuenberger)
- Coffee-based beverages (by Schuchmann)
- Laundry-powder components (by Boerefijn)
- Agrochemical products (by Frank)

In 2013, Bröckel et al. [20] extended the product design and engineering series even further with a focus on the formulation of gels and pastes. Again, many examples from food, personal care, and ionic liquid applications are discussed, together with the fundamentals of rheology of dispersed systems, and how to apply these fundamentals for the formulation of gels and pastes products.

8.5 Relations between product and process modeling

8.5.1 Causal flow of information on process – product modeling

The purpose of any manufacturing process is to convert the given feeds into products with target performances. For product and process design, one can make models of the feeds, the process, and the products. The product state is an output from a behavioral model of the product. This state is also important to predict the performance metrics of a product by means of a performance model of a product. This can be done on the basis of the physical-chemical state of the product as well as the processing conditions that prevail under a product application. An issue in product and process modeling is how to consistently link the behavioral models of products and processes. The behavioral models of the feeds, the process, and of the products must match, that is, be mutually consistent. For *simple* feeds and products, the situation is simple indeed. A simple product has a single phase with a homogeneous composition. This fully matches with the common characterization of a stream in a process flowsheet model by its composition and flow rate. However, a complex product (and/or a feed) has a single-phase product of a distributive nature (e.g., a crystal size distribution) or it has a multiphase composite structure, possibly also with one or more phase with distributive properties. This was covered already in the preceding paragraph (Section 8.4). This section will focus on connecting the models of the product and its manufacturing process.

(a) For product design and improvement, it is desirable to have clear, causal quantitative links between: (a) the structure and composition of a product at the (supra) molecular-scale, resulting from the product manufacturing process,

(b) The macroscopic physical properties of a product.

(c) The qualities of a product and associated performance for users.

Therefore, one needs three models in a cascade, to generate this flow of information from the product ingredients in the feeds via the product manufacturing process to the product properties and the customer qualities in product use. This is the causal flow of information (physical cause – > effect), shown in Fig. 8.9.

Each of these models will be highlighted. The positioning of the three models and the flow of information between the models is in line with [21].

8.5.1.1 The product manufacturing = > product structure and state model

The structure and the physical state of a product come into existence in a manufacturing process, starting from product ingredients or precursors in its raw feed streams. A model of a product manufacturing process has to be able to describe the structure and states of the process as well as the physical state of its product(s). Hence, a mathematical representation of the matter in process streams and unit operations should allow for the formation and transport of a structured product. Making a proper representation of product structure and the state(s) is to be considered as an integral part of process modeling. However, the current process modeling practice is a bit reluctant in this respect, dealing with simple (single phase) products only. Most commercial process flowsheet simulation programs do not offer facilities to model and simulate the formation of structured products, except as an add-on. Dedicated models of prod-

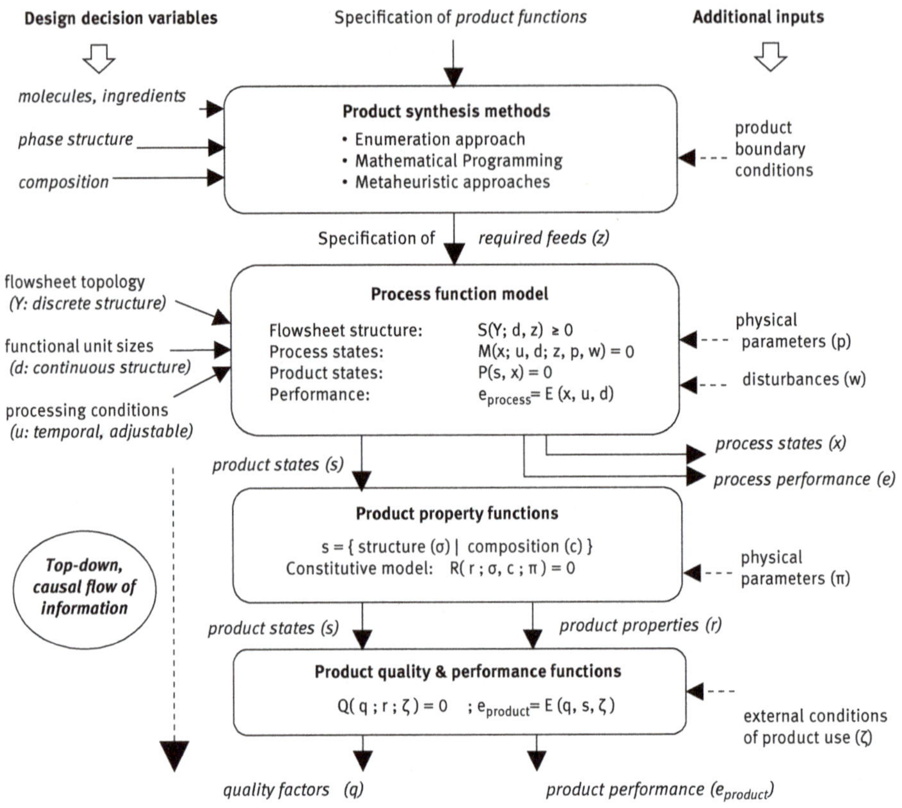

| Design decision variables | Specification of *product functions* | Additional inputs |

molecules, ingredients

phase structure

composition

Product synthesis methods
- Enumeration approach
- Mathematical Programming
- Metaheuristic approaches

product
boundary
conditions

Specification of ↓ *required feeds (z)*

flowsheet topology
(Y: discrete structure)

functional unit sizes
(d: continuous structure)

processing conditions
(u: temporal, adjustable)

Process function model

Flowsheet structure: $S(Y; d, z) \geq 0$
Process states: $M(x; u, d; z, p, w) = 0$
Product states: $P(s, x) = 0$
Performance: $e_{process} = E(x, u, d)$

physical
parameters (p)

disturbances (w)

process states (x)

process performance (e)

product states (s)

**Top-down,
causal flow of
information**

Product property functions

$s = \{ \text{structure } (\sigma) \mid \text{composition } (c) \}$
Constitutive model: $R(r; \sigma, c; \pi) = 0$

physical
parameters (π)

product states (s) *product properties (r)*

Product quality & performance functions

$Q(q; r; \zeta) = 0 \quad ; e_{product} = E(q, s, \zeta)$

external conditions
of product use (ζ)

quality factors (q) *product performance ($e_{product}$)*

Fig. 8.9: Causal flow of information through function models, from the manufacturing process to product qualities and performance.

uct formation may be interfaced with a process simulator. Equation-oriented model building software is well suited to include such models (e.g., gPROMS [22] and JACO-BIAN [23]). Dedicated product-process modeling tools are becoming available for classes of structured products, for example, gPROMS|FORMULATE is suitable for dealing with particle populations with distributive properties [24].

8.5.1.2 A product structure and state = > product properties model

This model derives physical and chemical product properties of the (bulk of the) product (rheological properties, heat capacity, conductivity, pH, etc.) from its underlying physical (micro) structures and molecular compositions. Such structures involve two or more thermodynamic phases, their distributive properties, and the interfacing between the phases. The chemical composition per thermodynamic phase must be known in order to derive the macroscopic properties from suitable constitutive equations. The latter links the molecular information to macroscopic transport properties.

In recent years, very nice progress has been made through the development of several product modeling case studies by the research groups of Prof. R. Gani, Prof. K.M. Ng, and Prof. M. Eden. This has recently led to the first "chemical product simulation and design tool" (ProCAPD), which is offered and supported by the new company 'PSE for SPEED' started by Gani in 2017 [25]. Product properties can be simulated and optimized based on the products' molecular composition and structure. The tool uses very extensive databases and has been demonstrated for design analysis of products as single molecular products (solvents, refrigerants), mixtures/blends (fuels, lubricants), and liquid-formulated products (cosmetics, paints). Each year, TU Delft hosts a workshop by Prof. R. Gani for PhD students and EngD trainees (see Section 14.3.4.1).

8.5.1.3 The product property = > customer qualities (attributes) model

This model links customer-related qualities and performance indicators with the physical properties of a product. Customer qualities are almost always empirical observable quantities, including the conditions under which a product is used or applied. Thus, the conditions in a product application process should be taken into account in a customer qualities model. Thus, such qualities are rooted in both product properties and product application conditions. Rigorous models, based on physical first principles of product use, are hard to derive or simply too expensive. Therefore, a customer qualities model is mostly an empirical, data-driven, black box model, either a statistical one or an artificial neural network – already discussed in Section 8.4.2. The model creates a data-supported bridge between the product properties, the conditions of (external) use of a product, and the product qualities and performance metrics, as appreciated by customers.

8.5.1.4 Interconnectivity between product and process models

The causal flow of information through these three models assumes that information on the required ingredients for the product(s) in the raw feeds to the manufacturing process is already available. This has to be generated by a preceding product synthesis effort. It is stressed that the design of a product and its manufacturing process run conceptually in opposite directions. First, one identifies the desired product qualities and the associated performance metrics (SHEETS-related). Then, one works towards a suitable set of product properties and thereon to product composition and structure. This allows for the identification of suitable product ingredients. Then, one can think of a conceptual design of a manufacturing process. The flow from product qualities to product ingredients and a process design generates a tremendous increase in information. This increase comes from a vast range of design decisions, particularly on the process manufacturing side:

- Process topology: It involves the selection and ordering of processing functions along with their streams connections in a process flow sheet. These are discrete decisions, initially man-made, sometimes leading to superstructures over which one can optimize a design.

- Type and sizes (volumes, exchange surfaces, length, etc.) of processing units, which have to be derived from specified functional targets for the units (e.g., degree of conversion, degree of separation, heat transfer duties, and ratio of pressure increase).
- Operational conditions in the process units (pressure, temperature, preferred flow regimes, phase ratios, etc.).

8.5.1.5 Mathematical model-based product-process optimization

Many product–process design cases are supported by repetitive process–product model simulations in which human designers vary design decisions while aiming for the target product qualities. Such iterative approaches often rely on serial design decision-making. Such iterative, serial approaches lead to feasible designs but not necessarily optimum ones. Therefore, more comprehensive optimization approaches are developed. These approaches rely on a computing-driven automated search over the joint spaces of the design decision variables for an optimum performance. The objective is to optimize some SHEETS performance metrics of the process and product. The search for a feasible optimum should remain within the validated domains of the process models, particularly taking care of staying within the regions of applicability of constitutive equations.

Model-based optimization of product-process designs poses difficulties of four generic kinds:

- The model equations do not comply with the mathematical requirements for the existence of an extremum (e.g., convexity), undermining a theoretical guarantee for an optimum.
- Even if a model would satisfy all conditions for the existence of an optimum, there may be multiple ones and the true extremum may be hard to find within the set of local extrema.
- Exhaustive computational efforts are needed during the search for the optimum due to the high dimensionality of the search space as well as nonlinearity of the process and product models.
- In addition to using real-valued continuous variables (e.g., P, T, x) it can be needed to use discrete variables with binary $\{0, 1\}$ or integer $\{0, 1, 2, 3, .\}$ values. Such integers may enumerate a class of chemical groups, for example, $\{ 1, -CH_2-; 2, CH_3-; 3, -CH = HC-; 4, HO-\}$.

To cope with the hardships in a comprehensive optimization of product and process designs, one can apply some form of problem decomposition. Two kinds of decompositions are mentioned:

– First, optimize the product structure, composition, and performance. One can put this outcome as a tight set of product-related constraints to be obeyed in a process optimization. However, the outcome is likely to be an expensive process, resulting in making the perfect product at a negative profit. Or, the process is even infeasible. One may work to a compromise by relaxing the product optimization and sacrificing slightly on product qualities that tend to make a manufacturing process less expensive. By means of repeated simulations, one may find a feasible solution for both product and process designs. It is hard to say if the combined solution is the optimum one.

– Adopt a sequential approach to product design to ease the computational burden. In computer-aided molecular design (CAMD), discrete variables occur to represent chemical groups that can be synthesized into myriads of different molecular structures for the product. Based on a molecular structure, continuous, real-valued product properties can be determined. The search over such discrete set of variables for molecular structure creation is too demanding, in combination with a process synthesis optimization. Then, an approximation would be to work only with continuous product variables, creating a hypothetical product with optimized product properties and qualities. Then, the first cycle is to jointly optimize the designs of the process and (hypothetical) the product. The next cycle of optimization is to search in the discrete space of chemical groups for a combination that forms a (real) product with properties and qualities close to those of the optimized hypothetical one. Having found the closest real product with near-optimal properties and qualities, one can re-optimize the process design for this real product. This two-step approach is developed and applied to the design of a solvent-based process in [26], and it is also an approach that is applied in case studies by Gani [25]. First, the designs of the process and of a hypothetical solvent (for internal use in the process) are jointly optimized. Such optimization is possible by using only continuous variables in the process and the product models. Once having identified the ideal hypothetical solvent, a search is made for a small set of real solvents, closely matching the identified ideal solvent properties. From this small set of near-optimum real solvents, the best performing candidate is selected by re-optimizing the process for each solvent. This design approach accounts for the interactions between the process and solvent designs in an affordable optimization procedure.

8.5.2 Notes on first-principles models of formation of a structured product

Structured products are made up of multiple thermodynamic phases. The continuous phase of the product may contain one or more dispersed phases of gas bubbles or liquid droplets or solid particles such as crystals. The particulates of a dispersed phase are often distributed in size (nonuniform) and may have internally different physical and mechanical states and chemical composition. Furthermore, a product will have a geometric extension in up to three dimensions. The distribution of the phases and their properties can vary with the spatial position. The structure of a product involves the distributions of phase properties as well as the spatial distributions. Last but not least, the structure of a product may change over time due to internal physical and chemical events in the phases and between the phases, such as chemical reactions, heat and mass transfer, and flow of phases. A visual abstraction of a structured product and an outline of a generic mathematical model of a structured product are given in Fig. 8.10.

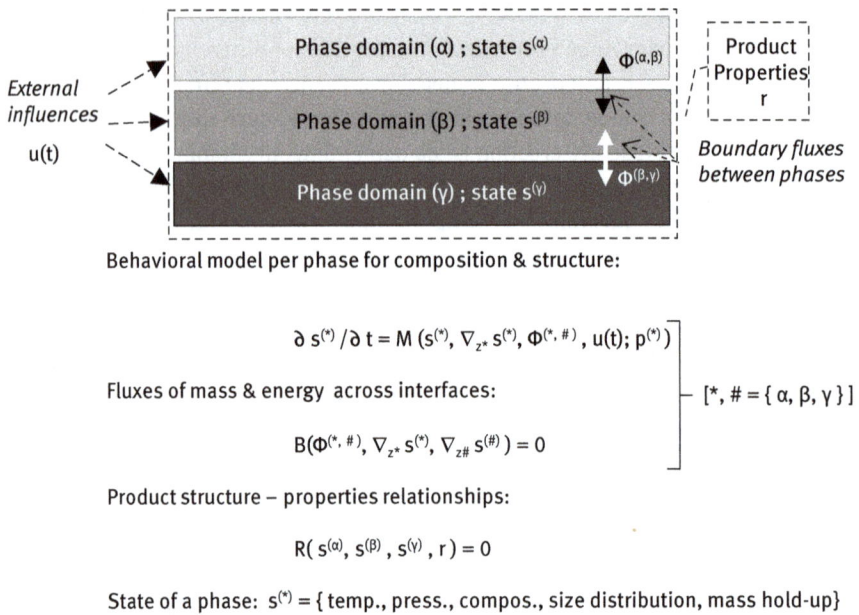

Behavioral model per phase for composition & structure:

$$\partial s^{(*)} / \partial t = M\left(s^{(*)}, \nabla_{z^*} s^{(*)}, \Phi^{(*, \#)}, u(t); p^{(*)}\right)$$

Fluxes of mass & energy across interfaces:

$$B\left(\Phi^{(*, \#)}, \nabla_{z^*} s^{(*)}, \nabla_{z\#} s^{(\#)}\right) = 0$$

$$[*, \# = \{\, \alpha, \beta, \gamma \,\}]$$

Product structure – properties relationships:

$$R\left(s^{(\alpha)}, s^{(\beta)}, s^{(\gamma)}, r\right) = 0$$

State of a phase: $s^{(*)} = \{\,$ temp., press., compos., size distribution, mass hold-up $\}$

Fig. 8.10: Summary of a generic model of a multiphase, structured product.

In a structured product, each phase has its own internal physical state(s) that evolves in time due to internal events in the phase domain as well as to interactions with the neighboring phases by fluxes (φ) and with the external world (u). Heat, mass, and momentum fluxes between adjacent phases are governed by the states of those phases. The properties of a multiphase product (r) are defined by the integral effect of the states (s) of the phases domains $\{\alpha, \beta,$ etc.$\}$. Each of the phases in a structured product has to be premanufactured in an upstream part of the process as a "simple" product, involving a single phase, characterized by its physical state (temperature, pressure, and composition) and physical properties (density, heat capacity, viscosity, thermal conductivity, and surface tension). The various simple products will go as ingredients into a product formation unit, such as an extruder or a colloid mill, to form the desired structured product.

Process modeling and simulation support the design and analysis of manufacturing and product application processes. For simple products, there is a coherent body of modeling knowledge for a broad class manufacturing processes. This knowledge is captured in the common commercial flowsheet simulators for processes dealing with locally uniform fluid phases (gases and liquids). Such generic modeling is far less common for processes that handle structured products. This is due to the complexity and diversity of structured products and the (labor) cost of modeling. Products leaving a manufacturing process and entering a product application process will also set the level of complexity for the associated process modeling. This complexity can be effectively dealt with by means of first principles and equation-oriented modeling, combined with the numerical power of modern equation solvers. Such modeling platforms are under development and are already offered for formulated products and their manufacturing processes [24, 25].

Example 8.4: Freezing ice cream

Let us have a look at a challenging example, the modeling of freezing ice cream. Ice cream has a microstructure, consisting of ice crystals, air bubbles, liquid fat, and a sugary solution (see Fig. 8.4). Starting from liquid feed and air, ice cream is formed in a freezer. The freezer is an externally cooled extruder with a scraped surface heat exchanger (Fig. 8.11). Such units and models are topics of study in chemical engineering literature: [5, 27, 28].

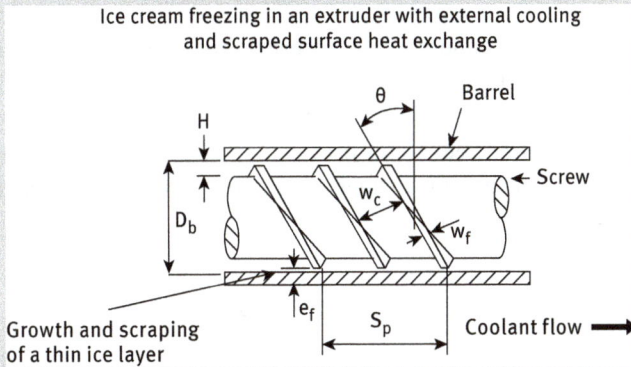

Ice cream freezing in an extruder with external cooling and scraped surface heat exchange

Fig. 8.11: Ice cream freezing in an extruder [5, 28].

The water in the sugary solution freezes at the cooled surface of the barrel, resulting in the growth of a layer of ice on this surface. This layer is periodically scraped off by rotating blades and dispersed as a population of ice crystals of different forms and size in the liquid–air mixture. The mixture of liquid/air/ice mixture is axially transported by the rotor. Fundamental modeling of the ice cream formation is of a staggering complexity. It is staggering because. *in principle*, one has to deal with:

– Eight phases: ice crystals, ice layer, air bubbles, fat droplets, aqueous sugar solution, rotor, wall, and coolant

- Three spatial coordinates (even a spiraling one) and time
- Distributions of particulate size (or mass) for the ice crystals, air bubbles, and fat droplets
- Kinetics of growth, agglomeration, breakage of crystals, bubbles, and droplets
- Mass, energy, and momentum balances for the phases in the ice cream mixture
- Energy balances for rotor, wall, and coolant as well as a momentum balance for the rotor
- Physical transport phenomena (mass, energy, and momentum) in and transfers between phases
- Thermodynamic equilibrium conditions at the interfaces of phases
- Physical properties for thermodynamics and transport (with non-Newtonian flow).

In view of this overwhelming complexity, current ice cream freezing modeling efforts are directed at pragmatically developing very much simplified, yet realistic models of this process unit for design and control [27, 28].

Another parallel modeling effort is required for the application process of the structured product. This is even more challenging. For instance, the "application" process of ice cream is human consumption and digestion. Currently, mechanistic models of the mouth-feel, melting behavior, flavor development and sensing, digestion of ice cream in intestines, and of the impact of ice cream ingredients on body condition and long-term health are hard to find. Yet, new developments in food engineering are on the way by studying food digestion in the human gastrointestinal track as a multiscale process, analogous to multiscale chemical processes [29].

In summary to the relation process-product modeling and computing, one can notice distinct developments toward models of (micro-)structured products, based on first principles as well as data-driven. First-principle product models have to account for much structural information (spatially and distributive properties), leading to complicated sets of (partial) different equations on multiple domains with changing boundaries. This is in contrast to much of the current process modeling and computing, where large sets of nonlinear algebraic equations prevail. Combining and integrating product and process modeling is needed for further advancements in chemical engineering applications.

8.6 Product models: overview

For product design in the (bio)chemical engineering field, the different types of models that are created and used in designing and discussed in this paragraph are listed in Tabs. 8.5 and 8.6. The focus in this table is on the first levels of the Delft design map to show the large types and number of models created and used in a design process.

Tab. 8.5: Schematic overview of various model types created and used in the DDM's project framing and product concept-consumer wants (quality function) design levels (see also Section 4.3.2).

Design levels	Models (physical, schematic, verbal, and mathematical (examples only, not complete list)
PF: project framing	– Business context graphs – *Development/prediction in time: market volume, sales volume, price level product, price level raw materials, energy, population growth, etc.* – Supply chain block diagram – Value chain diagram – Product's life cycle diagram – Schematic, showing design driver and design type – BOSCARD table – Project organization chart (members and accountabilities) – Project team – "Belbin roles" diagram – Project and time planning chart – . . . . . .
CW-PC: consumer wants – product concept: (quality function)	– Product application/use process block diagram – Product life cycle block diagram – Brain-writing collage or table with identified stakeholders from all life cycle phases – Story board showing various consumer interactions with the product – Story board showing other stakeholders interactions with the product – House of quality (1) diagram – *Consumer wants to product (quantitative) attributes* – Product attributes quantitative specifications table – . . . . . .

Tab. 8.6: Schematic overview of various model types: created and used during the DDM's product concept (product) property function design levels (see also Section 4.3.2).

Design levels	Models (physical, schematic, verbal, and mathematical (examples only, not complete list)

PC-PF:
Product concept: (product)
property function
	– Schematic drawing representing alternative product structures
	   – *Schematic drawings of the phases (solid/liquid/gas), their size (distribution), volume fraction, spatial distribution in the product, and composition of each phase)*
	   – *See Fig. 8.4 for Ice cream product*
	– Pure component properties table (See Tab. A13.3)

Tab. 8.6 (continued)

Design levels	Models (physical, schematic, verbal, and mathematical (examples only, not complete list)

	– Mathematical models of alternative products and states
	  – *composition, structure and physical properties (crystal size and distribution, droplet size and distribution, volume, specific surface area, particle density, powder bed density, and viscosity; see Section 8.3.2*
	  – *Steady-state but may be extended to the change of product state over its life cycle (manufacturing, storage, transport, application/ use, and end-of-use) or dynamic models*
	– Block diagram showing product application/use processes
	  – *Application/use of transportation fuels, lubricants, paint, laundry detergents, medication administering, and ice cream consumption*
	– Schematics showing key mechanism of the application/use processes: systems splits (Section 8.3.2)
	  – *Mass transfer: mixing, dispersion, dissolution, heat transfer, (bio) chemical reaction*
	– Time constant equations of the product application/use process mechanisms
	– Artificial neural network models
	– Mathematical models of the product's (components') behavior during the product's application/use
	  – *Also, in storage, transport, collection, and recycling*
	  – *(Bio)chemical reactions (stoichiometry, by-products, waste products production*
	  – *Physical changes: phase changes, dispersion, dissolution, and heat & mass transfer (energy use in product application/use)*
	  – *Simulating effects of different product alternatives*
	– House of quality (2) diagram
	  – *Product (quantitative) attributes translated to product properties (composition/structure, application/use conditions)*
	– Brain-writing diagram of alternative (bio)chemical components
	– Molecular formula/structure
	– Reaction equations
	  – *Stoichiometry, by-products, waste products produced in application/use*
	– Supply chain block diagram update
	– Preliminary "SHEETS" performance models (verbal and mathematical).
	  – *SHEETS: safety, health, environment, economy, technology, social*
	  – *Safety, health, environmental: effects of raw material and products transport and logistics, during waste production and during product application/use*
	  – *Economy: raw material costs, energy use in product application/ use, etc.*

8.7 Modeling for "safety, health, environment, economy, technology, social (SHEETS)"

For schematic and mathematical modeling of products in relation to the "SHEETS" evaluation criteria, the reader is referred to:
- Safety Chapter 11
- Health Chapter 11
- Environment Chapter 12
- Economy Chapter 10
- Technology This chapter and Chapter 9 (for process modeling)
- Social Chapter 12

8.8 Exercises

8.1 QFD tutorials

Go to www.youtube.com and watch:

https://www.youtube.com/watch?v=Mn_Hd9ZtIGw

https://www.youtube.com/watch?v=u9bvzE5Qhjk

8.2 House-of-quality (HoQ) practice

Go to the site: https://www.qfdonline.com/qfd-tutorials/

Train yourself by going through the tutorials

Symbols

A	(m^2)	Surface area (e.g., for heat transfer or mass transfer)
a	(m^2/m^3)	Specific surface area
c	$(*)$	Generic composition variable in product property model
C	$(kmol/m^3)$	Reactant concentration
C^*_g	$(kmol/m^3)$	Reactant concentration in gas corresponding to C^*_l
C^*_l	$(kmol/m^3)$	Saturation reactant concentration in the liquid corresponding to C^*_g
c_p	$(J/(kg.K))$	Specific heat capacity
d	$(*)$	Generic size (length, area, and volume) parameter of a process unit
D	(m^2/s)	Diffusion coefficient
$e_{process}$	$(*)$	Generic process performance indicator in process model
$e_{product}$	$(*)$	Generic product performance indicator in product quality model
F	(m^3/s)	Volume flow rate
k	$(kmol/(m^3 s))$	Zero-order rate constant
k	$(1/s)$	First-order rate constant
k_l	(m/s)	Liquid side mass transfer coefficient
L	(m)	Travel length, and diffusion length

m	(–)	Equilibrium concentration ratio (gas/liquid)
p	(*)	Generic parameter in a process/product model
q	(*)	Generic quality variable in a product quality model
r	(*)	Generic product property variable
r_c	(kmol/(m^{3}s))	conversion rate
s	(*)	Generic state variable of a product model
t_c	(s)	Conversion time
t_d	(s)	Diffusion time
t_{ht}	(s)	Heat transfer time
t_{mt}	(s)	Mass transfer time
u	(*)	Generic input variable to a process/product model
U	(W/(m^{2}K))	Overall heat transfer coefficient
i	(m^3)	Volume
v	(m/s)	Travel velocity
w	(*)	Generic disturbance variable of a process/product model
x	(*)	Generic state variable in process model
Y	(–)	Binary variable
z	(*)	Variable of a feed stream to a process model
α_I	(W/(m^{2}K))	Internal heat transfer coefficient
α_e	(W/(m^{2}K))	External heat transfer coefficient
$\varepsilon^{(\alpha)}{}_\beta$	(–)	Phase mass ratio (of phase β dispersed in continuous phase α)
$\varepsilon^{(\alpha)}{}_\gamma$	(–)	Phase mass ratio (phase γ dispersed in continuous phase α)
$\varepsilon^{(\beta)}{}_\alpha$	(–)	Phase mass ratio (phase α dispersed in continuous phase β)
$\varepsilon^{(\beta)}{}_\gamma$	(–)	Phase mass ratio (phase γ dispersed in continuous phase β)
$\varepsilon^{(\gamma)}{}_\alpha$	(–)	Phase mass ratio (phase α dispersed in continuous phase γ)
$\varepsilon^{(\gamma)}{}_\beta$	(–)	Phase mass ratio (phase β dispersed in continuous phase γ)
$\varepsilon^{overall}{}_\alpha$	(–)	Overall phase mass ratio of phase α in total system
$\varepsilon^{overall}{}_\beta$	(–)	Overall phase mass ratio of phase β in total system
$\varepsilon^{overall}{}_\gamma$	(–)	Overall phase mass ratio of phase γ in total system
ζ	(*)	Generic external condition of use in product quality model
η	(kg/(ms))	Viscosity
λ	(W/(mK))	Heat conductivity
π	(*)	Generic physical parameter for a product model
ρ	(kg/m^3)	Specific density
σ	(*)	Generic structure variable in a product property model
φ	(*/(m^2.s))	Generic flux of a physical quantity
τ	(s)	Travel time, residence time

Glossary

3D	Three-dimensional
BOSCARD	Background, objectives, scope, constraints, assumptions, reporting, deliverables
CAD	Computer-aided design
CAMD	Computer-aided molecular design
CW-PC	Consumer wants – product concept

DDM	Delft design map
DOE	Design of experiments
EFCE	European Federation of Chemical Engineers
HAZOP	HAZard & OPerability study
HoQ	House of quality
P&ID	Piping and instrumentation diagram
PBS	Process block scheme
PC-PF	Product concept – property function
PF	Project framing
EngD	Professional doctorate in engineering
PFD	Process flow diagram
PFS	Process flow scheme
QFD	Quality function deployment
SHEETS	Safety, health, environment, economy, technology, social
TAME	tert-Amyl methyl ether

References and further reading

[1] Overveld CWAM V. PowerPoint Slides guest lecture in EngD course Advanced Principles of Product and Process Design (ST6064) at TU Delft (April 2010).

[2] Skyttner L. General Systems Theory – Problems, Perspectives, Practice, 2nd edn, Singapore: World Scientific Publishing Co. Pte. Ltd, 2005, 97–99.

[3] De Haan AB, Swinkels PLJ, de Koning PJ. Instruction Manual Conceptual Design Project CH3843 – From Idea to Design, Delft, Netherlands: Department of Chemical Engineering, Delft University of Technology, 2016, Appendix–18.

[4] De Haan AB, Swinkels PLJ, De Koning PJ. Instruction Manual Conceptual Design Project CH3843 – From Idea to Design, Delft, Netherlands: Department of Chemical Engineering, Delft University of Technology, 2016, Appendix–17.

[5] Almeida-Rivera C, Bongers P, Zondervan E. A structured approach for product-driven process synthesis in foods manufacture. In: Martin M, Eden MR, Chemmangattuvalappil NG (eds). Computer Aided Chemical Engineering Vol.39 Tools for Chemical Product Design – From Consumer Products to Biomedicine, Amsterdam, Netherlands: Elsevier BV, 2017, 417–441.

[6] Zondervan E, Almeida-Rivera C, Camarda KV. Product-driven Process Design, Berlin, Germany: De Gruyter, 2020, Vols. 104–108 133–137. 148–150.

[7] QFD Online Quality Function Deployment tools and information for real time application, free QFD Templates 2007–2010. (Accessed October 10, 2017, at www.qfdonline.com/templates/.)

[8] Wesselingh JA, Kill S, Vigild ME. Design & Development of Biological, Chemical, Food and Pharmaceutical Products, Hoboken NJ, USA: John Wiley & Sons Inc, 2007, 27–38.

[9] Oosterhuis NMG, PhD Thesis, Delft University of Technology, 1984. (Accessed August 19, 2017, at https://repository.tudelft.nl/islandora/object/uuid:03a887b7-8c20-4052-8d6b-7fe76918d7ec/datastream/OBJ.)

[10] Sweere APJ, Luyben Kch AM, Kossen NWF. Regime analysis and scale-down: Tools to investigate the performance of bioreactors, Enzyme and Microbial Technology, 9, July 1987, 386–398.

[11] Wesselingh JA, Kill S, Vigild ME. Design & Development of Biological, Chemical, Food and Pharmaceutical Products, Hoboken NJ, USA: John Wiley & Sons Inc, 2007, Vol. 38 64–66. 77–80, 95–7, 119–120, 170–2.

[12] Janssen LPBM, Warmoeskerken MMCG. Transport Phenomena Data Companion, Delft: VSSD, Delftse Uitgevers Maatschappij, 1987.

[13] Cussler EL, Moggridge GD. Chemical Product Design, New York, USA: Cambridge University Press, 2001, 75–85.

[14] Bongers PMM. Model of the product properties for process synthesis. In: Braunschweig B, Joulia X (eds). Computer Aided Chemical Engineering Vol.25, 18th European Symposium on Computer Aided Process Engineering – ESCAPE 18, Amsterdam, Netherlands: Elsevier BV, 2008, 55–60.

[15] Dubbelboer A, Zondervan E, Meuldijk J, Hoogland H, Bongers PMM. A neural network application in the design of emulsion-based products. In: Lockhart Bogle ID, Fairweather M (eds). Proceedings of the 22nd European Symposium on Computer Aided Process Engineering, London. Amsterdam, Netherlands: Elsevier BV, 17–20 June 2012, 2012 692–696.

[16] Edwards DA, Brenner H, Wasan DT. Interfacial Transport Phenomena and Rheology, Boston MA, USA: Butterworth-Heinemann, 1991.

[17] Bröckel U, Meier W, Wagner G (eds). Product Design and Engineering – Best Practices, Volume 1: Basics and Technologies, Weinheim, Germany: Wiley VCH Verlag, 2007.

[18] Wesselingh JA, Kill S, Vigild ME. Design & Development of Biological, Chemical, Food and Pharmaceutical Products, Hoboken NJ, USA: John Wiley & Sons Inc, 2007, 259–293.

[19] Bröckel U, Meier W, Wagner G (eds). Product Design and Engineering – Best Practices, Volume 2: Raw Materials, Additives and Applications, Weinheim, Germany: Wiley VCH Verlag, 2007.

[20] Bröckel U, Meier W, Wagner G (eds). Product Design and Engineering: Formulation of Gels and Pastes, Weinheim, Germany: Wiley VCH Verlag, 2013.

[21] Bernardo FP. Integrated process and product design optimization, Ch. 12. In: Martin M, Eden MR, Chemmangattuvalappil NG (eds). Computer Aided Chemical Engineering Vol.39 Tools for Chemical Product Design – From Consumer Products to Biomedicine, Amsterdam, Netherlands: Elsevier BV, 2016, 347–371.

[22] Process Systems Enterprise (UK). gPROMS: Advanced process modelling platform. (Accessed October 5, 2017, at https://www.psenterprise.com/products/gproms.)

[23] Process Systems Engineering Laboratory, Massachusetts Institute of Technology (USA), Process modelling & computing software JACOBIAN. (Accessed October 5, 2017, at https://yoric.mit.edu/Jacobian.)

[24] Process Systems Enterprise (UK). gPROMS|FORMULATE: GPROMS Formulated Products, platform for integrated digital design of robust formulated products and their manufacturing processes. (Accessed October 5, 2017, at https://www.psenterprise.com/products/gproms/formulatedproducts.)

[25] PSEforSPEED Company Ltd., Computer-aided application of process systems engineering (PSE) for sustainable product-process engineering, evaluation & design (SPEED). (Accessed October 5, 2017, at http:www.pseforspeed.com.)

[26] Bardow A, Steur K, Gross J. Continuous-molecular targeting for integrated solvent and process design, Industrial & Engineering Chemistry Research, 49, 2010, 2834–2840.

[27] Arellano M, Benkhelifa H, Alvarez G, Flick D. Coupling population balance and residence time distribution for the ice crystallization modeling in a scraped surface heat exchanger, Chemical Engineering Science, 102, 2013, 502–513.

[28] Dorneanu B, Model reduction in chemical engineering. Case studies applied to process analysis, design and operation. PhD Thesis, Delft University of Technology, 2011. (Accessed October 5, 2017, at https://repository.tudelft.nl/islandora/object/uuid:afecbacd-0ac1-4a01-b7ab-db1ba0c42f50?collection=research.)

[29] Bornhorst GM, Gouseti O, Wickham MSJ, Bakalis S. Engineering digestion: Multiscale processes of food digestion, Journal of Food Science, 81, 2016, R534–543.

[30] Ng KM, Gani R, Dam-Johansen K (eds). Chemical Product Design: Towards a Perspective through Case Studies, Computer Aided Chemical Engineering – Volume 23, Amsterdam, Netherlands: Elsevier BV, 2007.

[31] Khalloufi S, Kharaghani A, Almeida-Rivera C, Nijsse J, Dalen G, Tsotsas E. Monitoring of initial porosity and new pores formation during drying: A scientific debate and a technical challenge, Trends in Food Science & Technology, 45, 2015, 179–186.

[32] Khalloufi S, Almeida-Rivera C, Janssen J, Van der Vaart M, Bongers P. Mathematical model for simulating the springback effect of gel matrixes during drying processes and its experimental validation, Drying Technology, 29, 2011, 1972–1980.

[33] Almeida-Rivera C, Jain P, Bruin S, Bongers P. Integrated product and process design approach for rationalization of food products. In: Pleşu V, Şerban Agachi P (eds). Computer Aided Chemical Engineering – Volume 24, Amsterdam, Netherlands: Elsevier BV, 2007, 449–454.

[34] Linke P, Papadopoulos AI, Seferlis P. Systematic methods for working fluid selection and the design, integration and control of organic rankine cycles – a review, Energies, 8, 2015, 4755–4801.

[35] Papadopoulos AI, Stijepovic M, Linke P, Seferlis P, Voutetakis S. Toward optimum working fluid mixtures for organic rankine cycles using molecular design and sensitivity analysis, Industrial & Engineering Chemistry Research, 52, 2013, 116–133.

[36] Schilling J, Tillmanns D, Lampe M, Hopp M, Gross J, Bardow A. From molecules to dollars: Integrating molecular design into thermo-economic process design using consistent thermodynamic modeling, Molecular Systems Design & Engineering, 2, 2017, 301–320.

Part C: **Design optimization**

9 Process modeling and optimization

Synopsis: This chapter deals with modeling in support of three successive innovation stages: concept, feasibility, and development. Linear process modeling is associated with applications in the concept stage. Nonlinear process modeling is linked with the feasibility and development stages. Elaborate process simulations will find applications in the development stage, while the feasibility stage also benefits from optimization applications that scout the entire design space and optimize over it.

Special attention is paid to the use of flowsheet simulators in specified ways for each of these three stages. There are also brief sections on the use of software packages for computational fluid dynamics and life cycle assessment for each of these three stages.

The modeling approach chosen here has a practical inclination. Many academic textbooks on modeling cover the common modeling steps, such as development of model equations, mathematical model analysis, and numerical means for computing solutions. Therefore, these steps are not treated here. Instead, the contextual and application sides of modeling in process engineering gets attention, for example, how does one get to a relevant model for a process engineering problem, how to analyze and evaluate the outcomes from computational model applications, and how to derive clues for process innovations? Also, aspects of sensitivity analysis, uncertainties, risks, and identification of technological constraints from model-based solutions will be discussed.

9.1 Justification and objectives of process modeling

Modeling objectives for process synthesis, analysis and optimzation

The purpose of process modeling and computing is to offer effective, quantitative means to synthesize, analyze, and evaluate process designs. The main goal of this modeling and computing chapter is to bring out the conceptual aspects of modeling, simulation, and optimization for process design applications. In view of the limited text space, this emphasis will go at the expense of the underlying mathematical and numerical aspects of model building and equation solving. Here, we will simply assume that:

- Models will be properly formulated from mathematical and computational perspectives.
- Their domain of applicability (containing feasible physical solutions) is well understood.
- An applied numerical solution procedure is consistent with the model and sufficiently accurate.

https://doi.org/10.1515/9783110782127-009

The focus will be on the context and practice of developing model applications for the concept, feasibility, and development stages in support of product and process designs; see also Chapter 4. Attention will also be given on how to extract relevant information from model applications for the purpose of improvement and innovation. There is an established industrial practice of supporting process design case studies with commercial steady-state process flowsheet simulators. This vast area of computer-aided process simulations has been thoroughly covered in [1] and will be skipped here. Also, dynamic modeling and model reductions for process control and operations are omitted.

Further background reading on process modeling for simulation and optimization in process development, design, and operations is offered in a number of high-quality text books regarding modeling [2–8], flow-sheeting simulations for design [1, 9–12], and optimization [13–16].

9.1.1 Conceptual representation of a process: *network features*

A process is an open system that internally consists of a network of interconnected and interacting unit operations. The system is in interplay with its physical, economic, and social surroundings. The system performs specified production functions, i.e., turning physical feeds into products with a chemical functionality. The production is guided by process and product control information. The system may operate in different modes over time. Each mode is characterized by a distinct internal state of the process.

A process is connected to its surroundings by:
– Feed, product and waste streams (the latter to waste treatment facilities)
– Control set points for its operational conditions and modes, as supervised by plant operators
– Environmental conditions
– Cash flow it generates for the process owner

The behavior and performance of a process is co-determined by the interactions between a process and its surroundings. This includes the imposed modes of operation. Each mode can be seen as the realization of an external *scenario*, under which the process must make product(s) while performing well enough. Development of scenarios is part of the design; these scenarios are used for testing candidate processes under design. Consequently, the set of process models must be able to handle all scenarios. This explains why next to steady-state models, dynamic models are also required.

9.1.2 Sharing a generic view on process modeling and computing

Design models carry a lot of process information on behavior and performance. The common steps in the modeling work procedure are outlined in Tab. 9.1. These steps are presented in a sequential fashion. This is NOT how it works in practice. There is a lot of going forward and backward over the steps during the modeling activity. One may also visualize these activities as placed on the perimeter of a circle, with easy crisscrossing between activities.

Tab. 9.1: Common steps in a modeling work procedure.

Step	Activity
1	Select the **object of modeling** and justify a modeling effort from **anticipated benefits**.
2	Agree on **goal(s) of a model** and the intended application domain(s) for future model use.
3	Find a suitable **conceptual representation** of the object to be modeled.
4	Develop a **mathematical model** of the object for its structure/behavior/performance.
5	Code the mathematical model in **software,** with **verification** of correctness of coding.
6	Select a **numerical solver** with enough numerical precision for intended applications.
7	**Validate** the coded model by comparing the computed answers with experimental data.
8	**Select case studies** based on **interaction scenarios** between the object and its surroundings
9	**Apply the model** to **case studies** and **analyze** the computed results on the achieved accuracies.
10	**Evaluate** the computed results **on significance** for goals and new insights.
11	**Report** the model, case studies, numerical results and the evaluation of results.

The next sections will deal with the nature of model applications in the three design stages: concept, feasibility, and development. The successive stages of process development and design call for models of different functionalities and capabilities. Three classes of models are considered:

- Linear models for process input-output blocks, for use in the *concept* stage
- Nonlinear models for process simulation, for use in the *development* stage
- Non-linear models for process optimization, for use in the *feasibility* stage

For each stage, the main modeling aspects generic to that stage will be discussed, involving:

- Consistent model formulation with proper degree of freedom analysis
- Significance of solutions and sensitivity analysis
- Analysis of active inequality constraints in the case of process optimization
- Identification of technological limitations in optimized solutions

Having dealt with the generic aspects, a number of (small scale) examples are presented as illustrations of certain modeling aspects. These examples are small-scale, on purpose, to illustrate a way of thinking and acting in modeling. There is a good reason to keep them small. The many facilities of process modeling software, with the associated computing power, often tend to overwhelm novice modelers. Therefore, it is preferred to develop basic model synthesis and analysis skills first, by means of small-scale problems. These problems can even be solved by hand; they serve as vehicles to transfer ways of thinking in model synthesis, analysis, and evaluation. Each example is concisely presented in this chapter by means of motivation and objective, modeling approach, and main outcomes. Details of the examples and results are collected in Appendix A9.

Before embarking on modeling for the three design stages, it is expedient to remark about a restrictive meaning of some terms in this chapter. The terms *behavior* and *performance* will be used in a more focused way than commonly done in chemical process engineering. The term *behavior* of a process or product will relate to patterns of change in its (bio) chemical and physical quantities. The term *performance* is intentionally reserved for factors that express some kind of value and appreciation by human society, including customers of a product, operators and owners of a process, and legal authority. Such performance factors are related to safety, health, economy, and ecology.

9.2 Contributions to a *concept* stage with linear modeling

9.2.1 Modeling for a concept stage

Let us move from a general modeling aspect to the specifics of modeling for process development and design. In this chapter, a distinction is made between three generic main items of any process model:

–   A submodel, describing the physical/chemical/biological *behavior* of a process. This submodel deals with changes in physical and chemical resources (species mass, energy, momentum, electrical charge, etc.) as they go through a process.
–   Another submodel, quantifying some indicators for process *performance*. These indicators reflect the degree of appreciation by process owners of what the process (design) achieves in terms of economy and ecology under safe operation.
–   *Scenarios*, defining pertinent influences of the surroundings of a process on the process itself. One must specify external scenarios under which a process is supposed to perform its function(s) in order to compute and analyze process *behavior* and *performance*.

A process interacts with its environment and this feature must be represented in a process model. The common, causal flow of information to and from a process model is shown in Fig. 9.1.

It represents the flow of physical quantities from process inputs to its outputs. The inputs cover the physical feed streams, the selected operating conditions (u), and physical parameters (p) for the process units as well as other known external influences (w) acting on the process. The fourth class of inputs involves the geometric design variables (d) of the process units (e.g., lengths, areas, and volumes). Their numerical values will

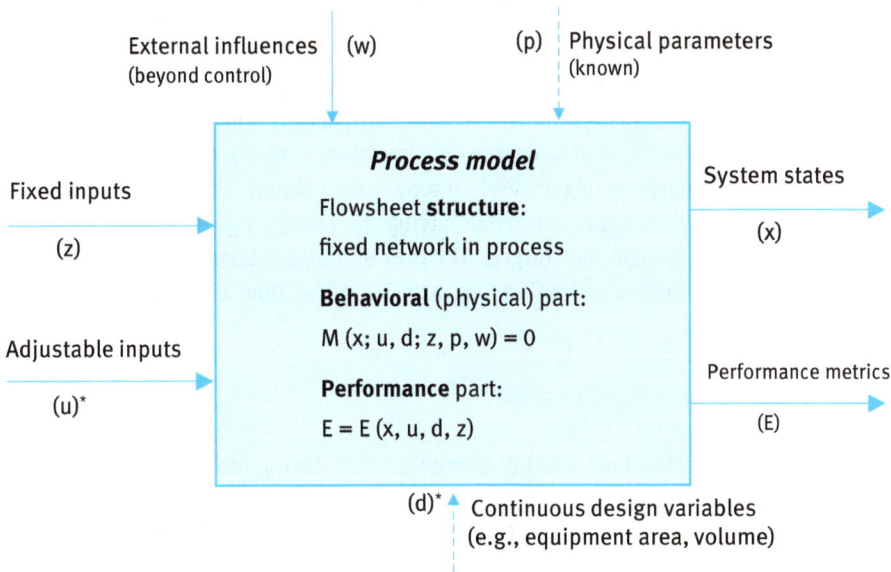

External influences (w) (p) Physical parameters
(beyond control)   (known)

Fixed inputs

(z)

Process model

Flowsheet **structure:**
fixed network in process

Behavioral (physical) part:

$M(x; u, d; z, p, w) = 0$

Adjustable inputs

(u)*

Performance part:

$E = E(x, u, d, z)$

System states

(x)

Performance metrics

(E)

(d)* Continuous design variables
(e.g., equipment area, volume)

Model based analysis
of a process design:

Given a **scenario:**   ==>   Compute:
{z, p, w} and {u*, d*}   {x, E} and sensitivities

{u*, d*} are given design decision variables

Fig. 9.1: Information flow through a process model for analysis of a process design.

have been established in the preceding design phase. The outputs (x) are the collective set of physical states of units and streams in the process. In this figure, it is assumed that the types of unit operation and the ways in which such units can be connected to an internal network structure have been decided upon by the preceding conceptual design decisions. Fig. 9.1 mentions the three main submodels: behavioral, performance, and the external scenario(s). Here, the structure part is assumed to be fixed. This assumption needs to be relaxed in the case of modeling for process synthesis studies that specifically deal with alternative process structures. Consequently, the model setup in Fig. 9.1 is suitable only for the *analysis of a design*. All these inputs must be given by the designer/modeler; there is no automatic adjustment of the inputs to target for process states or to optimize process performance metrics.

9.2.2 Design leads to a reversal of information flow to a model

In actual design situations, the geometric design variables will be unknowns rather than given quantities. These unknowns add to the degrees of freedom in a design situation. So, to take away the additional degrees of freedom, the model need to be augmented by as many design specifications as there are unknown geometric design variables. These new design specifications are often given as *specified targets for some key physical outputs of the process*, e.g., nominal production rate and product qualities as well as constraints on waste streams (sizes and waste content). There can also be specifications of the duties of process units, such as degree of conversion, selectivity in a reactor, and the desired degree of separation or purities in a separation column. The general rule is that targets on physical outputs are set, while having the required geometry of the equipment as an outcome to meet such targets. The flow of design information from targets to geometric features can be a reversal to the causal (physical) flow of information.

9.2.3 Modeling for the Concept stage

A Concept stage is focused on creating alternative, competing product concepts. In the background, though, one must also keep in mind the basic processing principles for eventually making such a product from the available feeds. It is prudent to think about a process block diagram that outlines the key processing steps to make a product. The viability of a product concept will also be determined by its manufacturing complexity and costs. At this Concept stage, process models and computations should be kept simple, addressing only the main features of processing functions. For this reason, it suffices to use *linear* models of the processing steps in a process block diagram for a first exploration. A model is linear when each equation in the model is linear in all variables. There are five common forms of deterministic models summarized in Appendix A9. The application examples in this section deal with linear algebraic and differential equations with real variables. The exception is the modeling of the structure of a block diagram / flowsheet involving a linear algebraic model with binary variables. Six examples are given to illustrate some aspects of modeling.

9.2.4 Examples for Concept stage modeling

Example 9.1: Exploring the stoichiometric space **for an innovative** chemical conversion

Motivation and objective:
Chemical processes have a conversion section in which the main chemical species in the feed streams are converted to target species for the product(s). A conversion is characterized at the chemical level by (a) the reaction stoichiometry; (b) associated thermodynamic reaction equilibrium; (c) reaction kinetics; and (d) level of catalytic activity and catalyst stability over time. The considerations of the associated physical transport phenomena, phase contacting patterns, and hydrodynamic flow regimes are part of the subsequent physical reactor selection and design. The reaction kinetics can be seen as an option for the optimization of product yield over the reaction conditions and over the ratios for reactant feed concentrations. Stoichiometry is often assumed to be given and fixed. Yet, stoichiometry can also be a degree of freedom in the design of a process concept. A simple example is of a search over a

stoichiometric space to find the best reaction serving a particular conversion goal. In this case, the objective is the chemical conversion of carbon-dioxide as a greenhouse gas into carbon-monoxide as a chemical reactant in syngas. The latter is, among others, a feed for methanol synthesis and Fischer-Tropsch synthesis for liquid hydrocarbon fuels. The reuse of carbon in chemical products helps in closing the carbon cycle.

Approach:
Linear algebra analysis of the stoichiometric matrix is applied to extend the range of dry-reforming reactions.

Outcome:
The outcome of this reaction modeling example shows how a systematic exploration of the space of reaction stoichiometry can lead to an optimized yield of a target product. The derived stoichiometric model enables one to explore for potentially interesting and useful reactions in a given species space. The super-dry reforming reaction is the highest performing reaction for CO_2 to CO conversion. An experimental study of this reaction and its inventive underlying catalytic scheme was reported in [17]. Stoichiometric analysis also revealed the high-dry and medium-dry reforming reactions as interesting side-catches. In principle, these reactions may occur simultaneously with super-dry reforming. In a reflection on uncertainties and limitations playing a role in this modeling, one should consider that the outcome space is determined by the chosen set of chemical species. One could argue that solid carbon (C_{solid}) should be added to the chemical species set to allow for more reactions, such as the Boudouard reaction ($2\ CO \leq\ > CO_2 + C_{solid}$).

Example 9.2: A conversion – separation – recycle **structure using lumped units**

Motivation and objective of modeling:
When moving from concept to feasibility stage, one must have some ideas about the use of species mass resources in a future manufacturing process for the product under consideration. Quite a few product options may have to be explored; so simple process modeling and fast computing are preferred. The simplest mathematical models in chemical engineering are linear ones that can be formulated and solved easily. As an example, a linear algebraic model of a common reactor-separator-recycle processing structure is derived and solved. This example may serve as a leading example on how to tackle such cases in general.

Approach:
A process block diagram is generated with a unique labeling of the process units, the streams connecting the units, as well as the chemical species in the streams. The units in the process are treated in a lumped, input-output way. The functional task(s) of a unit are simply characterized by means of model parameters, such as the degree of conversion and selectivity, and separation factors. Linear component mass balances will be generated over all process units and integrated to balances over the entire process. A degree of freedom analysis reveals how many external conditions need to be specified as part of a scenario under which a process operates.

Outcome:
The main outcome is a process block diagram (Figure A9-2), represented by a square linear algebraic model **A**. $\underline{x} = \underline{b}$. The vector **x** represents the species mass flows in the process. Hints are given on performing sensitivity analyses regarding how these mass flows depend on the chosen values for the process unit parameters. An energy balance is also of importance but it is bi-linear (mass flow multiplied with temperature). However, knowing the species mass flows and by adding some reasonable physical approximations, the energy balance is reduced to a linear form and easily solved.

Example 9.3: Specification of connectivity **in a block diagram by use of a** structure model

Motivation and objective of modelling:
The units in the process block diagram of example 9.2 are connected by streams; see Figure A9-2. The models of the process units include the species flows in the connecting streams. If units A and B are directly connected by stream (l), the models for unit A and B make use of the same flow variables. The behavioral models for the units account for the connections as well. There is no separate model to describe the connectivity of units by means of streams. The behavioral and structural models have been lumped together. However, suppose one would want to rearrange a process block diagram with different connections between process units using alternative streams, the model equations of the units involved in a reshuffle need to be rewritten because incoming and outgoing streams are relabeled. Such rewriting is error-prone and may involve quite some bookkeeping effort. Furthermore, it is conceptually unpleasant that a change in the connectivity of units must result in changes of the model equations describing unit behavior. It would be better to instead see a change in the connections by streams. In other words, a separation between the behavioral models of the units and a process flow sheet structure model is preferred.

Approach:
Each process unit is given slots for the incoming and outgoing flows, where each slot gets its own set of flow variables. This explicit definition of slot-related input and output flow variables allows for a formulation of the process unit model that is independent of the actual connections occurring in a process block diagram. The flow sheet structure is used to link stream slots of the process units. Each process stream must connect an output slot of a unit with an input slot of another unit.

Outcome:
A decoupling arises between the behavioral models and structural connectivities of units by streams in the flow sheet. This preferred way of modeling can be realized at a slight expense of having more linear connectivity equations in the model. The full set of the connectivity equations forms the *structure model* of the block diagram. To cover optional alternative connections between process units, a formal extension is made to include binary variables to switch on/off such connections. The price of the additional computing efforts due to the solving of additional linear connectivity equations is very small; it pales, compared to the benefit of transparent flow sheet models.

Example 9.4: Optimization of process performance by Linear Programming **(LP)**

Motivation and objective of modeling:
The species balances over the process units in a process block diagram give rise to a linear model with some residual degrees of freedom. The latter is taken away by specifying an external scenario under which the process must function. Such a scenario can set some intake rates of the feeds or targets for the product output rates. When the design problem is fully specified, the model can be solved. One often adds a performance metric for the design, e.g., an economic potential function and/or chemical element efficiency can be computed. The latter indicates how much of a key element (say carbon) in the feed stream ends up in the products; it is often expressed as a fraction.

A full specification of a design problem gives a "point" solution only. Practically, it is more relevant to explore the design-decision space for the best performance of a design. To stay in the linear domain, the performance metric must also be linear in the (unknown) process variables.

Approach:
For linear process design, optimization one has to do five things.
a) Open up some degrees of freedom in the design model for optimization
b) Select a linear performance metric
c) Introduce linear inequality constraints on process variables to demarcate the physical domain, in which a process design remains physically feasible
d) Select a reference production scenario and compute its performance
e) Compute the optimum solution and assess its relevance over the reference solution

Outcome:
The example of the conversion-separation-recycle process is used to illustrate an optimization study. An economic profit function is added to the design model as well as a target output for the main product, with an upper bound on the maximum amount of side product that can be sold. The optimized outcome is compared with the reference (nominal) situation. The gain in profit can be explained and visualized in a feed-intake diagram. Furthermore, sensitivities of the profit with respect to key processing parameters (selectivity and a separation factor), and a sales parameter are determined. This sensitivity analysis shows that a future profit enhancement would benefit most from an improvement in the selectivity factor among these three parameters. While this numerical example is fictitious, it is the way of analysis of the optimization outcome that is practically relevant.

Example 9.5: Inverse use of linear models for process targeting and estimation

Motivation and objective of modeling:
The use of a square linear process model, $A\,x = b$, has been covered in the preceding modeling examples. The process features are parametric-captured in matrix A and vector b, while the unknown process variables in vector x can be computed, that is, if matrix A is nonsingular. However, there are more complicated situations of practical relevance in process design and operations where linear process models play a role.

Let u represent a vector of inputs to a process and y the process outputs. These are related by a linear process model $A\,u = y$. The number of inputs and outputs will be unequal, in general. Such a nonsquare linear model can be used in three different ways:
1) Forward: Given A and u, compute y. This case is straightforward and will be skipped.
2) Targeting: Given A and target for y, compute u.
3) Estimation: Given a sequence of experimental observations of inputs $\{u_1, u_2, ., u_K\}$ and outputs $\{y_1, y_2, ., y_K\}$, find the connecting process parameters that are elements of matrix A.

Targeting and estimation problems will become complicated if the numbers of inputs and outputs are different. The objective is to offer approximate solutions to the targeting and estimation problems.

The practical question is, when can these methods be applied in chemical product and process engineering? In terms of product engineering, one can think of the outputs y as representing the empirical variables for product performance testing and u as representing the internal chemical and phase composition variables (e.g., for fuels and lubricants and food and health care products). In terms of process engineering, one can think of the relations between the inputs to a process unit and its outputs. One may also think of an identification of a process control matrix, linking steady-state relationship between inputs and outputs around an operational working point. The practical restriction is in the assumption of linear relationships between the inputs and outputs.

Approach:
One wants to find a solutions match in some best possible way with the linearity of the model. The "best possible way" has to be defined more specifically. This can be done by means of an additional function that aims for minimization of deviations from model linearity. For this purpose, one often uses quadratic functions over the model residuals: $\underline{\varepsilon} = \mathbf{A}. \underline{u} - \underline{y}$. The benefit of using quadratic expressions as fitting functions in both targeting and estimation is that one can derive closed-form linear analytical expressions as answers.

Outcome:
Linear analytic expressions are derived for the following cases:
(a) Targeting, when there are more adjustable inputs $\underline{u}_N$ than output targets $\underline{y}_M$, (M < N)
(b) Estimation, assuming for measurement errors in outputs $\underline{y}$ only
(c) Estimation, accounting for random measurement errors in both inputs and outputs

The full expressions are given in the Appendix to chapter 9. These expressions are analytical in form, yet computationally demanding due to repeated matrix and vector operations. These are better done numerically, e.g., by the use of Matlab™.
   Regarding the resulting uncertainties in the computed model variables, it is also possible to derive estimates for their covariance matrices, though the expressions for such covariances become quite involved.
   Statistical targeting and estimation are complicated subjects and it is not hard to get into conceptual traps. It is recommended to consult a book on engineering statistics, e.g. [18], or expert statisticians. They can advise on suitable error distributions, fitting functions, and the interpretation of the computed results.

Example 9.6: Notes on the use of other types of linear models

Motivation and objective of modeling:
Highlight the existence of some other types of linear process models beyond the algebraic ones.

Approach:
Offer an overview of the analytical expressions for each type of model.

Outcome:
Linear models are given for:
(a) Dynamic evolution of process state variables;
(b) Dynamic process models with spatially distributed states in one dimension;
(c) Discrete time process dynamics: development of a FMCG warehouse inventory
(d) Linear equations with integer coefficients and binary variables

This overview completes the set of applications of deterministic linear models in the Concept stage. These models are supposed to always contain a fair degree of knowledge of what is supposed to happen inside a process (grey box model). If it is relatively cheap and easy-to-collect experimental data, there is an alternative of using fully data-driven linear black box models. Such a model can be generated and used in an experimental optimization procedure. Data-driven modeling is covered as an optimization example in section 9.4.

9.3 Nonlinear process model simulations for the *Development* stage

There are many introductory academic text books available on the principles of process modeling, associated mathematical analysis of the solutions, and numerical simulations in the context of process design and flow sheeting. For background reading, one is referred to text books [1–16]. Much of this material is about looking inside the model, and how to structure and solve the model equations. This section aims at looking more outwardly at process modeling from a somewhat higher vantage point of model use. First, some generic structural aspects of process models and their use will be reviewed. This review is supplemented with a few examples highlighting aspects of the use of nonlinear process models.

9.3.1 Process representation

Any process model is based on a conceptual representation of a process (to be). A process is a network consisting of multiple nodes interacting with each other by exchange of physical and information resources. Each node represents a unit operation. Within a unit operation, several physical and chemical tasks can be carried out in parallel, e.g., reactions and separations in a reactive distillation column. The connections between the units are specified as physical streams and as signals (for measurement and control) in a process flow sheet. Also, utility streams servicing process units will be considered as part of the process system.

Looking on a more detailed scale inside each unit operation, it is made up of a subnetwork consisting of interconnected compartments. Within these compartments, the states of the resources change because of physical and (bio) -chemical phenomena (e.g., reactions, phase changes, fluid and particles flows, and heat and mass transfers). There can be active agents facilitating the changes of states of physical entities, such as catalysts and mass separation agents, including membranes. The active agents for signals in process control and automation are sensors (physical quantities = > information signal) and actuators, e.g., valves (signal = > physical quantity). The process control software using dedicated control laws act as "unit operations" for processing signals.

Obviously, a process is an open system having connections with the outer world:
- Physical streams entering the process as feeds and leaving as products and wastes
- Physical disturbances acting on the process units, such as heat losses to the environment
- Information flow from the process sensors to the plant operators
- Process control set points imposed on the process, collectively defining an operating mode

The interactions between process and its outside world are partially defined by the operational scenario imposed on a process as well as by the external disturbances, and partly by the resulting state of the process.

9.3.2 Product modeling

In this subsection, the focus will be on the modeling and simulation of a process without particular attention to the structural complexities of the product(s). The process stream delivering the product has no special status, i.e., in the sense of having some different modeling attributes. The product stream, such as any process stream, is made up of one or two coexisting fluidic thermodynamic phases. No provision is made for distributive properties of the product, such as a size distribution of the particulates in a phase or a chain length distribution of polymers. Also, it is assumed that the properties (e.g., chemical composition, geometric shape, porosity, etc.) of the active agents in a process (catalyst, solvents, membranes, etc.) are known and can be used in the process model. One may want to account for distributive properties of thermodynamic phases in the process units and in the formation of products. This capability requires a high level of integrated product and process modeling sophistication (see chapter 8). Such a level is still beyond the capabilities of the current generations of process flow sheeting software.

9.3.3 Process equipment modeling

A process plant is made up of process equipment items in which processing operations take place. In other words, a process equipment piece hosts one or more nodes of the processing network. Regarding the function of process equipment, there is much more to it than just the process events happening inside. The complementary aspects are the integrities of its mechanical condition, its operational software for safety and control, and its availability for the production function the equipment was designed for. The latter aspects (its state of availability for normal production conditions, state of mechanical integrity, and safety system aspects) are not part of conventional process modeling in chemical process engineering. These aspects are modeled separately under the common assumption that actual physical state of the matter inside a process equipment item has rarely an immediate influence on its mechanical state and availability. Having outlined these aspects of process representation, the focus will narrow to the modeling of the physical and chemical events and states inside a process, along with the flow of resources (physical, information).

9.3.4 Scope of a process model

A process model is an amalgamation of models of the various units in the process as well as of the streams connecting these units, including utility streams. In turn, the unit operations models are an ensemble of compartment models with submodels of the behavior of the active agents and of the transfer of the physical and chemical resources between compartments. It is practical to introduce a formal compact representation of the process model. Let process model [M] represent the process system [PS] at a suitable scale of detail, for design purposes:

$$PS => [\,M\,] \qquad\qquad M(\underline{x};\,\underline{u},\,\underline{d};\,\underline{p},\,\underline{w}) = 0$$

M: Set of all equations describing the physical and chemical behavior inside the process
$\underline{x}$: State variables (physical, information) of the process units and internal and exit streams
$\underline{u}$: Input variables associated with *incoming* streams that can be manipulated
$\underline{d}$: Geometric design variables (fixed structures and sizes)
$\underline{p}$: Physical parameters, having some uncertainty, characterized by a covariance matrix
$\underline{w}$: Disturbances, acting more or less randomly on the process, from outside or from inside

The process model [M] is deterministic, while most of its model equations are nonlinear in the process state variables. The model equations are continuous in the variables and (at least once) differentiable in its variables.

The above model formulation [M] may seem to suggest that this model structure is restricted in its validity to continuous processes with lumped process units, resulting in a set of nonlinear algebraic equations. What about distributed process units having an internal geometric coordinate, such as a tubular reactor? Or a batch process unit, where time is imminent and such a unit is charged and discharged at specific times? Rather than having a continuous, steady flow in and out, it will be a time-varying flow or even a big chunk of mass transferred at once from one process operation to the next one. Fortunately, the same model format and reasoning applies. The only change is in a broader interpretation of the state variable x. In a distributed process unit, a state variable x may have a spatial derivative in direction z: $\nabla_z \underline{x} = \partial \underline{x}(z,t) / \partial z$. Similarly, there can be a time derivative of a state variable: $\underline{x}' = \partial \underline{x}(z,t) / \partial t$. In these cases, one can extend a single state $\underline{x}$ to its triple $\{\underline{x},\ \nabla_z \underline{x},\ \underline{x}'\}$ and insert this triple in the overall state vector. Then of course, the process model must include the corresponding boundary and initial conditions for these derivatives.

The process model [M] describes the internal physical-chemical behavior of the process. But the modeling problem is still open-ended, leaving some degrees of freedom because the process model itself does not yet account for the external conditions in its surroundings for which it has to be solved to find the process states $\underline{x}$. This open-ended nature is evident from the number of unspecified variables in the model $\{\underline{u}, \underline{d}; \underline{p}, \underline{w}\}$. A set of specifications for these variables is called a scenario. A scenario reflects the design and external conditions under which one wants to determine the physical and (bio) chemical process behavior. It is practical to make a distinction between the relatively simple process analysis cases and the more complicated process synthesis cases.

9.3.5 Process analysis scenarios

In such cases, a design of the process is available and the design variables d are known. For instance, one may want to analyze how the state variables x (of internal process units and streams) will vary under variations of inputs (u) and design variables (d). For that purpose, the process model [M] must be complemented with an **analysis scenario**. A typical base case analysis scenario contains the following numerical specifications:

$$
\begin{array}{llll}
[S_{\text{Design}}] & : \underline{d} - \underline{d}_{ref} & = 0 \\
[S_{\text{Input}}]: & \underline{u} - \underline{u}_{ref} & = 0 \\
[S_{\text{param}}]: & \cdot \underline{p} - \underline{p}_{ref} & = 0 \\
[S_{\text{disturb}}]: & \underline{w} - \underline{w}_{average} & = 0
\end{array}
$$

Here, each spec quantity represents a particular numerical value, applicable in the base case. So, given specifications for all variables $\{\underline{u}; \underline{d}; \underline{p}, \underline{w}\}$, except the states $\underline{x}$, which can be solved from M, the combination $\{\underline{x}; \text{M and S}\}$ is often a very large set of algebraic variables and equations in equal number that needs to be solved numerically. In the case of a dynamic process system, one must specify the trajectories of the inputs and disturbances as functions of time.

9.3.6 Process performance evaluation metrics

The process model enables computation of its internal states, and this model has an inward focus on what is happening in the process. There is also a complementary outward look on the performance of a process from a societal perspective. The process under design can be evaluated by means of different, complementary metrics. Here, three kinds of metrics are distinguished, representing different values to the process operator, owner, and society.
- Technical metrics: $E_{\text{tech}} = P_{\text{te}}(\underline{x}, \underline{u}, \underline{d})$
- Economic metrics: $E_{\text{econ}} = P_{\text{en}}(\underline{x}, \underline{u}, \underline{d}; c_I, c_O, p_P)$
- Ecological metrics: $E_{\text{ecol}} = P_{\text{ec}}(\underline{x}, \underline{u}, \underline{d}; e_I, e_O)$

The technical metrics can cover product yields over feed intake(s), energy efficiency, element efficiency, and risks (DF&EI). The economic ones account for size-related investment costs (c_I), throughput-related operating costs (c_O), and the proceeds from product sales (p_p). The time value of money can be accounted for by deriving the Net Present Value profiles. Ecological metrics will address sustainability issues. For instance, a process design can be assessed by second law analysis to determine its (irreversible) exergy losses. Once having solved a process model under a specified scenario with a process simulator, it is relatively a minor effort to compute the performance metrics as well and perform sensitivity analyses.

9.3.7 Sensitivity analyses

One can compute how much of small changes in inputs $\underline{u}$ and in design variables $\underline{d}$ will influence the performance metrics (E) through the state variables $\underline{x}$. This can be done by determining the scaled sensitivities of the performance metrics with respect to the states: $S_{Ex} = \{\partial\ln(\underline{E})/\partial\ln(\underline{x})\}$ and of the process states with respect to the process inputs ($\underline{u}$): $S_{xu} = \{\partial\ln(\underline{x})/\partial\ln(\underline{u})\}$ and with respect to the design variables ($\underline{d}$): $S_{xd} = \{\partial\ln(\underline{x})/\partial\ln(\underline{d})\}$. Process simulators offer such facilities for such local sensitivity analyses. One may rank these sensitivities in order of decreasing magnitude (i.e., using absolute values of computed sensitivities) and see which inputs and design variables affect the process behavior most strongly. Similarly, one may compute and rank-order the sensitivities with respect to the physical model parameters. The latter is particularly relevant if some of the model parameters have a substantial uncertainty margin in their determination from experimental data. An overview of sensitivity analyses for chemical systems is offered in [19]

Simulations for process analysis are conceptually simple as a process model computes the states directly from the inputs and the design variables. It is a forward flow of information in the process model [M]. The situation becomes more complicated when a few of the key process states are fixed as design targets for process performance. Then, one must search for those values of the design variables ($\underline{d}$) to match the specified process performance. This is a so-called "inverse" problem as the flow of information from targets on a few states goes "backward" to the required design variables that generate these targets. For process models consisting only of algebraic equations, this reverse flow of information can usually be handled well without excessive computational efforts. For models containing (partial) differential equations, an inverse problem is much tougher to solve than the forward problem. That is because the integration of the equations follows the causal chain of events along the geometric and time coordinate(s) and there is no way of reversing it. Consequently, the models have to be repeatedly solved from inputs to outputs, searching for an input that hits a target output. This iterative procedure can be computationally very demanding.

9.3.8 Process design and synthesis cases with targets

In a design situation, one needs to find values for the unknown process design variables d. Hence, there are more unknown variables, both x and d. The degree of freedom in model [M] is now equal to the number of design variables, assuming there is
an external scenario for setting the remaining variables $\{u; p; w\}$. This degree of freedom must be taken away by setting some targets $\{t\}$ for the performance of a design.
The number of targets must be equal to the number of process design variables, otherwise, the design is under- or over-specified. These targets can be imposed both on
important performance indicators as well as on the geometric design variables. For
instance, the yield of a product, relative to its raw material, its production rate, and
its product quality can be design targets. Rather than giving direct specifications of
geometric design variables, certain heuristics for their proportions are applied, e.g.,
fixing the aspect ratio (diameter/length) of a vessel. The specifications can be set up in
three categories: technical process performance, operational conditions of process
units, and constraints *on* geometric design variables. However, the total number of
specs cannot exceed the number of design variables in the process model; it must exactly match.

$$[S_{perf}]: \quad T_{perf}\,(\underline{x}) - \quad \underline{t}_{perf} \quad = 0$$
$$[S_{oper}]: \quad T_{oper}\,(\underline{x}) - \quad \underline{t}_{oper} \quad = 0$$
$$[S_{design}]: \quad T_{design}\,(\underline{d}) - \quad \underline{t}_{design} \quad = 0$$

T_{***}: Performance target expressed as a function of process states or design variables
t_{***}: Numerical value for a performance target

The technical process performance variables express the effectiveness of the process to
turn inputs (feeds and utilities) into outputs (product qualities and yields, and amount of
waste). These performance variables must consistently relate to the Key Performance Indicators (KPIs) used in the higher management levels in a manufacturing organization.

Special attention is warranted to the occurrence of different modes of operations.
(a) Normal continuous production over a time period to make the main product grade(s)
(b) Normal continuous production over a time period to make another product grade(s)
(c) Dynamic transitions between two successive product grades
(d) Transitions between different production output rates
(e) Repetitive cleaning (e.g., for food production) and regeneration operations of pro-
    cess units
(f) Startup from standby (with low hold-up and internal circulation) to nominal
    production
(g) Shutdown from high production to standby level

Apart from the steady-state production modes, all other modes of operations will require dynamic modeling and simulation.

9.3.9 Additional specification of an external scenario for a design

Having set the targets, one must impose additional specifications for the other variables $\{\underline{u};\, \underline{p};\, \underline{w}\}$ in the process model [M]. A process design scenario consists of specifications for the external inputs that can be manipulated – the model parameters and the external influences (beyond manipulation):

$$[S_{input}]: \quad \underline{u} \; - \; \underline{u}_{ref} \quad = 0$$
$$[S_{param}]: \quad \underline{p} \; - \; \underline{p}_{ref} \quad = 0$$
$$[S_{disturb}]: \quad \underline{w} \; - \; \underline{w}_{average} \; = 0$$

$\underline{u}_{ref}$: reference values for the process model input variables
$\underline{p}_{ref}$: reference values for the process model parameters
$\underline{w}_{average}$: reference value for the disturbances

The input specifications will address the flow rates, compositions, temperatures, and pressures of the raw material streams as well as the conditions of the available utility streams. The amounts of utility consumed or produced can be made part of the process state vector as utility amounts depend on the inner workings of the process.

It may happen that a designer faces a situation where more targets need to be set than there are design degrees of freedom. The excess number of targets going beyond the number of design specifications can be handled by removing as many specifications on the inputs. The inputs that are set free will be manipulated in the model computations to meet the excess number of targets.

In the case of a dynamic mode of operation (e.g., a switching between production modes), the input specifications will be dynamic ones. These dynamic scenarios may challenge the design performance and call for readjustment of the geometric design decision variables, such as lowering the volumes and holdups to increase the rates of transition (within safe bounds, of course).

The parameters in the process model will in general be the statistical averages resulting from a parameter estimation procedure using experimental data. In favorable circumstances, one may even have a covariance matrix (V_{param}) describing the extents and patterns of uncertainties in the parameters. Specifying the external random disturbances can be conceptually hard. It often involves random (noisy) phenomena with poorly specified or unknown probabilistic distributions. The disturbances can be, for instance, climate-related quantities such as the environmental temperature and humidity, affecting the processing states. It may also be a random variation or a gradual decline in the activity of a process agent, such as a catalyst. In design situations, one often takes the expectation value (averaged value of a disturbance variable taken over a long time interval). It is already a boon to know the spread in the variation per disturbance variable; getting a full covariance matrix ($V_{disturb}$) is practically ruled out.

9.3.10 Consistency in model formulation

When formulating a process model [M], there is a challenge to be consistent in the selection of variables and equations within a physically feasible domain. In other words, one must avoid redundancies and contradictions in the model. For instance, the design specification imposed on a process design model should not violate thermodynamic feasibility and stability. For example, it may happen that a process unit model in a flow sheet has two thermodynamic phases and the evolution of the states of the two phases in the model causes one of the two phases to lose its stability. In physical reality, such phases would merge into a single phase. In the model, such spontaneous merging will not happen. Instead, the physical properties of the two thermodynamic models will start to coincide, possibly leading to linear dependencies between equations, with a resulting loss of model solvability.

The computational solving of a model [M] requires it to be well posed:

– All equations are differentiable with respect to all variables $\{\underline{x}; \underline{u}, \underline{d}; \underline{p}, \underline{w}\}$.
– The finite derivatives of equations are required to avoid numerical singularity of the Jacobian matrix J_M (= $\partial M/\partial x$).
– Model [M] has as many linear independent equations as state variables x. Linear dependent equations are combinations of other independent equations and so, redundant.

It is assumed that all model equations can be solved simultaneously by means of a suitable numerical solver. In process simulation practice for novice modelers, it happens that complaints arise about slow convergence to a weird solution or even total failure of a numerical solver to converge to a solution. The first resort is to attribute the responsibility for ill-performance to the numerical solver. Upon closer inspection of the model, it often appears there is a mathematical or physical inconsistency in the model. Such an inconsistency will challenge any solver to converge to an unphysical or nonexistent solution. The first example in this section (Example 9.7) is a simple illustration of a locally ill-conditioned model, violating the rule of bounded derivatives and causing trouble in model simulations. Example 9.8 deals with the consequences of a design specification for a reactor violating thermodynamic reaction equilibrium.

9.3.11 Process model computations

Having the set the design targets {t} as well as the complementary design scenario specifications for {$\underline{u}$, $\underline{p}$, $\underline{w}$}, one can jointly solve the process model [M] and the target specifications [S] equations for $\underline{x}$ and $\underline{d}$. The solution that one obtains for a chosen reference scenario is called the base case solution. One can compute solutions to variations on this base case by varying the specified reference values for the inputs ($\underline{u}$), physical parameters ($\underline{p}$) and the external disturbances ($\underline{w}$). Using the solutions in the perturbed cases in comparison with the reference solution, one can compute the sensitivities. These sensitivities can be ranked in order of (absolute) magnitude to find out which of the variables going into the process model determine the computed states the strongest. Figure 9.2 offers a visual overview of a simulation problem for process design, using design targets for process states.

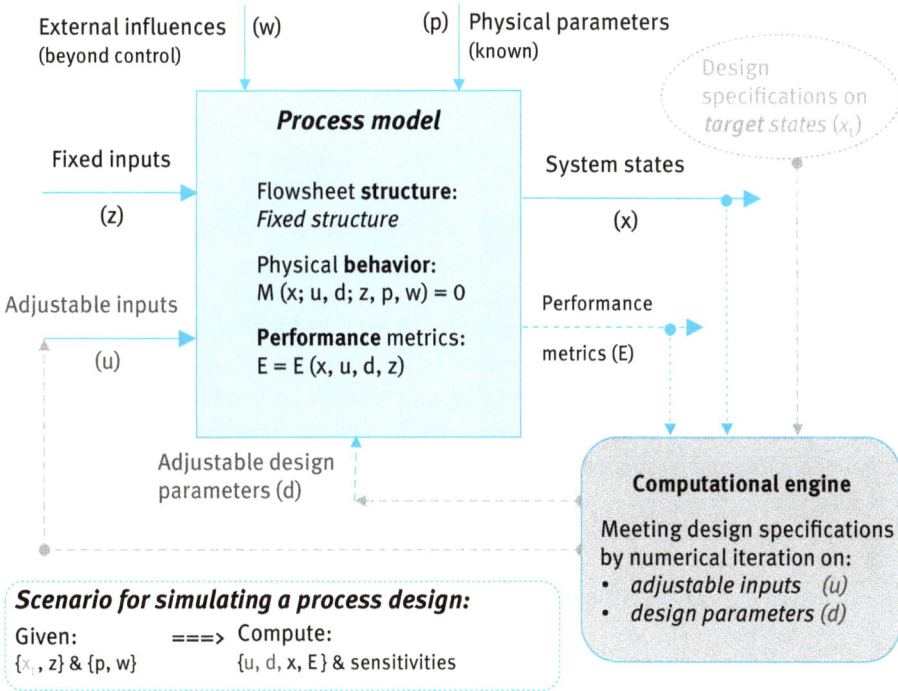

Fig.9.2: Overview of process simulation for design.

9.3.12 Analysis of the solution obtained from a model

What kind of intricacies of solutions can one expect when having solved a process model?

Though rather rare, one may occasionally find computationally multiple solutions. Theoretically, this can never be ruled out for nonlinear models. Remember that every polynomial of degree N has N mathematical solutions, some of which are in the complex domain. Fortunately, in many chemical engineering systems, there is often only one real solution among the mathematical set of (complex) solutions. Even if there would be a few real solutions, only one of these has the better physical stability characteristics, i.e., being more robust against external disturbances. Usually, it is this solution that is preferred as *the* design solution.

Having found more than one solution may also turn out to be a boon. The unexpected new solution may have a better performance than the anticipated solution. Moreover, it will make the designers aware that the multiplicity of solutions has implications for the dynamics and stability of the future process, e.g., there is a possibility of disturbance-driven transitions between steady states.

9.3.13 Process performance evaluation metrics

Three kinds of metrics are considered, representing different values to the process operator, owner, and society:

- Technical metrics: $E_{tech} = P_{te}(\underline{x}, \underline{u}, \underline{d})$
- Economic metrics: $E_{econ} = P_{en}(\underline{x}, \underline{u}, \underline{d}; c_I, c_O, p_P)$
- Ecological metrics: $E_{ecol} = P_{ec}(\underline{x}, \underline{u}, \underline{d}; e_I, e_O)$

Examples of technical metrics are product yields over feed intake(s), energy efficiency, element efficiency, risks e.g. by DF&EI. The technical metrics will be closely linked with the earlier introduced design performance functions (T_{perf}) and targets ($\underline{t}_{perf}$) for the process design. The quantities $\{c_I, c_O, p_P, e_I, e_O\}$ in the economic and ecological metrics are model parameters, related to investments (I), cost of operation (O), and prices of products (P).

9.3.14 Using uncertainty information in a sensitivity analysis

Having obtained a base case design, one has still to consider the impact of the uncertainties in a design situation on the design outcome and the evaluation metrics. There can be variability in the specifications for operational and input variables, for instance, when there are different production modes. Usually, these uncertainties are of a deterministic nature. A designer may know there can be multiple production modes with different feed intakes. These variations in the operational and input variables are often handled by defining a number of likely production modes, as part of the specified external scenarios. The resulting designs for these modes can be compared and possibly reconciled to arrive at a robust master design that can handle all production modes. Apart from having distinct production modes, there can be uncertainty about the proper targets for some of the operational variables as well as about the input variables and streams. One can analyze the impacts of these uncertainties by means of model-based sensitivity analysis. If necessary, a design can be adapted to attenuate the sensitivities. Fortunately, there is also a way of coping with remaining sensitivities to disturbances after completion of a design and having put the process plant into operation. Then, one can still compensate for the many arising additional uncertainties by adjusting the operational and input variables through process control to meet set performance targets.

In the case of a rigorous model development with experimental validation and model parameter estimation, the uncertainties in the system parameters (p) can be co-estimated. These uncertainties are expressed by a covariance matrix of the parameter estimates. Computationally, this covariance matrix comes for free along with the mean values of the parameter estimates as obtained in a statistical maximum likelihood parameter estimation procedure from experimental data. For further model uncertainty analysis, one may sample parameter values from the space spanned by the covariance matrix. Given such a set of sampled parameter values, one can compute and analyze the resulting variances in process design variables (d), states (x) and performance metrics (E).

Uncertainties in external disturbances are often harder to cope with. There is often hardly any information on the distributions of their magnitudes, if anything at all. These uncertainties are often handled after completing the base case design by means of:

-   Application of experience-based overdesign factors to process equipment: There is a risk of this approach by applying overly conservative factors. Historical factors may not account yet for the increased accuracy in equipment design that is possible by model-based design simulations.
-   Some form of a mathematical uncertainty analysis, based on probabilistic distributions of the uncertainties and reducing the chance that a design cannot cope with disturbances to a low acceptable level. Such global system analysis facilities are offered by advanced simulators, e.g., gPROMS process modeling platform [20].

The above outline completes the story on process modeling for simulation purposes. The application of process models for optimization of a process design is part of the next subsection.

9.3.15 Examples for development stage modeling

Three modeling examples are presented. The first two examples will highlight two common pitfalls in modeling, while the third example offers a plea for model-supported operability analysis in the design phase.

Example 9.7: Note on ill-conditioned constitutive equations in design models

Motivation and objective of modeling:
Highlight the detrimental effect of an ill-conceived constitutive equation on the solvability of a process (unit) model.

Approach:
Use a continuous-flow well-mixed tank reactor (CSTR) model with a commonly used reaction rate expression that has a local physical anomaly.

Outcome:
The model has a region in which the solution will numerically fail due to an unbounded first derivative of the model equation. The implication of this example for modeling practice is that it pays to carefully check all constitutive equations in a design model on their asymptotic behavior before embarking on any process engineering computation.

Example 9.8: Consistency of design specifications with thermodynamics

Motivation and objective of modeling:
In modeling for process design, it is rather easy to overlook the limits imposed by thermodynamics on setting feasible design performance targets. If such design targets are set beyond thermodynamic equilibrium, the computations with the process model will fail. The failure can be a lack of convergence in the computations or finding unphysical design outcomes after "successful convergence". The purpose of this example is to show such a situation through a reactor design with an analytical solution.

Approach:
The design case is the computation of the required residence time to achieve a target conversion in a continuous flow, well-stirred tank reactor (CSTR) with a single, reversible first-order reaction A = > B at isothermal conditions.

Outcome:
Having specified a degree of conversion above the reaction equilibrium, the reactor design equations generate a numerical solution having a negative residence time as a displeasing feature.

   The resulting design heuristic is obvious enough: *Physical design targets should be located within a thermodynamically feasible domain. The preferred distance between the design target and the thermodynamic equilibrium boundary is negotiable in design.*

Example 9.9 Model-based analysis of operability aspects in process design

Motivation and objective of modeling:
The operational success of a process plant depends to large extent on the quality of its design. The operational characteristics of process should be taken into consideration in process design and modeling. The conventional model-based process analysis is suitable to study the effect of the dominant phenomena on the behavior and performance of a process at steady states under normal operational modes. Operability analysis is needed to assess the effects of major deterministic and random changes outside and inside a process. Operability analysis of a process has become an extensive subfield of process engineering, linking process design, control, and operations. It relies much on dynamic process modeling, away from the usual steady states. Here, the aim is to create awareness of the main aspects of operability while giving some references to pertaining literature for further study.

Approach and outcome:
Following Marlin [21], eight topics of operability are briefly covered from the perspective of teaching such topics in process design. Dynamic process models are applicable in all of the above operability cases, except for reliability and safety. Reliability is an exception as it requires probabilistic models for equipment failure rates. Safety is another exception because it often involves process events that go far beyond the normal operating window. The conventional process models are not tuned to such exceptional conditions beyond the normal operating window. Safety requires dedicated models to study the physical behavior of a process unit under abnormal conditions, such as run-away or flow reversal effects.

9.4 Nonlinear process model optimization for a feasibility stage

9.4.1 Nonlinear modeling and optimization

An explanation is in order as to why modelling in the Feasibility stage is covered only after modeling for the development stage. Commonly, in process design, it is the feasibility stage that precedes the development stage. The reason to switch the order here is rooted in the additional complexity of model use for optimization and its associated computational procedure. In the feasibility stage, one may perform an optimization over a range of process design options using relatively simple process models to search for an optimized performance over a wide space of design options. A few of the more promising process options that are identified in the optimization procedure will be transferred to the subsequent process development stage. There, such options will be analyzed in more detail, requiring a more extensive process model along with strong process simulation capabilities. A process model in the feasibility stage may contain less detail and be simpler, in general, than the model for development. The paradox is that the computational framing of the models in the feasibility stage is seen as being more complicated due to its optimization features, relative to simulation in the development stage.

The main benefit of optimization over simulation in design is that optimization offers a search over the entire solution space for the best outcome. A simulation results in a single point in the solution space. When additionally, a simulation-based global system analysis is applied [22], some more light is shed on the behavior in the neighborhood of that single solution point. In system optimization, one can explore a predefined space for a point of optimum performance. It is because of this capability to explore a wider space that one wants to apply optimization in the feasibility phase to identify the more promising options within the entire solution space (for process/products). Furthermore, advanced optimization methods can also be used for the synthesis of process and/or product structures from a predefined set of building blocks (e.g., unit operations for a process and phase building blocks for a product). In process simulation, the process structure is fixed a priori. Therefore, it is the combination of (i) creation of structure, (ii) the computing and simulation of physical behavior, and (iii) optimization of system performance that makes optimization applications attractive in the Feasibility phase. Yet, there are two restrictions to be accounted for. The search over a solution space is always a heavy computational task. Therefore, the models of the building blocks for product / process synthesis must remain relatively simple and should focus on core features only. A similar restriction on models may emerge from a relative lack of process and product knowledge in the Feasibility stage. The details of the mechanistic functioning of the building blocks may still be lacking and thus approximate model descriptions are used.

The price to pay for the enhanced capabilities by optimization is a much more extensive description of the design problem, relative to a simulation case. It is necessary to describe the boundaries of the search space in an optimization as well as the domain of validity of the process model. In addition to the usual model equations, one has the following attributes in an optimization problem:

- Structure-related: discrete (binary) variables (Y) to characterize the presence of building
  blocks, the types of building blocks, and the connectivity of blocks
- Structure-related: equations (equalities and inequalities) to model flow sheet structure
- Domain-related: valid domain of model application in the optimization
- Feasible design targets: ranges for variation of process, design, and performance variables

Objective function(s): One performance metric or two metrics for a Pareto trade-off.

9.4.2 Overview of common elements in an optimization frame

The elements of a design optimization problem are given:

Process variables:		$\{x, u, d, Y; p, w\}$		
Structure model	$[S]$:	$S\,(Y, x)$		≥ 0
Process model	$[M]$:	$M\,(x; u, d, Y; p, w)$		$= 0$
Model application domain	$[A_M]$:	$A_M\,(x, u, d)$		≥ 0
Design co-targets:	$[S_{design}]$:	$T_{design}\,(d)$	$-\quad t_{design}$	≥ 0
Process design scenario:	$[S_{input}]$:	u_{fixed}	$-\quad u_{ref}$	$= 0$
	$[S_{param}]$:	p	$-\quad p_{ref}$	$= 0$
	$[S_{disturb}]$:	w	$-\quad w_{average}$	$= 0$

Process performance metrics for optimization:
- Technical metrics: $\quad\quad$ max. $E_{tech} = P_{te}(x, u, d)$
- Economic metrics: $\quad\quad$ max. $E_{econ} = P_{en}(x, u, d; c_I, c_O, p_P)$
- Ecological metrics: $\quad\quad$ max. $E_{ecol} = P_{ec}(x, u, d; e_I, e_O)$

Degrees of freedom: $\quad\quad$ dim $\{d\}$ + dim $\{Y\}$ + dim $\{u_{adjust}\}$

One may wonder about the role of the design cotargets. Why not optimize all design variables at once? It may happen that a design optimization is part of a retrofit of an existing process, where the design of certain parts of the process should be retained. The use of design cotargets allows keeping a portion of the design variables fixed. In a similar fashion, if one does not want to reinvest in new equipment, one can keep the existing equipment, fix the design variables, and still optimize the process by varying the operating conditions such as pressure, temperature, and internal recirculation flows. There is often some trade-off possible between making changes in the geometric design variables or in the temporal ones, such as operating conditions {pressure, temperature, and flow rates}.

9.4.3 Mixed integer nonlinear programming format

When the physical meaning of the variables in a process model is ignored, one can write the design optimization in its usual compact mathematical form as a mixed integer nonlinear optimization problem (MINLP):

$$\text{Min.}_{\{x,\ Y\}} F(\underline{x},\ Y)$$

Subject to:
$$\underline{g}(\underline{x},\ Y) = 0$$
$$\underline{h}(\underline{x}Y) \geq 0$$
$$Y \in \{0,\ 1\} \text{ and } \underline{x} \in \mathfrak{R}^N$$

The mathematical conditions for solving such a MINLP problem and the various numerical algorithms are covered in text books on optimization in chemical engineering [11, 13–16].

9.4.4 Flow of information in the computational process

In a process model, the causal flow of physical information runs from physical and information inputs (feed streams, preset operating conditions, and design parameters) to physical process outputs. When targets are set on the outputs, being some states and performance metrics, one must start adjusting some inputs and design parameters to meet these targets. In view of the nonlinearity of a process model as well as (numerical) stability considerations, it is not possible to invert the model, as done easily with linear algebra equations: $A.\ \underline{x} = \underline{b} \Rightarrow \underline{x} = A^{-1}.\ \underline{b}$. Stability considerations play another important role. For instance, if one solves a model of a tubular reactor by integrating the differential equations in the model from entrance to exit, both the analytical and numerical solutions are stable. Going the reverse way, from exit, back to entrance, it is observed that the same solutions exhibit unstable behavior, effectively prohibiting such a route. Consequently, process models must be solved in a forward manner. When aiming for targets on process states or searching for the optimum performance, the process model must be solved in iterative moves many times, toward the targets or the optimum performance. The causal flow of information in a process design optimization procedure is shown in Fig. 9.3.

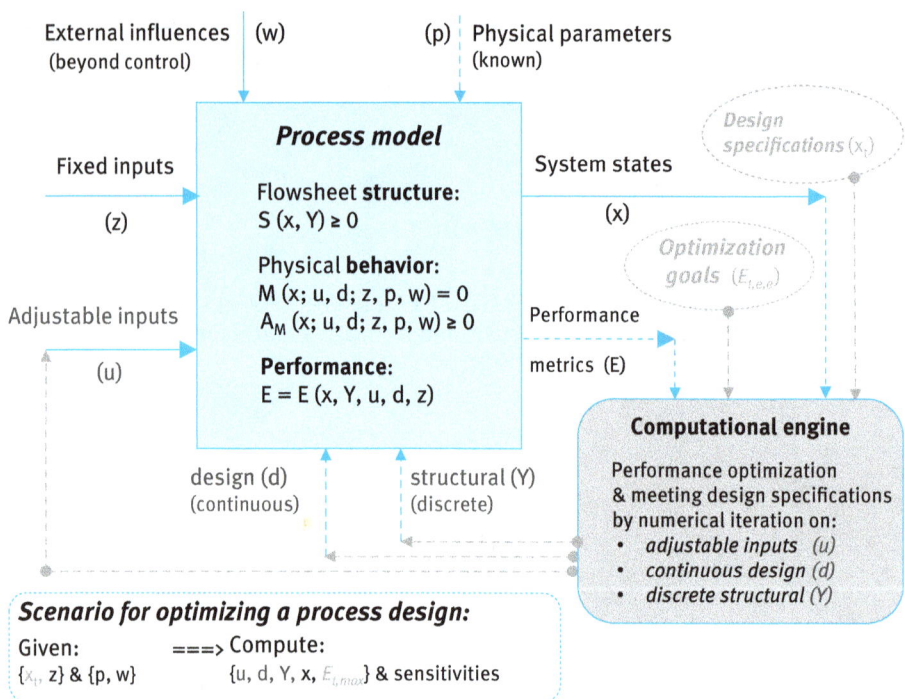

Fig. 9.3: Design optimization by solving an inverse problem.

For a proper evaluation and interpretation of the solution of an optimization problem, it is strongly recommended to obtain an basic understanding of the role of the Lagrangian function, of the necessary conditions for the existence of a local extremum (be it a minimum or a maximum) and of the sufficiency conditions for a local minimum. The necessary theory is presented in optimization text books focused on chemical processes [11, 13–16].

What are the practical issues arising in the application of MINLP to process optimization?

a) *Establishing the domain of model validity*

The derivation of the process model equations is similar to process simulation. The hardship is in getting a good set of inequalities to bound the solution and so prevent fictitious results outside the domain of model applicability. In practice, it is often harder than expected to collect relevant information on the bounds of the domain of model applicability. The latter is the domain in which the model and, particularly, its constitutive equations can be applied with sufficient confidence. Commonly, one is well able to retrieve the original documentation of a constitutive equation from which theoretical considerations and experimental data was derived. But it is far less common to find along with this equation an explicit specification of the domain in which it can be applied with a particular degree of confidence. Rarely, the delineation of such a domain is given, e.g., as a set of inequality constraints. Without such a specification, the numerical optimizer may move into regions where the performance seems to "improve," while one or more constitutive equations in the process model have lost their significance. Such a loss yields a fictitious optimization result. Often, the model developer has to go back to the original literature and look into the original sets of experimental data and infer the ranges over which the experimental conditions were varied, with the resulting ranges of the response variables.

b) *Finding a representation of model uncertainty*

Similarly, it can be hard to find proper documentation of the statistical properties of the estimated parameters in the constitutive equations beyond their mean values, such as covariance matrices. Having such an uncertainty description would be helpful in assessing the significance of the computed optimization results, in relation to the underlying uncertainties in the process model.

c) *Numerical conditioning of the model*

Scaling of process variables and equations is advised due to a wide range in magnitudes. Enthalpy flows and terms in an enthalpy balance may be of the order of 10^6, while molar fractions of trace components in mass balances can be of the order 10^{-6}. Applying some scaling will help to stabilize the numerical computations by the optimizer. It will also facilitate any internal local scaling by the optimizer.

d) *Initialization of model variables to enhance convergence of iterations on model equations*

The assignments of starting values to the process model variables needs due attention in order to facilitate a smooth convergence of the model equations to a solution. Inconsiderate assignments of initial values may play havoc with the magnitudes of the nonlinear terms in the model equations. Such nonlinearity is often found in rate expressions (e.g., power-law reaction kinetics) and in thermodynamic constitutive equations (e.g., for activity coefficients). To get started with model initialization, it is found helpful in practice to artificially tone down the impact of the nonlinear rate terms in the balance equations. It is convenient to introduce into the model, a common artificial scalar "inhibition factor" as a multiplier, to all strongly nonlinear terms in the model equations. The purpose of such an inhibition factor is to slow down the rates of change. Assigning initially a small value to this inhibition factor, a nonlinear model tends to behave as close-to-linear. Then, it is easier to achieve initial convergence. Having obtained such a converged in-between solution, one can use it as a starting point for the next round with a gradual ramping-up of this "inhibition factor." Stepping up the inhibition factor to unity will restore the original nonlinear features of the model.

e) *Use of suitable modeling platforms for the optimization of processes*
Modern process model building and computing platforms, based on equation-oriented computations, offer compatible links to advanced MINLP optimization facilities. Examples are gPROMS [20] and AIMMS [23]. The conventional sequential, block-oriented process flow sheet simulators are more restricted and offer some optimization facilities for process operating conditions (e.g. control set points). They do not offer optimization of flow sheet (super) structures.

These remarks complete the review of modeling for process (design) optimization. The next section will deal with four examples to illustrate some aspects of optimization applications.

9.4.5 Examples for feasibility stage modeling

Four examples are covered that highlight aspects of:
- Constraint analysis to identify options for technological innovations
- Pareto trade-offs in process design between two performance criteria
- Synthesis of gas-to-liquid (GTL) processes
- Experimental process optimization with data-driven black box models

---

Example 9.10: Constraints analysis to identify options for technological innovations

Motivation and objective of modeling:
Realistic process models should reflect current limitations of the technologies underlying process unit operations in a process (conversion, separations, compression, etc.). In process design optimization studies, such limitations should be expressed as inequality constraints in the process model. These inequalities delineate the operational domain in which a particular technology (e.g., catalyst, membrane) can trustfully be applied (e.g., validity range of reaction kinetics, maximum allowable temperature, and maximum allowable pressure differences). The purpose of constraint analysis after an optimization of a design is to identify those active inequality constraints that predominantly limit process performance. Knowing how inequality constraints relate to underlying technologies helps one find directions for improving process performance by means of innovations in the limiting processing technologies.

Approach:
Model-based process optimization delivers optimized process conditions and design variables. In addition, it will also yield an active set of inequality constraints constraining the process performance. The relative importance of these inequality constraints can be identified by means of their associated Kuhn-Tucker multipliers, coproduced as a result of the optimization. One can put the Kuhn-Tucker multipliers in order of decreasing magnitude. In that way, one obtains the order by which active inequality constraints are restraining the objective function. This approach of finding the relative impact of active inequalities is explained by means of small reactor design example.

Outcome:
The result of the reactor design optimization shows that one inequality constraint is strongly dominating the optimum and could be a candidate for relaxation, while another inequality constraint has only a marginal influence on the performance in the optimum. An analysis of the active inequality constraints and their Kuhn-Tucker multipliers in an optimization outcome can help to find incentives and clues for further technology development and to enhance process performance.

In the same vein, an extensive, deep modeling study was reported in [24] for the conceptual design of a steam cracking reactor. This design was reoptimized with respect to olefin yields by lifting the current material-induced temperature constraint by 100 K to a new anticipated level of 1,300 K. The result is a very significant potential improvement in olefin yields (55 = > 66% mass yield of ethylene).

Example 9.11: A simple example of a Pareto trade-off between two performance metrics

Motivation and objective of modeling:
Virtually, every process design has to consider multiple performance criteria. The ranges of all design decision variables together form a high-dimensional process design space. Each point in this design space represents the complete array of design-decision variables. In these points, one can compute the process performance metrics. This is a low dimensional space of two or three dimensions. The overall objective of a design is to improve the performance criteria as much as possible. In certain regions of the design space, one can move a design from one point to another point and all performance criteria improve indeed. However, often there appears a transition to another region, where opposition between criteria arises. Making changes in a design can improve one performance metric while making anther metric worse. There is no way of improving both. The set of all designs where one cannot improve a performance metric without making at least one other metric worse is called a (non-inferior) *Pareto set of design solutions*. If one takes all performance criteria equally serious, it is impossible to say which design within this Pareto set is the better one. All are equally qualified as "best" designs.

Working with multiple performance functions in a design conceptually gives rise to multiobjective nonlinear programming (MONLP). In process engineering practice, one wants to keep using a normal single objective mathematical programming format and the associated NLP solver. Then, there are three options to deal with multiple performance criteria. The assumption is that all criteria have to be minimized.

1) Set a threshold level for the performance variable of each criterion, except one: The latter one is called the dominant criterion. The purpose of having thresholds is to relax from the minimization and stay well above the attainable minimum level of performance for each criterion. Each threshold will appear as an additional inequality constraint in the optimization. Next, the design is optimized with respect to the dominant criterion, while making sure that all other criteria meet their threshold levels. When a criterion is at its threshold level (i.e. becoming an active inequality), its Kuhn-Tucker multiplier indicates the sensitivity of the dominant criterion with respect to the criterion at threshold. The outcome may give reason to adapt the thresholds and repeat optimization till the results are "satisfactory" to the problem owners of a design. In other words, "satisfactory" depends on a reflective judgement by process engineering management, depending on societal and company cultures as well as on local site conditions where the process will be erected.

2) Define a new master criterion as a linear weighted sum over all separate criteria: Often, economic performance is used as the dominant criterion and all other criteria must be translated into monetary values. The weight factors in the linear sum expression represent the relative economic weights of the various criteria. The approach under 1) can help to establish reasonable estimates for such weight factors (≈ using Kuhn-Tucker multiplier). One can then minimize the new master criterion. If the results are not satisfactory to the problem owner of a design, one may start iterative improvements in adjusting the relative weights in the master criterion.

3) Create a Pareto set of design solutions, where each design in the set will have different performance values: Such a Pareto set can be generated by means of approach 1) where the selected threshold levels (initially high and relaxed) are systematically stepped down in severity, till the level for that particular criterion is so low that optimization with respect to the dominant criterion becomes infeasible.

Approach:
A Pareto optimization is demonstrated by means of a small-scale reactor design problem with two performance criteria, profit and mass yield.

Outcome:
The results indicate there is a region of design in which both performance criteria simultaneously improve. Adjacent is another region with a Pareto front – improvement of one criterion makes the other criterion worse. A similar result is obtained in an optimization of the design of a reactive-distillation column for MTBE synthesis [25]. An economic performance indicator is placed in a trade-off with an ecological indicator, the exergy efficiency of this column.

Example 9.12: Optimization of designs of gas-to-liquid processes

Motivation and objective of modeling:
The main purpose of GTL processes is to turn natural gas (NG) into liquid fuels. GTL processes tend to serve huge gas fields and are truly large scale. The capital investment is very high (up to several billions US$). There are many design decisions with interacting outcomes on process performance. Also trade-offs emerge between capital investment, carbon, and energy efficiencies. Therefore, GTL processes are prime candidates for model-based optimization process synthesis.

Approach:
The feasibility of such an approach is shown in an industrial case study on model-based, optimization-driven synthesis of a large-scale GTL process [26], using profitability and carbon efficiency as performance measures. The process flowsheet is represented by means of a block diagram with reactors and unit operations. The flowsheet is generalized to cover many possible synthesis options. The reactor models and unit operations are reduced to their core features to keep the model relatively small. An extensive set of inequality constraints is implemented to make sure the conditions in the reactors and the unit operations will stay in validated domain during optimization.

Outcome:
An industrial case study [26] shows a design region in which both performance metrics can be improved simultaneously. However, after reaching the top in profitability, there is a Pareto frontier with a trade-off: an improvement in carbon efficiency leads to a decrease in profitability.

Few other industrial case studies of (MI)NLP optimizations for process synthesis are found in the scientific literature. This is compensated for by a growing number of academic publications. Floudas and coworkers have applied optimization comprehensively to synthesize XTL processes. In XTL, the symbol X stands generically for either biomass or coal or natural gas or even all feeds in parallel. Also, in these studies, the product slate has been extended from liquid hydrocarbon fuels to chemicals, including methanol, DME and olefins [27]. It is even possible to extend optimizations to the synthesis of a national supply chain (for the USA) for the best locations for new GTL processes, given local sources of natural gas and main customer markets for liquid fuels [28].

Example 9.13: Process performance optimization by model and experiments

Motivation and objective of modeling:
Some new products as well as the processes to make these products are conceived experimentally in the concept stage. Some further optimization is warranted before moving to the next stages. Such an optimization is part of the Feasibility stage. If a product or a process is complicated and only poorly understood, effort to mechanistically model its behavior and performance could be daunting and costly. In such situations, an experimental optimization of the product / process performance with the help of data-driven models can be more effective. Data-driven process model development is sometimes cheaper and quicker than developing a (semi-)mechanistic model. However, it is required that one can indeed measure all input and output variables of a product / process, relevant to the characterization of its behavior and performance.

Approach:
A data-driven modeling and optimization procedure is presented that applies equally well to the optimization of process and product performance. This story will focus on the conceptual outline of experimental process optimization, using a response surface approach.

Outcome:
The outcome is a procedure, leading to a parametric process model and its experimental validation. The model is repetitively improved though the following sequence:
(a) making a computational prediction of the conditions of optimum performance;
(b) experimental validation of this point;
(c) Adaptation: if there is still a significant deviation between the model predictions and the experimental outcomes, the most recent experimental data are used to refine model parameters.

In the end, the model has parameter estimates accurate enough to closely predict the optimum performance point. The strength of this successive approach of gradual refinement of parameter estimates is that model outputs and experimental data should become consistent in the region of optimum performance.

Another, similar response surface approach is reported in [18], chapter 20.5, on the modeling and optimization of a coating process. A design of experiments as well as the successive response surface modeling (regression) are extensively discussed and analyzed for statistical significance of the results. This application is an adaptation from a training exercise in E.I. du Pont de Nemours and Company (USA). It is possible to extend the response surface approach to dynamic process systems [29], such as (semi-)batch reactors. One applies a dynamic design of experiments. Upon collection of the corresponding experimental data, a black box or grey box model can be fitted to the data and applied for process operations (finding optimal trajectories) and control.

9.5 Flowsheet simulators and their usages

9.5.1 Flowsheet simulators

The previous sections explain the details of mathematical process modeling in combination with process design in the innovation stages. This section is about the use of flow sheet simulators, which solve process model equations. These simulators can be purchased and by a little training, can be used directly.

These simulators calculate the input and output compositions of each process step or unit operation by solving the steady-state mass balances. In addition, these simulators have physical property models by which, for each temperature and pressure, physical properties, such as transport properties and phase equilibria, are calculated. Also, enthalpy balances are solved, so that the amount of heat needed or produced, and the temperatures of the unit operations are calculated.

9.5.2 Usage

In the ideation stage, a simulator can be used to show the input and output stream sizes and compositions of each process step. In the ideation stage, the reactor type and the separation type may not be known or not yet decided. Still, the simulator can be used in several ways. A stoichiometric reactor model, a yield model, or a chemical equilibrium model can be chosen.

For separations, simple split models can be used with a split factor as an input variable. Such a simulator model reveals input and output stream compositions and the sensitivity to the degree of conversion and split factor on the input and output compositions.

In the concept stage, when reaction kinetics have been determined, more advanced reactor models, such as plug-flow, or back-mixed, or combinations can be explored on obtaining a high product yield on feed. Optimizer options of the simulator can be used to optimize for temperature and residence time, and residence time distribution, for instance, by optimizing for the number of back-mixed reactors in series for maximum product yield on feed. This can also be optimized, in combination with a separation unit operation and recycle of unconverted feed.

Some simulators also have cost calculation options. These can be used to optimize the function to optimize the whole process for minimum cost. However, care should be taken to make sure that these optimizations are executed for the same yield of product on feedstock, as for most cases, feedstock cost is the overriding cost factor. So, it is better to first optimize for maximum product yield on feed and then optimize for minimum investment cost.

Advanced designs such as reactive distillation can also be treated with simulators and with various degrees of sophistication, such as using equilibrium trays or using mass transfer and reaction kinetics to calculate and optimize the design. Several simulators have models for batch reactors

In the feasibility stage and also in the development stage, detailed process equipment design of all unit operations, including process control can be simulated using dynamic simulation options.

9.5.3 List of steady-state and dynamic simulation packages

Here is a shortlist of simulation packages with their application areas. It is certainly not complete. It serves to show that for many application areas, packages are available.

AspenTech's product portfolio (https://www.aspentech.com/en/products/full-product-listing)**: Aspen HYSYS®, Aspen Plus®,, Aspen Custom Modeller®, Aspen Plus Dynamics**

Application and Use:
- Process simulation software widely used in the chemical, petrochemical, and biochemical industries for process design, process analysis, optimization, and dynamic simulation

AVEVA™ PRO/II™ (AVEVA https://www.aveva.com/en/products/pro-ii-simulation/)**:**

Application and Use:
- Applied in the simulation of various processes, including those in the bioprocess and food industries
- Provides a user-friendly interface and robust capabilities for process optimization and troubleshooting

BioSolve Process (Biopharm Services, https://www.biopharmservices.com/biosolve-software/biosolve-process/)**:**

Application and use:
- Excel spreadsheet-based; applied in the bioprocess and pharmaceutical industry for the optimization and analysis of bio-production processes
- Provides tools for process understanding and economic evaluation

ChemCAD (Chemstations™: https://www.chemstations.com/CHEMCAD/)

Application and use:
- Widely used in the chemical engineering industry for process simulation, design, and analysis
- Known for its intuitive interface and versatile capabilities

ChemSep™ (http://chemsep.com/program/index.html)

Application and use:
- Primarily developed and used for simulating distillation, absorption, and extraction processes
- Particularly valuable for engineers working on separation process design

gPROMS (Siemens): https://www.siemens.com/global/en/products/automation/industry-software/gproms-digital-process-design-and-operations.html

Application and use:
– Applied in the pharmaceutical, chemical, bioprocess, and food industries for process optimization and design
– Allows rigorous modeling of complex processes, including detailed kinetics and thermodynamics

Honeywell UniSim Design Suite (Honeywell, https://www.honeywellforge.ai/us/en/products/industrial-operations/honeywell-unisim-design-suite**):**
   Application and use:
– Applied in various industries, including chemical, refining, and petrochemical
– Provides a platform for rigorous process simulation and optimization

MATLAB Simulink (Mathworks: https://nl.mathworks.com/products/simulink.html
   Application and use:
– Widely used across engineering disciplines, including chemical and biochemical engineering
– In chemical and biochemical engineering, used for dynamic system modeling and simulation
– Allows for the design and analysis of control systems in real-time

Mobatec Modeler (Mobatec: https://www.mobatec.nl/web/products/mobatec-modeller/**)**
   Application and use:
– Applied in various industries, including chemical, petrochemical, and pharmaceutical
– Focuses on dynamic simulation and provides tools for process optimization, control strategy analysis, and operator training

SuperPro Designer (Intelligen, Inc. https://www.intelligen.com/products/superpro-overview/**)**
   Application and use:
– Widely used in bioprocess and food industries for process development, design, optimization, and dynamic simulation
– Provides a comprehensive platform for modeling, simulating, and optimizing bioprocesses

9.5.4 Risks in using flow sheet simulators

Some major risks in using flow sheet simulators are:
– Trusting the simulator results without experimental validation
– Trusting distillation results for components with hydrogen bonding, such as water and alcohol mixtures using available physical property models with no experimental validation

- Not investigating the error report of the simulation results
- Trusting the simulations of a process with one or more recycle streams, without checking the long-term steady-state mass balance

9.6 Life cycle analysis packages and their usages

9.6.1 Life cycle analysis packages

Life cycle analysis (LCA) packages calculate environmental impacts over the entire life cycle, cradle-to-grave of products. The input needed is the functional unit product composition, decided by the designer [30]. The packages can be used in the concept stage, feasibility stage, and development stage. Their reliability and accuracy depend entirely on the databases used. Large, accurate, and validated data bases can be purchased from a few institutes.

Here are some well-known software packages, namely, SimaPro, OpenLCA, Gabi, Umberto obtained from Silva, who discusses these and others [31].

9.6.2 Usage

In the concept stage, an LCA package can be useful with a standard database, when a reference case is used and only conclusions are drawn on the environmental impacts of the design case, relative to the reference case, and the design case has far lower impacts than the reference case.

In the feasibility stage and the development stage, reliable and accurate impacts are needed to decide to pursue the new product development or not. To this end, specific databases should be purchased. Some large companies have their in-house LCA group and do all research to get the LCA data themselves.

9.7 Computational fluid dynamic (CFD) packages and their usages

9.7.1 CFD packages

Computational fluid dynamic (CFD) packages are used to calculate fluid flows in complex configurations. The packages can calculate single-phase mixing, residence time distribution, mass transfer between phases, and heat transfer. Many packages are available on the market. They vary a lot in their specifics. Therefore, a list is not provided here.

9.7.2 Usage

CFD packages are mostly used in the development stage of a new complex process step, where the precise configuration is important for the performance, such as in furnaces, complex reactors, and mixing equipment.

CFD is also used to improve existing processes. An example is the horizontal crossflow gas-liquid reactor of the ethylbenzene hydroperoxidation reactor, operated at commercial scale by Shell. By making a CFD model of the liquid and gas bubble flow patterns, in combination with reaction kinetics, areas of low oxygen concentration could be discovered by simulation. With tracer experiments in the commercial scale reactor, the CFD model could be validated on flow pattern aspects. By modifying the gas spargers, using the CFD simulations, the oxygen starvation areas could be prevented. When the new gas spargers were installed, the product yield and the production capacity increased [31].

9.8 Concluding remarks

This chapter has highlighted approaches to model development and the applications related to the Concept, Feasibility, and Development stages of product and process development and design. The emphasis here has been on making effective and inventive use of models for process innovation and optimization, illustrated by means of small-scale examples. This amounts to making useful interpretations of the modeling and computing results. The balance in the use of models has been in favor of steady-state (algebraic) models, at the expense of modeling of process dynamics for control and (dynamic) optimization, though the core thinking patterns in analyzing models and their results remain the same.

List of symbols, subscripts, and abbreviations

Symbols

A	Real (non-)square matrix (real values as entries)
b	Real vector
c	Cost-related parameters in economic model expressions
d	Symbol for design parameters (continuous) in a process model
E	Set of quantitative performance metrics for a process design.
N	Dimension of a vector
p	Generic symbol for a set of physical parameters in a process model
p_P	Proceeds from product sales, in economic model equations
P	Performance function
t	Target value for a design quantity or performance function in process design

T	Function expressing a metric (function) related to a target
u	Generic symbol for inputs to a process model
w	Disturbance variables in a process model
x	Generic symbol for physical states in a process model
y	Set of measured outputs of a process (model)
Y	Generic symbol for structural design parameters (discrete) in a process model
z	Generic symbol for a set of adjustable inputs to a process model
ε	Deviation between computed and experimental value of a process output variable.
$\Re$	Mathematical space of real variables
∇	Spatial derivative

Subscripts

Average	Averaged value of a quantity
design	Design related
ecol	Ecology related
ecom	Economy related
oper	Operations related
perf	Performance related
ref	Reference value of a quantity when making a comparison
tech	(Process) technology related

Abbreviations

CSTR	Continuous flow, well-stirred tank reactor
DF&EI	Dow Fire & Explosion Index
DME	Dimethyl ether
GTL	Gas to liquid
KPI	Key performance indicator
LP	Linear programming
MINLP	Mixed-integer nonlinear programming
MONLP	Multiobjective nonlinear programming
MTBE	Methyl *tertiary* butyl ether
NLP	Nonlinear programming
XTL	X = {Biomass, coal, natural gas} to liquid fuels

References and further reading

[1] Dimian AC, Bildea CS, Kiss AA. Integrated Design and Simulation of Chemical Processes, 2nd edn, Vol. 35, Elsevier, Computer-Aided Chemical Engineering, 2014.

[2] Aris R. Mathematical modelling techniques. In Research Notes in Mathematics 24, Pitman Advanced Publishing Program, 1979.

[3] Bequette BW. "Process Dynamics, Modelling, Analysis & Simulation", Prentice Hall, 1998.

[4]   Aris R. Mathematical modelling. In PSE Series, Vol. 1, Academic Press, 1999.
[5]   Hangos KM, Cameron IT. Process modelling & analysis. In PSE Series, Vol. 4, Academic Press, 2002.
[6]   Cameron IT, Gani R. "Product & Process Modelling A Case Study Approach", Elsevier, 2011.
[7]   Rice RG, Do DD. "Applied Mathematics & Modeling for Chemical Engineers", 2nd edn, Wiley, 2012.
[8]   Zondervan E. "A Numerical Primer for the Chemical Engineer", Taylor & Francis (CRC Press), 2014.
[9]   Rudd DF, Watson CC. "Strategy of Process Engineering", John Wiley &Sons, Inc, 1968.
[10] Westerberg AW, Hutchison HP, Motard RL, Winter P. "Process Flowsheeting", Cambridge University Press, 1979.
[11] Biegler LT, Grossmann IE, Westerberg AW. "Systematic Methods of Chemical Process Design", McGraw Hill, 1997.
[12] Seider WD, Seader JD, Lewin DR, Widagdo S. "Product and Process Design Principles. Synthesis, Analysis, and Evaluation", 3rd edn, John Wiley & Sons, Inc, 2010.
[13] Floudas CA. "Nonlinear and Mixed-Integer Optimization. Fundamentals and Applications", Oxford University Press, 1995.
[14] Floudas CA. "Deterministic Global Optimization. Theory, Methods and Applications", Kluwer Academic Publishers, 2000.
[15] Edgar TF, Himmelblau DM, Lasdon LS. " Optimization of Chemical Processes", 2nd edn, McGraw-Hill, 2001.
[16] Biegler LT. Nonlinear Programming. Concepts, Algorithms and Applications to Chemical Processes, SIAM, 2010.
[17] Buelens LC, Galvita VV, Poelman H, Detavernier C, Marin GB. Super-dry reforming of methane intensifies CO_2 utilization via Le Chatelier's principle, Science, 354(6311), 2016, 449–452.
[18] Ogunnaike BA. Random Phenomena. "Fundamentals and Engineering Applications of Probability and Statistics, CRC Press, Section, 20(5), 2010, 879.
[19] Varma A, Morbidelli M, Wu H. Parametric Sensitivity in Chemical Systems, Cambridge University Press, 1999.
[20] gPROMS process modelling platform of PSEnterprise (London, UK);https://www.psenterprise.com/products/gproms/platform
[21] Marlin TE. Teaching "operability" in undergraduate chemical engineering design education, Computers & Chemical Engineering, 34, 2010, 1421–1431.
[22] Global system analysis in gPROMS process modelling platform of PSEnterprise (London, UK); https://www.psenterprise.com/products/gproms/technologies/global-system-analysis
[23] AIMMS (Advanced Interactive Multidimensional Modeling System) of Paragon Decision Technology, Haarlem (NL) https://aimms.com/english/developers/resources/manuals/optimization-modeling/
[24] Van Goethem MWM, Barendregt S, Grievink J, Moulijn JA, Verheijen PJT. Model-based, thermo-physical optimisation for high olefin yield in steam cracking reactors, Chemical Engineering Research and Design, 88, 2010, 1305–1319.
[25] Almeida-Rivera CP, Grievink J. Process Design Approach for Reactive Distillation Based on Economics, Exergy and Responsiveness Optimization, Industrial & Engineering Chemistry Research, 47, 2008, 51–65.
[26] Ellepola JE, Thijssen N, Grievink J, Baak G, Avhale A, van Schijndel J. Development of a synthesis tool for Gas-To-Liquid complexes, Computers and Chemical Engineering, 42, 2012, 2–14.
[27] Baliban RC, Elia JA, Weekman V, Floudas CA. Process synthesis of hybrid coal, biomass, and natural gas to liquids via Fischer–Tropsch synthesis, ZSM-5 catalytic conversion, methanol synthesis, methanol-to-gasoline, and methanol-to-olefins/distillate technologies, Computers and Chemical Engineering, 47, 2012, 29–56.
[28] Elia JA, Baliban RC, Floudas CA. Nationwide, Regional, and Statewide Energy Supply Chain Optimization for Natural Gas to Liquid Transportation Fuel (GTL) Systems, Industrial & Engineering Chemistry Research, 53, 2014, 5366–5397.

[29] Klebanov N, Georgakis C. Dynamic Response Surface Models: A Data-Driven Approach for the Analysis of Time-Varying Process Outputs, Industrial & Engineering Chemistry Research, 55, 2016, 4022–4034.

[30] Silva DA, Nunes AO, Piekarski CM, da Silva Moris VA, de Souza LS, Rodrigues TO. Why using different Life Cycle Assessment software tools can generate different results for the same product system? A cause–effect analysis of the problem, Sustainable Production and Consumption, 20, 2019 Oct 1, 304–15. Harmsen J, Bos R. Multiphase Reactors: Reaction Engineering Concepts, Selection, and Industrial Applications. De Gruyter; Berlin, 2023.

[31] Harmsen J, Bos R. Multiphase Reactors: Reaction Engineering Concepts, Selection, and Industrial Applications, Berlin: De Gruyter, 2023.

10 Evaluating economic performance

10.1 Introduction

Before companies agree to invest large amounts of capital on proposed projects, their management must be convinced that the selected project(s) will provide a sound *return on investment* (ROI) compared to alternatives. Therefore, it is crucial that economic evaluations are made before initiating a project, at various stages in its development, and before attempting the design of a process and plant. These evaluations determine whether a project should be initiated, abandoned, continued, or taken to the scale-up and implementation stages. Furthermore, economic evaluations pinpoint those parts of the project (product formulation, manufacturing, and process) that require additional investigation. Even if the available technical information is insufficient for complete product/process design, economical evaluations must still be made to determine whether the project is economically (more profitable than competing projects) and financially (capital required for implementation can be raised) feasible. Although cost estimation is a specialized subject and a profession in its own, design engineers must be able to make rough cost estimates to support the decisions between project alternatives, and optimize the design.

The purpose of most chemical engineering design projects is to provide the input information from which estimates of capital and manufacturing costs can be made and subsequently, the profitability of a project can be assessed. Therefore, this chapter first introduces the principal methods used for comparing the *eco-*

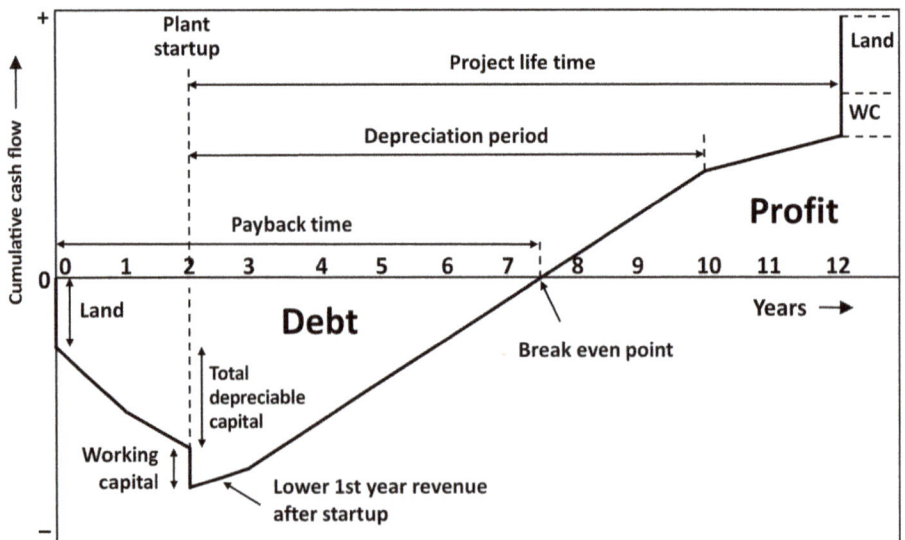

Fig. 10.1: Project cash flow diagram.

https://doi.org/10.1515/9783110782127-010

nomic performance and *profitability* of different projects. Subsequently, the components comprising the manufacturing costs and the methods used for capital costs estimating are introduced and discussed.

10.2 Economic project evaluation

10.2.1 Project cash flow

During any project, cash initially flows out of the company to cover the engineering costs, equipment procurement, plant construction, and plant startup. After construction is completed, the plant is put into operation and revenues from product sales start to flow into the company. The *net cash flow* (CF) is the difference between earnings and expenditures. A *cash-flow diagram*, such as that shown in Fig. 10.1, illustrates a typical cumulative net cash flow forecast over the life of a project. The cash flows are based on estimates of investment, operating costs, sales volume, and sales price for the project. The diagram can be divided into the following characteristic regions:

– Investment required to design, construct and provide funds for startup of the plant.
– Upturning cash-flow curve due to a positive net cash flow as income is generated from sales; the cumulative amount remains negative until the investment is paid off, which is known as the *break-even point*.
– Positive cumulative cash flow where the project is earning a *return on the investment*.

The *cash flow* (CF) summarizes all the incoming and outgoing cash flows through the operating year. Investments are considered a negative cash flow, while after-tax profits plus depreciation are positive cash flows. During the years of plant construction, the cash flow for a particular year n is:

$$CF_n = -\varphi\,TDC - WC - C_{Land} \qquad (10.1)$$

where φ is the fraction of the total depreciable capital TDC expended that year, WC the working capital, and C_{Land} the cost of land expended during that year. *Working capital* comprises the additional funds, on top of the costs to build the plant, to startup the plant, and keep it in operation. It flows in and out of an existing operation and is usually assumed to be completely recoverable at the end of a project without loss. As illustrated in Fig. 10.2, working capital is divided into two main categories:

1. *Current liabilities* that consist of bank loans and accounts payable (money owed to vendors for various purchases).

2. *Current assets* that comprise
    - available cash (salaries, raw material purchases, maintenance supplies, taxes);
    - accounts receivable (product shipped but not yet paid, extended credit to customers);
    - product inventory (material in storage tanks and bins);
    - in-process inventory (material contained in pipe lines and vessels);
    - raw material inventory (material in storage tanks and bins);
    - spare parts inventory.

Working capital can vary from as low as 5% of the fixed capital for a simple, single-product, process, with little or no finished product storage, to as high as 30% for a process producing a diverse range of product grades for a sophisticated market, such as synthetic fibers. A typical figure for general chemicals and petrochemical plants is 15% of the total permanent investment, as defined in Fig. 10.7.

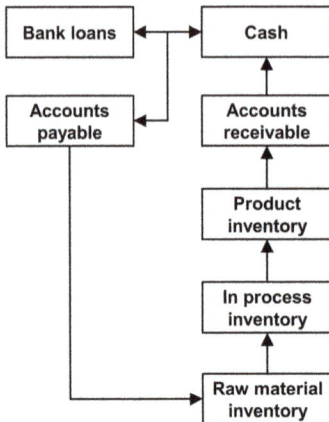

Fig. 10.2: Flow of working capital.

The cash flow for a particular year of plant operation is calculated from the *operating income* (OI), resulting from subtracting the operating expenses (variable and fixed costs) from the sales or revenue for that specific year:

$$OI_n = S - VC - FC \tag{10.2}$$

where S is the annual sales, VC the annual variable costs, and FC the annual fixed costs. The total annual product sales revenue is the sum of the unit price of each product multiplied by its sales quantity. When depreciation (capital costs) is subtracted from the operating income, annual *gross profit* (GP) is obtained:

$$GP_n = S - VC - FC - D \tag{10.3}$$

where D is the annual depreciation, calculated by dividing the total depreciable capital, C_{TDC} in Fig. 10.7, by the depreciation period of N years, often chosen equal to 10 years. The annual *net profit* (NP) represents the net result in a specific year and is calculated by subtracting income taxes from the gross profit:

$$NP_n = GP_n \times (1 - t) \qquad (10.4)$$

where t is the taxation rate. As depreciation is considered an internal expense that is retained in the company, the cash flow from plant and product manufacturing operations for year n equals the after tax earnings plus the depreciation:

$$CF_n = (1 - t)(S - VC - FC - D) + D \qquad (10.5)$$

The *cumulative cash flow* sums the calculated cash flow for each year up to the point of the calculation:

$$CCF_N = \sum_{n=0}^{N} CF_n \qquad (10.6)$$

where CCF_n is the cumulative cash flow for a period of N operational years, CF_n the cash flow for year n, and n is a positive integer for the operational year. The *discounted cash flow* (DCF) calculates the value of current money in the future, considering that the value of money earned in the future is less than money in the present. This calculation is dependent on the discount rate, which further decreases the value of money when it increases. The discount rate used is generally the appropriate weighted average cost of capital (WACC) that reflects the risk of the cash flows, by incorporating the time value of money and a risk premium. Through the time value of money (risk-free rate), investors are compensated for having to wait until their investment is paid back. The risk premium reflects the extra return investors demand because they want to be compensated for the risk that the cash flow might not materialize after all. As a result, the *discount rate* typically ranges from 10% for low-risk projects up to 20% for high-risk projects.

$$DCF_n = \frac{CF_n}{(1 + r)^n} \qquad (10.7)$$

where DCF_n is the discounted cash flow for year n and r the discount rate. The *cumulative discounted cash flow* (CDCF) calculation uses the calculated discounted cash flow for each year and sums them up to the point of the calculation. This number is indicative for the net result of a business case. When the cumulative discounted cash flow equals zero, the *internal rate of return* is equal to the discount rate.

$$CDCF_N = \sum_{n=0}^{N} DCF_n \qquad (10.8)$$

where $CDCF_n$ is the cumulative discounted cash flow for a period of N years.

10.2.2 Economic potential project evaluation method

Before even starting a design, it is crucial to verify that the target product(s) at least provide a value that is higher than the minimal consumption of raw materials. Such evaluation requires nothing more than first indications of possible sales prices, raw material prices, and targeted production volume, which are typically obtained from market data. For important chemicals, market data and prices are recorded and published by various organizations such as ICIS (www.icispricing.com), IHS (www.ihs.com/industry/chemical) that also publishes the Chemical Economics Handbook containing market overviews for > 300 compounds, Nexant (www.nexant.com/industries/chemicals), Orbichem (www.orbichem.com), and Intratec (www.intratec.us) or can be obtained via online brokers and suppliers (www.business.com/directory/chemicals, www.alibaba.com). For specialty chemicals and new products, it is often more difficult to obtain reliable price data. In such cases, consultants can be involved to provide economic and marketing information or estimates of attainable prices levels need to be made, based on envisioned product performance and the related/competing products in the targeted market(s).

Based on the obtained price data indications and the calculated raw material consumptions, a first estimate of the *economic potential* (EP) of a project can be made by simply subtracting the annual cost of the raw materials from the expected annual sales revenues:

$$EP = S - RMC \tag{10.9}$$

where S is the annual sales and RMC the annual raw material costs. Sales can constitute only the main product and known by-products. The main benefit of this first high-level evaluation is that nothing more than a black box *input-output diagram* and only an overall mass balance is required to assess the difference in economic potential of the various design alternatives. For most projects, this is a valuable approach as raw material costs are the major part (50–90%) of the total manufacturing costs. When required, this economic potential estimation can be refined to include utility costs, but then more detailed information on the energy balances is required.

10.2.3 Simple project evaluation methods

This section introduces some simple economic indicators that can be calculated quickly once the project investment and cash flows are estimated. These indicators are widely used for preliminary screening of project attractiveness. By dividing the total capital investment (fixed capital plus working capital, see Fig. 10.7) by the average annual cash flow, the *payback time* (PBT) is obtained:

$$PBT = \frac{C_{TCI}}{CF} \tag{10.10}$$

where C_{TCI} = the total capital investment and CF the average annual cash flow. This is not the same as the break-even point indicated by the cash-flow diagram, as it assumes that all the investment is made in year zero and revenues begin immediately. For most chemical plant projects, this is not realistic as investments are typically spread over one to three years and revenues may not reach 100% of design basis until the second year of operation. The simple payback time is strictly based on a cash flow, but for simplicity, taxes and depreciation are often neglected and the average annual income is used instead of cash flow. Another simple indicator of economic performance ROI is defined as

$$\text{ROI} = \frac{NP_n}{C_{TCI}} \times 100\% \tag{10.11}$$

which is generally stated as a percentage per year. The required ROI will typically depend on the degree of risk associated with a project. For low-risk projects, a ROI of 10% might be sufficient while for high-risk ventures (new products, new markets, and new technology), the required ROI might be as high as 50%.

10.2.4 Present value project evaluation methods

The above simple economic indicators do not capture the time dependency of cash flows during the project. Cash flow timing is crucial to investors because not all capital must be financed immediately and money earned can be reinvested to earn a return as soon as it is available. For that reason, money earned in the early project years is more valuable than that earned later. This *time value of money* can be accounted for by bringing the net cash flow in each year of the project to its "present value" at the start of the project through discounting it at some internally agreed discount rate, which is set at the weighted average cost of capital (WACC) in most companies.

The *net present value* (NPV) is one way to value a project in terms of cash flows and the value of money. It states how much the business is worth at a certain point of time (typically, a 10-year period is taken). It includes future revenues and expenses, and converts those cash flows into a positive or negative value.

$$\text{NPV}_N = \sum_{n=1}^{N} \frac{CF_n}{(1+r)^n} \tag{10.12}$$

The *internal rate of return* (IRR) is directly linked to both the discount rate and the net present value. It says something about the rate of growth that a project is expected to generate. By calculating the NPV at various interest rates, it is possible to find an interest rate at which the cumulative net present value at the end of the project is zero:

$$\text{NPV}_N = \sum_{n=1}^{N} \frac{CF_n}{(1+\text{IRR})^n} \qquad (10.13)$$

This particular rate is called the internal rate of return and is a measure of the maximum interest rate that the project could pay and still break even by the end of the project life. The more positive the IRR, the more favorable the project is in terms of ROI. Often, this value is used for the comparison of different projects, as it is relative easy to calculate and compare. Companies usually demand IRR levels higher than the internally agreed discount rate before they invest in a project.

10.3 Manufacturing costs

To determine the financial attractiveness of a project, the manufacturing (production) costs of a product are needed. Figure 10.3 separates the total production costs into three main categories – direct costs, indirect costs, and general costs. Direct (variable) costs are proportional to the production rate, whereas the indirect and general cost (fixed and plant overhead cost) tend to be independent of the production rate. General costs cover management of the firm, marketing the product, and research and development.

Feedstocks, catalysts, solvents	Raw materials	
Steam, electricity, fuel, refrigeration, water, waste treatment, landfill	Utilities	**Direct costs**
Supplies: operating, maintenance Operating labor, supervision Maintenance labor, supervision Laboratory charges Royalties	Operational costs	
General plant overhead Safety and protection Medical, salvage services Payroll overhead Packaging, shipping, storage facilities Restaurant, recreation facilities Control laboratories Maintenance facilities	Plant overhead costs	**Indirect cost**
Depreciation Property taxes, financing, insurance, rent	Fixed costs	
Executive, clerical, engineering, legal, communications	Administration	**General cost**
Sales (or transfer), advertising, product distribution, technical sales service	Marketing costs	
Research and development		

Fig. 10.3: Total production costs breakdown.

10.3.1 Direct costs

In calculating the *direct costs*, the annual cost of each raw material is found by multiplying the annual consumption by the price. Raw material costs often dominate the overall manufacturing costs. Consumables such as acids, bases, adsorbents, membranes, solvents, and catalysts are depleted or degraded over time and require replacement. Utilities comprise steam, electricity, fuel, cooling water, process water, compressed air, refrigeration, waste treatment, and landfill. Utility equipment is usually located outside the process area (OSBL) and may supply to several processes. Analogous to raw materials, the annual cost of each utility is determined by multiplying its annual consumption by the price.

Chemical plants require operating labor to produce a chemical, and maintenance labor to maintain the process. There is also indirect labor needed to operate and maintain facilities and services. When developing a new process, the quantity of operating labor can often be estimated in preliminary cost analysis from company experience with similar processes or published information on comparable processes. If a flowsheet or PFD is available, the *operating labor* can be estimated from the number of operations, based on previous experience:

$$\text{Annual labor cost} = \sum \left(\text{Equipment type} \times \frac{\text{operator/shift}}{\text{unit}} \right) \times (\text{number shifts})$$
$$\times \left(\frac{\text{cost/year}}{\text{operator}} \right)$$

$$(10.14)$$

Table 10.1 indicates some typical labor requirements for various types of process equipment. The number of shifts varies between countries but is generally 5 for 24/7 operating continuous plants and 4 for 24/5 operating batch plants. The required amount of operating supervision is closely related to the total amount of operating labor, and typically averages about 15% of the operating labor costs.

Maintenance costs consist of materials, labor, and supervision, and may sometimes be higher than the cost of operating labor. Although maintenance costs increase as a plant ages, for economical estimates, an average value over the plant life is assumed. The *maintenance supplies* (materials and services) are typically estimated as a fraction of the total depreciable capital investment, and range from 3.5% for a fluids processing plant to 5% for a solids processing plant, and 4.5% for solids-fluids processing. *Maintenance labor* wages and benefits are typically of the same magnitude as the maintenance supplies costs. The costs for engineers, supervision, and overhead are estimated at 30% of the direct maintenance labor. Custodial supplies, safety items, tools, column packing, and uniforms, which are not raw materials or maintenance supplies, are considered as *operating supplies*. A typical value for these types of supplies cost is 15% of the maintenance supplies.

In order to meet the specifications of the products, regular analysis of process streams is mandatory to determine and control their quality. As initial estimate, these *laboratory charges* for quality control typically range from 10 to 20% of the operating labor. The amount of *royalties* (licensing fees) depends heavily on the uniqueness of the process and can range from 0 to 6% of the total production cost or, alternatively, 0 to 5% of product sales. All these direct costs components are summarized in Tab. 10.1.

Tab. 10.1: Typical labor requirements for process equipment [9].

Equipment type	Operators/unit/shift
Compressors and blowers	0.15
Centrifuge	0.35
Crystallizer	0.20
Dryer	0.50
Evaporator	0.25
Filter (continuous/batch)	0.15/1.0
Heat exchanger	0.10
Vessel, tower	0.2–0.5
Reactor (continuous/batch)	0.5/1.0

10.3.2 Indirect and general costs

As depicted in Fig. 10.3, *indirect costs*, consisting of fixed and plant overhead costs, change only little or not at all with the amount of production (plant capacity) because they are indirectly related to the production rate.

Fixed costs include depreciation, property taxes, insurance, interest, and rent. These expenses are a direct function of the capital investment and financing structure. Rent is usually not taken into account in a preliminary estimate. The initial investment is paid back by charging *depreciation* as a manufacturing expense. Although there are several depreciation methods, it is commonly accepted to use a constant yearly depreciation rate for a fixed period in preliminary economic evaluations. A manufacturing plant or individual equipment has an economic, a physical, and a tax life. The economic life is the period until a plant becomes obsolete, the physical life is when maintenance of a plant becomes too costly, and the tax life is fixed by the government. The plant life is usually set between ten and twenty years. Depreciable capital includes all incurred costs for building a plant up to the point where the plant is ready to produce, except land and site development. The magnitude of local *property taxes* depends on the particular location of the plant and may be initially estimated at 2% of the total depreciable capital investment. Property insurance rates, typically 1% of the total depreciable capital per year, depend on the actual operation carried out in the manufacturing plant and the extent of protective measures taken. Interest is only a cost when it is necessary to borrow the capital needed to invest in the plant. However, in most cases, the capital originates from the company and/or investors, making project profitability calculation, the main evaluation objective. Therefore, *interest* is normally excluded from the cash flow. *Plant overhead* is the costs involved for services and facilities to make the complete plant function as an efficient unit. These costs are closely related to the costs for all labor directly connected with the production operation, and amount about 22% of the total direct labor costs (operating, maintenance, and supervision).

In addition to manufacturing costs, other *general costs* are associated with management of a plant. Main categories to be included are administrative, marketing and sales, and research and development. *Administrative costs* are those expenses connected to executive and administrative activities. For a preliminary estimate, they may be approximated as 2% of the sales. *Marketing and sales* cost vary widely for different types of manufacturing plants. For larger volume products, 3% of the product sales is considered a reasonable estimate. When the product is only transferred "over the fence," 1% of sales is a suitable approximation.

Finally, the process and product need to be continuously improved. Therefore, the cost of *research and development* should be added to the production cost. In the chemical industry, these costs range from 1 to 10% of the total sales, depending on the business. Table 10.2 summarizes the indirect and general cost components.

Tab. 10.2: Breakdown of direct operational and indirect costs [4].

Operational costs	
Operating labor	See eq. (10.14)
Operating supervision	15% of operating labor
Operating supplies	15% of maintenance supplies
Maintenance supplies	
– Fluid handling	3.5% of total depreciable capital (C_{TDC})
– Fluid-solids handling	4.5% of total depreciable capital (C_{TDC})
– Solids handling	5.0% of total depreciable capital (C_{TDC})
Maintenance labor, supervision	120% of maintenance supplies
Laboratory charges	15% of operating labor
Royalties	4% of total production costs without depreciation
Indirect costs	
Overall plant overhead	22% of operating + maintenance labor and supervision
Fixed costs	
– Depreciation	10% of total depreciable capital (C_{TDC}, chosen lifetime dependent)
– Property taxes	2% of total permanent investment (C_{TPI})
– Insurance	1% of total depreciable capital (C_{TDC})

Tab. 10.2 (continued)

Indirect costs	
– Interest	Excluded from cash flow when profitability calculation is the main objective
– Rent	0% of total permanent investment (C_{TPI})
General costs	
– Administration	2% of sales
– Marketing and sales	3% of sales (transfer 1% of sales)
– Research and development	5% of sales (2–10% depending on business)

10.3.3 Cost sheet

For economic project evaluation, accuracy is crucial in estimating total production cost. It is most important to ensure that all costs associated with making and selling the product are included. These are included in the cost sheet depicted in Fig. 10.4. In this cost sheet, the raw materials and utilities comprise the so-called *bill of materials* (BOM), indicating exactly what is needed as input to produce one kilogram or metric ton of product, and therefore a main means of communication between the technical and financial people within a company.

10.4 Estimation of capital costs

10.4.1 Capital cost components

The *total depreciable capital* (TDC) involves the total cost for the design, construction, and installation of a new plant or the revamp of an existing plant. It also includes the associated modifications required for the preparation of the plant site. A new manufacturing plant may be an addition to an existing site or a *grassroots* plant without nearby auxiliary facilities. It is customary to separate the directly associated processing equipment from the auxiliary facilities by an imaginary border, named *battery limits*. Utilizing this division, the total capital investment is made up of direct costs:

1. *Inside battery limits (ISBL)* investment for the plant itself
2. *Outside battery limits (OSBL)* investment auxiliary facility/site modifications and improvements

and indirect costs:

3. Engineering and construction
4. Contingency charges

Bill of Materials			Annual production		xxxxx ton/year
Raw materials	Consumption (unit/kg product)	Price (€/unit raw material)	Cost (€/kg product)	Annual consumption (unit/year)	Anual costs (€/year)
Reactant A					
Reactant B					
Catalyst					
Solvent					
Carbon					
...					
Total raw materials					
Utilities					
Steam (ton)		25			
Electricity (kWh)		0,07			
Natural gas (GJ)		1,9			
Process water (m³)		0,35			
Cooling water (m³)		0,05			
Waste water (m³)		1			
Landfill (ton)		20			
Other waste (ton)		300			
...					
Total utilities					
Total variable costs					
Operational cost					
Operating cost					
Labor					
Supervision					
Supplies					
Maintenance cost					
Supplies					
Labor, supervision					
Laboratory charges					
Royalties					
Total operational cost					
Indirect cost					
Plant overhead					
Depreciation					
Other					
Total indirect cost					
Manufacturing cost					
General cost					
Administration					
Marketing & sales					
R&D					
Total general cost					
Total production cost					

Left margin labels: BOM (Bill of Materials); + Direct cost; + Indirect cost; + General cost.

Fig. 10.4: Cost sheet outline.

As summarized in Fig. 10.5, the *ISBL investment* includes the purchasing and installation costs of all the equipment items that make up the new plant. These *direct plant costs (DPC)* consist of:
- *Purchased equipment costs (PEC)* for all major process equipment, including field fabrication and testing, if necessary
  - vessels, reactors, and columns
  - furnaces, heat exchangers, and coolers
  - pumps, compressors, motors, fans, and turbines
  - filters, centrifuges, and driers
  - . . . .

I **S** **B** **L**	Major equipment, spare parts, surplus equipment freight charges, taxes, insurance, duties, startup allowance	**Delivered equipment**		**Direct costs**
	Placing equipment, paint, insulation, foundations, structural steel	**Equipment installation**		
	Instrumentation purchase, calibration, installation	**Instrumentation**		
	Process piping, pipe hangers, fittings, valves, insulation	**Piping**		
	Electrical equipment, materials, installation	**Electrical**		
O **S** **B** **L**	Process buildings, maintenance shops, warehouses, buildings for services, garages, steel structures, laboratories, medical, cafeteria	**Buildings**		
	Site development, site clearing, grading, roads, walkways, railroads, fences, parking, landscaping	**Yard improvements**		
	Utilities, waste treatment, receiving, shipping, packaging, storage, lighting, communications	**Service facilities**		
	Property cost, surveys, fees	**Land**		
	Administration, process design, general engineering, cost engineering, drafting, purchasing, expediting, inspection, supervision, reproduction, communications, travel	**Engineering**		**Indirect cost**
	Temporary facilities, offices, roads, parking, railroads, electrical, piping, fencing, equipment, supervision, accounting, purchasing, timekeeping, expediting	**Construction**		
		Legal expenses Contingency Contractor's fee		

Fig. 10.5: Components of the total depreciable capital (TDC) for a chemical manufacturing plant (Adapted from [2]).

– *Installation Costs* to integrate all equipment items into an operating plant
    – foundations, structures, roads, piling, process buildings, and sewers
    – piping, valves, insulation, and paint
    – wiring and instrumentation
    – solvents and catalysts
        – Installation labor and supervision, including construction costs such as construction equipment rental, temporary construction (rigging, trailers, etc.), temporary water and power, and construction workshops
    – . . .

It is extremely important to define the ISBL scope of a project carefully in the early stages of a project, as other project costs are often estimated from the ISBL cost. With a poorly defined ISBL scope, the overall project economics can be badly miscalculated. After discussing the other capital cost components and the various, the following sections of this chapter provide several methods, from very coarse to highly detailed, for estimating the ISBL costs.

Utilities

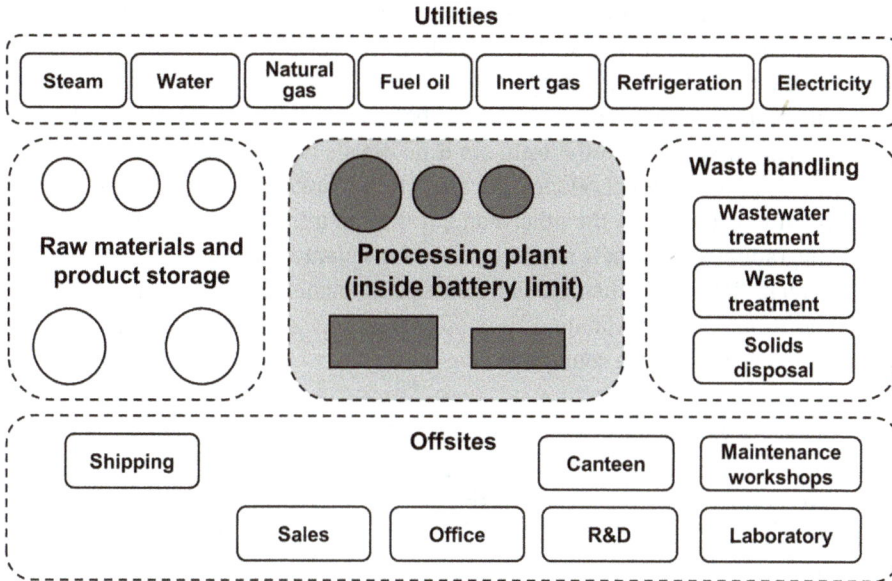

Fig. 10.6: Layout, including auxiliary facilities (OSBL), of a chemical manufacturing plant.

OSBL investment includes the costs of all additions, improvements, and modifications that must be made to the site infrastructure to accommodate adding a new plant or revamp an existing plant. Figure 10.6 illustrates typical offsite auxiliary facilities that are associated with the construction of a grassroots plant. Depending on the extent of needed/available offsite facilities, they can be a significant contribution to the total depreciable capital. Typical offsite investments include:

- Power generation plants, turbine engines, standby generators, electric main substations, transformers, switchgear, and powerlines
- Boilers, steam lines, condensate lines, boiler feed water treatment plant, and supply pumps
- Cooling towers, circulation pumps, cooling water lines, and cooling water treatment
- Water supply pipes, demineralization plant, waste water treatment, and site drainage and sewers
- Air separation plants to provide site nitrogen for inert gas and nitrogen lines
- Dryers and blowers for instrument air and instrument air lines
- Pipe bridges, feed and product pipelines, tanker farms, loading facilities, silos, conveyors, docks, warehouses, railroads, and lift trucks
- Laboratories, offices, canteens, changing rooms, central control rooms, and maintenance workshops
- Emergency services such as firefighting equipment, fire hydrants, and medical facilities
- Site security, fencing, gatehouses, and landscaping

OSBL investment is typically estimated as a proportion of the ISBL investment in the early stages of design. Off-site costs are usually in the range from 10% to 100% of ISBL costs, depending on the project scope and its impact on site infrastructure. For typical chemicals projects, off-site costs are usually between 20% and 50% of ISBL cost, and 40% is usually used as an initial estimate if no details of the site are known. For an established site with well-developed infrastructure ("brownfield" sites), off-site costs will generally be lower. On the other hand, if the site infrastructure is in need of repair or upgrading to meet new regulations, or if the plant is built on a completely new site (a "green- field" site), then off-site costs will be higher. Table 10.3 provides some guidelines to make approximate estimates of OSBL investment as a function of the process complexity and site conditions.

Tab. 10.3: Approximate OSBL investment as percentage of ISBL investment [1].

Process complexity	Existing site	New site
Typical bulk chemical	30%	40%
Specialty chemical	30%	50%
Solids handling process	40%	100%

The *engineering and construction costs*, comprise the costs of detailed design of the plant and other engineering services required to carry out the project:
–   Detailed design engineering of process equipment, piping systems, control systems and off sites, plant layout, drafting, cost engineering, scale models, and civil engineering
–   Procurement of main plant items and bulks
–   Construction supervision and services
–   Administrative charges, including engineering supervision, project management, expediting, inspection, travel and living expenses, and home office overheads
–   Contractor's profit

In most cases, operating companies bring in one or more of the major engineering contracting firms to carry out all these activities. Initially, engineering and construction costs can be estimated as 30% of the total plant costs (ISBL plus OSBL) for smaller projects and 10% of the total plant costs for larger projects.

As all cost estimates are uncertain, *contingency charges* are added to the project budget to allow for variation from the cost estimate. In addition, contingency costs also help to cover minor project scope changes, price changes (e.g. steel, copper, catalyst, etc.), currency fluctuations, and other unexpected problems. A contingency charge can be thought of as an additional fee charged by the contractor to reduce the likelihood of losing money on a fixed-price bid. A minimum contingency charge of 10% of ISBL plus OSBL cost should be used on all projects. For uncertain technologies, higher contingency charges (up to 50%) can be used.

Abbreviation	Definition
PEC	Purchased equipment cost + delivery/installation/instrumentation/piping/electrical (ISBL)➔
C_{DPC}	Direct plant cost + buildings/site development/service facilities (OSBL) ➔
C_{TPC}	Total plant cost + engineering and construction➔
C_{DPI}	Direct permanent investment + contractor's fee, legal, contingency➔
C_{TDC}	Total depreciable capital + cost of land/plant startup➔
C_{TPI}	Total permanent investment + working capital (inventory, stock, reserves,...)➔
C_{TCI}	Total capital investment

Fig. 10.7: Connection between capital investment components.

As illustrated in Fig. 10.7, the direct plant costs, OSBL investment, engineering and construction, and contingency make up the *total depreciable capital (TDC)*. The cost of land, typically about 1-3% of the total depreciable capital, is not included because its value usually does not decrease with time. Additionally, capital for plant startup/modification expenditures (materials, equipment, maintenance, repairs) is commonly required before the constructed plant can operate at maximum design conditions. It should be part of any capital estimate because it is essential to the success of the venture. Normally, an allowance of 8–10% of the total depreciable capital is satisfactory, which is commonly included as a one-time-only expenditure during the first year of plant operation, and therefore also not included. Adding the cost of land and plant startup yields the *total permanent investment (TPI)*. As already discussed before, the owner needs to invest some capital in maintaining plant operations, in addition to the fixed capital investment that was used to design and construct the plant. The capital that is tied up in maintaining inventories of feeds, products, and spare parts, together with cash on hand and the difference between the money owed by costumers (accounts receivable) and the money owed to suppliers (accounts payable), is termed the *working capital* of the plant. By adding the working capital, the final *total capital investment (TCI)* of a project is obtained.

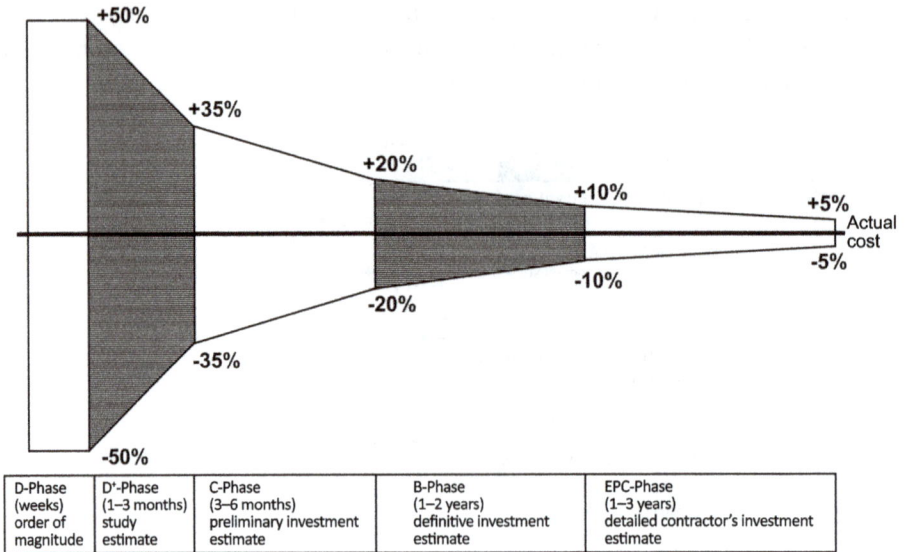

Fig. 10.8: Evolution of investment estimates.

10.4.2 Evolution and purpose of capital cost estimates

The accuracy of an estimate depends on the available amount of design detail, the cost data accuracy, and the time spent on estimate preparation. As illustrated in Fig. 10.8, the available information only justifies approximate estimates in the early project stages and the accuracy of the estimate improves as the project progresses. Commonly, according to their accuracy and purpose, the following capital cost estimate classification is used:

- *Order of magnitude estimates*, Class 5; ± 50% is usually applied for initial feasibility/screening studies, based on the costs of similar processes or functional units and requiring minimal-to-no design information.
- *Study estimates*, Class 4; ± 35% is used to make coarse choices between design alternatives, based on limited cost data and design detail.
- *Preliminary estimates*, Class 3; ± 20% is used to make detailed choices between design alternatives and first optimizations, based on more detailed design and more accurate cost data, including vendor quotes for critical equipment.
- *Definitive estimates*, Class 2; ± 10% is used to authorize funds to bring the design to the point where a more detailed, and thus accurate estimate can be made. With experience and cost data from similar projects, acceptable accuracy estimates can be based on the project flowsheet, a rough P&I diagram, and the approximate sizes of the major equipment items.

– *Detailed estimates*, Class 1; accuracy of ± 5%, which is used for project cost control and estimates for fixed-price contracts. Theses quality estimates require a completed front-end engineering design (FEED), including a (near) complete process design, firm equipment quotes, and a detailed construction cost breakdown and estimation.

To go beyond the preliminary estimate, accuracy requires a fairly detailed design of the plant. The cost of estimate preparation thus becomes the cost of process design and sizing the main equipment items. To improve the accuracy further, the contractor needs to establish the plot plan and plant layout to make accurate estimates of actual installation costs (piping, wiring, and structural steel). The cost of estimate preparation can thus range from 0.1% of the total project cost for ± 35% accuracy up to approximately 5% for a detailed ± 5% estimate.

Tab. 10.4: Plant capacity and capital investment for some chemicals (year = 2000) [4].

Product	Process	Reference capacity (10^3 tons/year)	C_{REF} (10^6 \$)	n
Acetic acid	Methanol + CO	10	8	0.68
Acetone	Propylene	100	33	0.45
Ammonia	Natural gas	100	29	0.53
Butanol	Propylene + CO + water	50	48	0.40
Ethylene	Refinery gas	50	16	0.83
Ethylene oxide	Ethylene oxidation	50	59	0.78
Methanol	Natural gas	60	15	0.60
Nitric acid	Ammonia	100	8	0.60
Sulfuric acid	Contact, sulfur	100	4	0.65
Urea	Ammonia and CO_2	60	10	0.70

10.4.3 Order of magnitude estimates

During early design stages or preliminary marketing studies, it is often desired to make a quick (order of magnitude) capital cost estimate without having to complete a plant design. Several shortcut methods using existing plant data or step counting have been developed to enable total plant investment estimates within ± 50% accuracy for preliminary studies. These methods are also suitable to perform a rough check on more detailed estimates, based on process equipment costs later in the process design.

10.4.3.1 Existing plant data
The quickest way to make an order-of-magnitude estimate of plant cost is to scale it from the known cost of an earlier plant that used the same technology or from pub-

lished data. This requires no design information other than the production rate, year the plant was built, and plant location. The capital cost of a plant can be related to capacity by the equation

$$C_{\text{TDC}} = C_{\text{REF}} \left(\frac{\text{Desired capacity}}{\text{Reference capacity}} \right)^n \tag{10.15}$$

where C_{TDC} is the total depreciable capital of the plant with the desired capacity and C_{REF} is the total depreciable capital for the reference capacity. As illustrated by Tab. 10.4, the capacity exponent n can range from 0.4 for small-scale processes (pharmaceuticals and specialty chemicals) up to 0.9 for processes that use a lot of mechanical work or gas compression (solids or gas processing). As the exponent n is always less than 1.0, constructing larger plants tends to cost less per unit production capacity, which is known as *economy of scale*. Averaged across the whole chemical industry, n is about 0.6, and hence eq. (10.15) is commonly referred to as the *"six-tenths rule"* and can be used to get a rough estimate of the total depreciable capital.

As all cost-estimating methods use historical data, and material/labor costs are subject to inflation. The old cost data have to be updated for estimating at the design stage, and to forecast the future construction cost of the plant. This is commonly done by using published composite cost indices published for various industries in the trade journals to relate present costs to the past costs. These are weighted average indices combining data for equipment categories, construction labor, buildings plus engineering, and supervision, in proportions considered typical for the particular industry. For the chemical process industry, the *Chemical Engineering Plant Cost Index (CEPCI)* is the most widely used composite index, which is based on price developments in the United States process industry, and published monthly in the journal Chemical Engineering. The estimated total depreciable capital can now be updated to the desired year by multiplying with the CEPCI index ratio between the desired and reference year:

$$C_{\text{TDC}} = C_{\text{REF}} \left(\frac{\text{Desired capacity}}{\text{Reference capacity}} \right)^n \left(\frac{\text{CEPCI}}{\text{CEPCI}_{\text{REF}}} \right) \tag{10.16}$$

Figure 10.9 depicts the development of the annual CEPCI indices since 1957–1959, when the plant costs were relatively stable and the index defined as 100.

Most plant and equipment cost data are available on a U.S. Gulf Coast (USGC) or North West Europe (NWE) basis, as these have historically been the main centers of the chemical industry. However, for other locations, the cost of building a plant will depend on local costs for labor, fabrication, construction, shipping, transporting, and importing (taxes) as well as currency exchange rates. In early-stage cost estimates, these differences are typically captured by using a *location factor (LF)*, relative to the USGC or NWE basis, to correct the total depreciable capital:

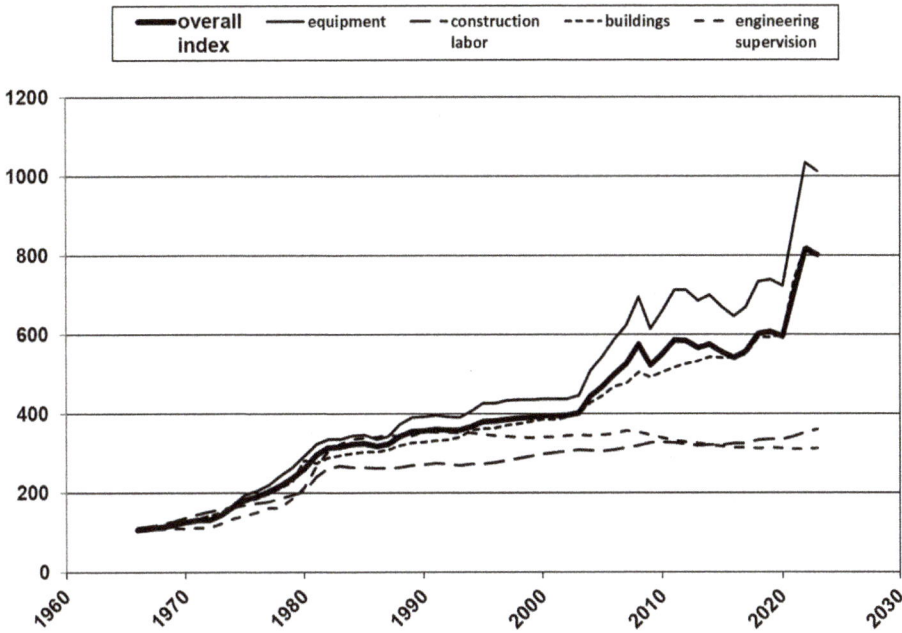

Fig. 10.9: Chemical Engineering Plant Cost Index (CEPCI).

$$C_{TDC}(\text{location}) = \text{LF} \times C_{TDC} = \text{LF} \times C_{REF} \left(\frac{\text{Desired capacity}}{\text{Reference capacity}} \right)^n \left(\frac{\text{CEPCI}}{\text{CEPCI}_{REF}} \right)$$

(10.17)

Location factors for international locations are a strong function of currency exchange rates and hence fluctuate in time. Table 10.5 provides example location factors in US dollars, based on 2003 data.

Tab. 10.5: Location factors in US dollars, based on 2003 data [1].

Country	Region	Location factor
US	Gulf Coast	1.00
Canada		1.00
Western Europe	France	1.13
	Germany	1.11
	The Netherlands	1.19
	United Kingdom	1.02
Mexico		1.03
Brazil		1.14
China	Imported	1.12
	Indigenous	0.61

Tab. 10.5 (continued)

Country	Region	Location factor
Southeast Asia		1.12
Japan		1.26
India		1.02
Middle East		1.07
Russia		1.53
Australia		1.21

10.4.3.2 Step counting methods

In case no process cost data are available, an order-of-magnitude estimate can be made by adding the contributions of the different plant sections, which are called *Functional Units*. Such an approach has been developed by *Bridgewater*, correlating plant cost against the number of processing steps for liquids and solids processing plants [1, 7]:

$$Q \geq 60,000: \quad C_{DPC} = 4,320\,N \left(\frac{Q}{s}\right)^{0.675} \tag{10.18}$$

$$Q < 60,000: \quad C_{DPC} = 380,000\,N \left(\frac{Q}{s}\right)^{0.30} \tag{10.19}$$

where C_{DPC} = direct (ISBL) plant cost in U.S. dollars (U.S. Gulf Coast, Jan 2010 basis, CEPCI = 551), Q = plant capacity (metric tons/year), s = reactor "conversion" (kg desired reactor product/kg total reactor input) and N = number of functional units. A functional unit is defined as:

- A significant process step, such as reactions, separations, or other major unit operations such as evaporators, compressors and blowers, except for low-compression ratio recycle compressors and blowers.
- A multistream operation is taken as one functional unit.
- Individual heat exchange and pumping are normally ignored as they are part of a functional unit, unless substantial costs are involved (i.e., main stream compressors, refrigeration systems, or process furnaces).
- Standard "in process" and raw material storage are ignored because they tend to be a constant process function, unless mechanical handling for solids is involved.
- Large raw material, intermediate or product storages are normally treated separately from "the process" in the estimate.
- Simple "mechanical" separation, without moving parts, is also ignored (cyclone, gravity settler, flash and reflux drums) as the cost is usually relatively insignificant.

The direct permanent investment, C_{DPI}, is now obtained by adding the OSBL factor, F_{OSBL}, as given in Tab. 10.7, to the Bridgewater direct plant cost that already includes engineering and construction for the ISBL part of the plant:

$$C_{DPI} = (1 + F_{OSBL}) \times C_{DPC} \tag{10.20}$$

Finally, the total depreciable capital, C_{TDC}, total permanent investment, C_{TPI}, and the total capital investment, C_{TCI}, (see Fig. 10.7) are obtained by including a contingency of 40%, additional 10% for costs of land, royalties and plant startup, and taking 15% of the total permanent investment as working capital:

$$C_{TDC} = 1.40 \times C_{DPI} \tag{10.21}$$

$$C_{TPI} = 1.10 \times C_{TDC} \tag{10.22}$$

$$C_{TCI} = 1.15 \times C_{TPI} \tag{10.23}$$

An alternative, as presented by Seider [3], is based on the method of *Hill*, based on the year 2006 (CEPCI = 500), a base production rate of 4,536 metric tons/year, carbon steel construction, a design pressure of less than 7 bar, and located at the US Gulf Coast. As the first step, the production rate factor, F_{PR}, is calculated using the six-tenths rule:

$$F_{PR} = \left(\frac{\text{main product flow rate (ton/year)}}{4,536}\right)^{0.6} \tag{10.24}$$

The production rate factor is used to calculate the module cost, MC, for purchasing, delivering, and installing each piece of equipment needed to perform/operate a significant process step, similar to the previously described functional units.

$$MC_i = F_{PR}\, F_M \left(\frac{P_{design}\,(\text{bar, if } > 7\,\text{bar})}{7}\right)^{0.25} \times \$160,000 \tag{10.25}$$

where F_M is a material factor, accounting for the differences in construction materials, as given in Tab. 10.6 and $160,000 is the reference price for a base functional unit with reference capacity of 4,536 metric tons/year, carbon steel construction, design pressure <7 bar, and located at the US Gulf Coast in the year 2006.

The individual module costs are summed, updated to the current CEPCI cost index, and multiplied by the process-type factor, F_{PI}, given in Tab. 10.7 to account for piping, instrumentation, controls, and indirect costs to obtain the direct plant costs, C_{DPC}:

$$C_{DPC} = F_{PI} \left(\frac{\text{CEPCI}}{500}\right) \sum_{i=1}^{N} MC_i \tag{10.26}$$

Tab. 10.6: Material factors [1].

Material	F_M
Carbon steel	1.0
Aluminum	1.07
Cast steel	1.1
Stainless steel 304/316	1.3
Stainless steel 321	1.5
Hastelloy C	1.55
Monel	1.65
Inconel and nickel	1.7
Titanium clad	2.0

By multiplying the direct plant costs with the engineering and construction factor F_C and OSBL factor F_{OSBL}, both given in Tab. 10.7, the direct permanent investment, C_{DPI}, is obtained:

$$C_{DPI} = (1 + F_C + F_{OSBL}) \times C_{DPC} \tag{10.27}$$

Finally the total depreciable capital total permanent investment and the total capital investment are obtained through eqs. (10.21)–(10.23) to include contingency, costs of land, royalties, plant startup, and working capital.

Tab. 10.7: Hill method factors [1].

F_{PI}	Process type
1.85	Solids handling
2.00	Solids-fluids handling
2.15	Fluids handling

F_C	Construction factor
0.15	Outdoor construction
0.40	Mixed indoor and outdoor construction
0.80	Indoor construction

F_{OSBL}	OSBL factor
0.10	Minor addition existing facilities
0.30	Major addition existing facilities
0.80	Grassroots plant

10.4.4 Study and preliminary estimates (factorial methods)

The more accurate study (±35%) and preliminary (±20%) estimates are based on estimating the purchase cost for all the major equipment items and adding the other
costs as factor(s) of these purchased equipment cost. Clearly, the accuracy of this approach will highly depend on the design stage reached when the estimate is made,
and on the reliability of the equipment costs data used. When detailed equipment
specifications are available and firm vendor quotes have been obtained, more accurate preliminary estimate can be obtained, but in general, this approach leads to the
study-estimate accuracy of ± 35%.

10.4.4.1 Estimation of purchased equipment cost

The major effort in estimating the depreciable capital cost is estimating the cost of equipment. There are three main sources of equipment cost data: current vendor quotations,
past vendor quotations, and literature estimates. Vendor quotations are the most accurate,
but the effort required to prepare detailed specifications and quotations are not usually
warranted in the early stages of a project. Thus, for quick estimates, one has to rely on
literature estimates and past quotations. Equipment costs reported in the literature are
either FOB, delivered, or installed cost, all usually given at some time in the past. Shipping
and delivery cost will typically amount to about 5–15% of the purchased cost. Before adding equipment costs, tall must be on the same basis, either FOB, delivered, or installed.

Usually, the literature contains equipment costs for capacities other than that required. To scale the purchased equipment cost to the required capacity, we usually
assume that its cost varies to some power of its capacity:

$$PEC_{Base} = PEC_{REF} \left(\frac{\text{Desired capacity}}{\text{Reference capacity}} \right)^n \tag{10.28}$$

where n is the *size exponent* that can vary from less than 0.3 to larger than 1, depending on the type of equipment. Some typical values are given in Tab. 10.8.

Tab. 10.8: Some typical size exponents
for equipment cost [4].

Equipment	n
Crystallizer	0.37
Compressor	0.69
Evaporator	0.54
Heat exchanger	0.44
Reactor	0.55
Tower	0.62
Tank	0.57

Tab. 10.9: Pressure and temperature factors [2].

Design pressure (atm)	F_P	Design temperature (°C)	F_T
0.005	1.3	−80	1.3
0.014	1.2	0	1.0
0.048	1.1	100	1.05
0.54–6.8	1.0	600	1.1
48	1.1	5,000	1.2
204	1.2	10,000	1.4
408	1.3		

Often, the cost literature contains equipment cost at base conditions, being at low temperature and pressure, carbon steel construction, and a specific design. To update for the actual equipment cost at other conditions, the base is multiplied by correction factors:

$$PEC = PEC_{Base}\, F_T\, F_P\, F_M \tag{10.29}$$

The material factor F_M has already been given in Tab. 10.6. Tab. 10.9 contains values for the temperature, F_T, and pressure, F_P, and correction factors. These factors depend on the type of equipment; thus using the same correction factors for all equipment is an approximation. For shell-and-tube heat exchangers, these factors should be used with care as the shell material may be different than the tube material. Also, it is a good practice to place the high-pressure on the tube side to reduce material cost.

Finally, the purchased equipment costs need to be updated to the desired year by multiplying with the CEPCI index ratio between the desired and reference years:

$$PEC_i^{year} = PEC_{REF}\, F_T\, F_P\, F_M \left(\frac{Desired\ capacity}{Reference\ capacity}\right)^n \left(\frac{CEPCI}{CEPCI_{REF}}\right) \tag{10.30}$$

Equipment costs information can be found in professional cost engineering literature, engineering textbooks, and on various websites. Useful textbooks are Towler and Sinnot [1], Lewin, and Widagdo [3], Peters, Timmerhaus, and West [4], Woods [6], Ulrich and Vasudevan [9], Dutch Association of Cost Engineers [10], Seider, Seader, Cooper, Penney, and Fair and Walas [11], In these textbooks, correlations given for preliminary estimates are typically of the form:

$$PEC = c_1 + c_2\, S^n \tag{10.31}$$

where c_1 and c_2 are cost constants, S a size/capacity parameter, and n the scale exponent for that type of equipment. A useful web-based costing tool is available at www.matche.com. Other sources to obtain first-estimates are resale websites such as www.equipnet.com and www.ippe.com or websites such as www.alibaba.com. Alternatively, when available, cost estimating programs like Aspen Process Economic Analyzer can be used. It is highly recommended to thoroughly crosscheck the estimates obtained from multiple sources to arrive at a reasonable reliability.

10.4.4.2 Overall installation (Lang) factors

To achieve study-estimate accuracy (±35%), it is typically sufficient to consider overall installation factors that relate the total depreciable capital (without working capital) and total capital investment (including working capital) of a plant to the total purchased equipment cost by the equations:

$$C_{\text{TDC}} = F_{\text{L(TDC)}} \sum_i \text{PEC}_i \qquad (10.32)$$

$$C_{\text{TCI}} = F_{\text{L(TCI)}} \sum_i \text{PEC}_i \qquad (10.33)$$

where ΣPEC_i is the total delivered cost of all the major equipment items and F_L the installation factor, widely known as the *Lang factor*. These factors depend on the type of plant being considered and include process equipment, Installation, instrumentation and control, piping, electrical, engineering, etc. The values and breakdown of the Lang factors are given in Tab. 10.10.

Tab. 10.10: Breakdown (percent of purchased equipment cost) of Lang factors [4].

Item	Process type		
	Fluid	Fluid-solid	Solid
Purchase cost major equipment	100	100	100
– Installation	47	39	45
– Instrumentation and control	36	26	18
– Piping	68	31	16
– Electrical	11	10	10
– Buildings and structures	18	29	25
– Yard improvements	10	12	15
– Service facilities	70	55	40
Total direct plant cost	**360**	**302**	**269**
– Engineering and supervision	33	32	33
– Construction expenses	41	34	39
Total and indirect plant cost	**434**	**368**	**341**
– Contractors fee and legal expenses	26	23	21
– Contingency	44	37	35
Total depreciable capital (TDC)	**504**	**428**	**397**
$F_{\text{L(TDC)}}$	5.0	4.3	4.0
– Working capital and land	89	75	70
Total capital investment (TCI)	**593**	**503**	**467**
$F_{\text{L(TCI)}}$	6.0	5.0	4.7

10.4.4.3 Individual factors

Further improvement in accuracy to achieve a preliminary estimate (±20%) can be achieved by using *individual installation factors* for different types of equipment. This requires the availability of much more detailed design information and is therefore only justified if reliable cost data are available and the design has been taken to the point where all the cost items can be identified and included. In Guthrie's book [8], the detailed cost estimating method includes the costs for installation, piping, and instrumentation, taken separately for each piece of equipment. As an illustration, some of his factors are given in Tab. 10.11. For more information about these individual factor methods, the reader is referred to Cooper [5], Gerrard [7], Guthrie [8], and Ulrich and Vasudevan [9], as this is beyond the scope of this book.

Tab. 10.11: Guthrie Bare module factors [8].

Equipment	Bare module factor
Shell-and-tube heat exchanger	2.19
Vertical pressure vessel	4.16
Horizontal pressure vessel	3.05
Gas compressor	2.15
Centrifuges	2.03
Crystallizers	2.06
Evaporators	2.45

10.5 Exercises

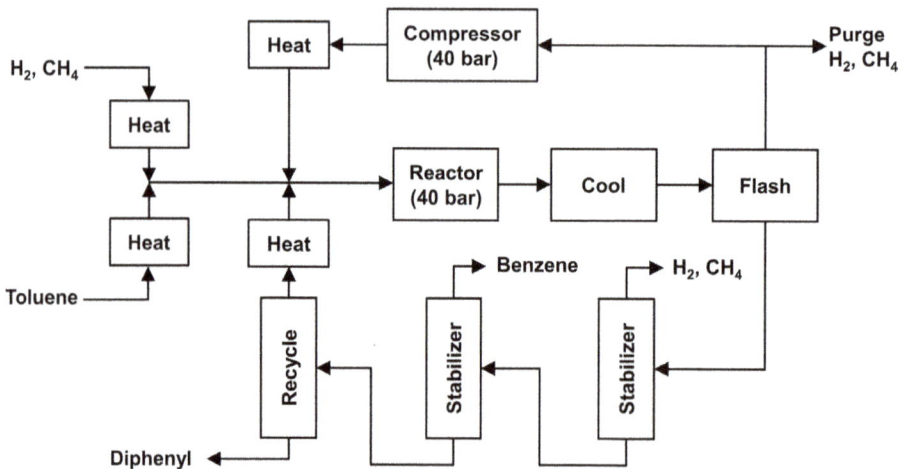

Fig. 10.10: Block scheme for the toluene hydrodealkylaton process.

10.1 Compare the order of magnitude estimates for the total capital investment, according to the Bridgewater and Hill methods to produce 100 kton/year benzene using the toluene hydrodealkylaton process (Fig. 10.10) in the year 2022.
  a. Determine the number of functional units.
  b. Make an order of magnitude estimate with the Bridgewater method.
  c. Make an order of magnitude estimate with the Hill method.
  d. Compare the results from an order-of-magnitude perspective.
  e. What would be the total capital investment when the plant would be located in Europe or China?

10.2 Estimate the total depreciable capital investment in the year 2022 for a plant located in Europe that produces 200 kton/year of ethylene oxide.

10.3 Make a first-estimate of the annual operating labor cost for the toluene hydrodealkylation process (Fig. 10.10) when operating 24/7.

10.4 Estimate the purchase cost of a Hastelloy C reactor of 50 m³ in 2022. In 2005, the purchasing department received a quotation of 70.000 euro for a similar SS316 reactor with a volume of 10 m³.

10.5 Estimate the economic potential (eq. (10.9)) for the following conversions:
  a) Sugar to biobutanol
  b) Bioethanol to bio-based butadiene
  c) Sugar via lactic acid and lactide to polylactic acid (PLA)

  Use literature to obtain the product prices, raw material costs, and conversion yields

Nomenclature

BOM	Bill of materials	
CF	Cash flow	Valuta/year
CCF	Cumulative cash flow	Valuta
CDCF	Cumulative discounted cash flow	Valuta
CEPCI	Chemical engineering plant cost index	–
C_{Land}	Cost of land	Valuta
C_{DPC}	Direct plant cost	Valuta
C_{DPI}	Direct permanent investment	Valuta
C_{REF}	Total depreciable capital reference capacity	Valuta
C_{TCI}	Total capital investment	Valuta
C_{TDC}	Total depreciable capital	Valuta
C_{TPC}	Total plant cost	Valuta
C_{TPI}	Total permanent investment	Valuta
C_1	Cost constant	Valuta
C_2	Cost constant	Valuta
D	Depreciation	Valuta/year
DCF	Discounted cash flow	Valuta/year
DPC	Direct plant costs	Valuta

EP	Economic potential	Valuta/year
FC	Annual fixed costs	Valuta/year
FEED	Front-end engineering design	
F_C	Construction factor	–
F_M	Material factor	–
F_L	Lang factor	–
F_{OSBL}	OSBL factor	–
F_P	Pressure factor	–
F_{PI}	Process-type factor	–
F_{PR}	Production rate factor	–
F_T	Temperature factor	–
FOB	Free on board	
GP	Gross profit	Valuta/year
IRR	Internal rate of return	–
ISBL	Inside battery limits	–
LF	Location factor	–
MC	Module cost	Valuta
n	Year	–
n	Capacity exponent, size exponent	–
N	Number of years	–
N	Number of functional units	–
NP	Nett profit	Valuta/year
NPV	Net present value	Valuta
OI	Operating income	Valuta/year
OSBL	Outside battery limits	–
P	Pressure	Bar
PBT	Payback time	Years
PEC	Purchases equipment costs	Valuta
PFD	Process flow diagram	–
PLA	Polylactic acid	
Q	Plant capacity	Metric tons/year
r	Discount rate	–
RMC	Annual raw material costs	Valuta/year
ROI	Return on investment	–
s	Reactor "conversion"	–
S	Sales	Valuta/year
S	Size/capacity parameter	m^3, m^2, ton/h
t	Taxation rate	–
TCI	Total capital investment	Valuta
TDC	Total deprecial capital	Valuta
TPI	Total permanent investment	Valuta
VC	Annual variable costs	Valuta/year
WC	Working capital	Valuta
WACC	Weighted average cost of capital	–
Φ	Fraction of total depreciable capital (TDC)	–

References and Further Reading

[1] Towler G, Sinnot R. Chemical Engineering Design: Principles, Practice and Economics of Plant and Process Design, 2nd edn, Amsterdam: Elsevier, 2013.
[2] Silla H. Chemical Process Engineering, Design and Economics, New York: Marcel Dekker, 2003.

[3] Seider WD, Seader JD, Lewin DR, Widagdo S. Product and Process Design Principles, 3rd edn, New York: John Wiley & Sons, 2010.

[4] Peters MS, Timmerhaus KD, West RE. Plant Design and Economics for Chemical Engineers, 5th edn, New York: McGraw Hill, 2004.

[5] Cooper JR. Process Engineering Economics, New York: Marcel Dekker, 2003.

[6] Woods DR. Rules of Thumb in Engineering Practice, Weinheim: Wiley-VCH, 2007.

[7] Gerrard AM. Guide to Capital Cost Estimation, 4th edn, London: IChemE, 2000.

[8] Guthrie KM. Process Plant Estimating, Evaluation and Control, Solano Beach, California: Craftsman, 1974.

[9] Ulrich GD, Vasudevan PT. Chemical Engineering Process Design and Economics: A Practical Guide, Process Publishing, 2004.

[10] Dutch Association of Cost Engineers. Price Booklet, 32nd edn, Nijkerk: DACE, 2017.

[11] Cooper JR, Penney WR, Fair JR, Walas SM. Chemical Process Equipment, 3rd edn, Oxford: Butterworth-Heinemann, 2012.

11 Evaluating for safety

11.1 Introduction

When designing new processes or products, it is important to properly identify and manage associated hazards and risks. The emphasis on systematically assessing these factors and the methodologies that have been developed for this purpose were largely influenced by a series of major industrial accidents. These incidents unveiled critical blind spots and deficiencies in the prevailing safety awareness within the industry. The practical lessons learned and regulatory fallout from these accidents resulted in improving safety and risk management in the design and operation of chemical and industrial processes and the design of products [1]. Conducting hazard and risk assessments should be methodical and timely during a design project's life cycle. The goal of this chapter is to provide readers an introduction to basic safety concepts and the tools to get started on performing risk assessments during early design stages. References are provided at the end of this chapter for further reading on this subject.

11.2 Hazard and risk

Before discussing the risk assessment and evaluation process, we must first define "hazard" and "risk."

Hazard: *"An inherent chemical or physical characteristic that has the potential for causing damage to people, property, or the environment [2]."*
     The keyword here is "potential": a hazard is something that has the potential for adverse effects such as harm, illness, death, pollution, loss, etc. For instance, natural gas is a flammable hydrocarbon, making it a fire and explosion hazard. Chlorine, being toxic when inhaled, presents a toxic hazard.

Risk: *"A measure of human injury, environmental damage, or economic loss in terms of both the incident likelihood and the magnitude of the injury or loss [2]."*
     Risk relates to the occurrence of undesired events (e.g., explosion), which lead to undesired effects (e.g., damage from an explosion) and the likelihood (probability or frequency) that these events and effects occur. A simplified expression of risk is to express it as a function of likelihood of occurrence times the consequences:

$$\text{risk} = f((\text{likelihood}) \cdot (\text{consequences}))$$

https://doi.org/10.1515/9783110782127-011

11.3 Process risk management during design

Risk management is the systematic application of management policies, procedures, and practices to the tasks of analyzing, assessing, and controlling risk in order to protect employees, the general public and the environment, as well as company assets while avoiding business interruptions [3].

There are various models available that describe how risks are managed during design, but the basic principle is that it is an iterative process in which hazards are identified and the risks evaluated until deemed acceptable. The following steps are general for risk reduction strategies, as also shown in Fig. 11.1:
1. Identify the hazards.
2. Assess the consequences.
3. Analyze and evaluate if the risks are acceptable.
4. Decide on (more) mitigating measures based on risk evaluation.
5. Implement in design.

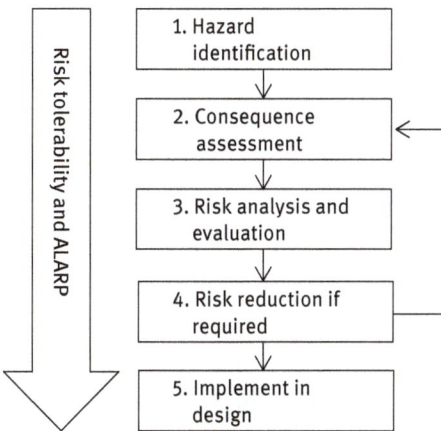

ALARP: risks As Low as Reasonably Practicable

Fig. 11.1: Risk reduction steps during design.

The principles applied to the risk management of process engineering can also be applicable to product engineering. The difference is more in where the focus of the assessment lies. When assessing the hazards and risks during product engineering there is a shift of focus towards the end user. The production of toys is an example. During the risk assessments of the process the focus is on risk exposure to the workers on the facility, surrounding population, local environment, etc.

The starting principle of a design should be to approach it from an inherent safety principle. This will be discussed in the next section.

11.3.1 Inherent safer design principle

The concept of inherent safer design was introduced by Trevor Kletz after the Flixborough disaster, and his expression sums up the basic principle: *"What you don't have cannot leak or burn"* [4]. There are four strategies for achieving inherent safer designs [44]:

1. Minimize: to reduce the quantity of material, energy, or energy density contained in a manufacturing process or plant.
2. Substitute: the replacement of a hazardous material or process with an alternative that reduces or eliminates the hazard.
3. Moderate: using materials under less hazardous or energetic conditions, or designing the plant to mitigate the impact of an incident arising from the hazard.
4. Simplify: designing and/or operating the process to reduce or eliminate unnecessary complexity in order to reduce or eliminate the chemical hazard.

Since no process can be considered inherently safe, the accurate term to use is "inherently safer design." Hence, it is always relative to another reference process. The principle of inherent safer design is that first measures should be taken to avoid the hazard altogether and that mitigation measures should only be taken when the other options are not reasonably practical. It is for this reason that identification of hazards and the assessment of mitigating measures should be done early during design stages when possible. The true inherent safer design approach focusses on avoidance and prevention, thus choosing different technologies or materials to avoid hazards all together. This is as opposed to the safety in design approach where additional safety systems are introduced to control and mitigate hazards.

There are some tools and techniques available that can help compare design options from an inherent safety perspective. An example of such a tool is the Dow indexes [5, 6] which, based on the properties of the chemicals intended for use as part of the design, can rank design options based on their fire, explosion, and toxic hazards. Processes can also be compared based on the consequences during incidents. The principle has been adopted in many codes, standards, design engineering practices, and regulations and is an important one to keep in mind during all design stages. The American Institute of Chemical Engineers has also published a checklist with questions that can be used to evaluate a process based on the inherently safer strategies (see reference [45]). Some companies choose to hold team-based inherent safety reviews with the objective of understanding the hazards and finding ways to eliminate or reduce them [44].

It is important to recognize that trade-offs are inevitable. An inherently safer solution for one hazard may be inherently less safe for another hazard. For example, a chemical might have low toxicity but be flammable. Another chemical might be nonflammable and nontoxic, yet require the process to operate at higher pressures. Trade-offs might also arise in terms of process economics, increased maintenance frequency, and other factors. A weighted scoring technique can assist in evaluating the various options, facilitating more informed decision-making.

11.3.2 Process risk management strategies

Inherent safety is one method of risk reduction if feasible but other strategies are also available to reduce the risk to an acceptable level. Process risk management is the term given to collective efforts to manage process risks through a wide variety of strategies, techniques, procedures, policies, and systems that can reduce the hazard of a process, the probability of an accident, or both [44]. In general, the strategy for reducing process risk, whether directed toward reducing the frequency or the consequences of potential accidents, can be classified into four categories [44], as shown in Tab. 11.1:

Tab. 11.1: Process risk management strategies, ordered from most robust to least robust [44].

Strategy	Description
Inherent	Eliminate or greatly reduce the hazard by changing the process to use materials and conditions that are nonhazardous or much less hazardous.
Passive	Minimize hazards using process or equipment design features that reduce either the likelihood or consequence of an incident without the active functioning of any device.
Active	Manage risk using process control systems, safety instrumented systems (SIS), mitigation systems such as sprinklers, and other active systems; these may prevent an incident or reduce the consequences of an incident.
Procedural	Use operating procedures, safety rules and procedures, operator training, emergency response procedures, and management systems to manage risks.

Some sources also have "spatial" designated as a separate category between inherent and passive, which defines the use of distance to reduce the effects of hazards, such as siting the process as far as possible from occupied buildings and outside receptors [44]. Analyzing the risks and reductions steps should ideally be done in a hierarchical manner where more robust measures are considered before less robust measures. The difference between the inherently safer design and the above mentioned process risk management strategies is that the inherent strategies aim to reduce the hazards at the source while the process risk management strategies aim to control the hazard by some other means.

Now that we recognize the importance of mitigating hazards at their source and comprehend the distinction between inherently safer design and other risk management strategies, the initial step in implementing these principles is the identification of hazards inherent in the proposed design.

11.4 Hazard identification studies

The first step of a systematic evaluation requires that the hazards be identified. This section provides an overview of some common hazard identification methodologies used during engineering projects. It is not intended as a complete overview of all identification techniques or detailed treatise about the methodologies, but more as an introduction into this field. For a detailed description of the methodologies the reader should refer to books such as Lees' Loss Prevention in the Chemical Industry [7] and the international standard IEC/ISO 31010 Risk management–risk assessment techniques.

11.4.1 Chemical reactivity hazards

The first item to evaluate in a product risk assessment is the physical and chemical properties of the product. Following the inherent safer design principle, one should strive to avoid or minimize the use of chemicals and/or materials that have hazardous properties. Substitution or phasing out of chemicals that are known to be hazardous is now becoming more a part of international legislation. EU Regulation No. 1907/2006 concerning Registration, Evaluation, Authorization, and Restriction of Chemicals (REACH) came into force in 2007. The goal of REACH is to improve the protection of human health and the environment by identification of the intrinsic properties of chemical substances and phasing out chemicals of very high concern. When developing a product in the EU or for use in the EU, this regulation is important to be aware of and its requirements should be taken into consideration from the feasibility stage.

But how does one discover what the hazardous properties of a product are? There are various sources of information available that contain data with respect to toxic, fire, and explosion hazards of chemicals such as but not limited to:
- Material Safety Datasheets (MSDS)
- Information contained in standards such as NFPA 49, 325, 432, 491
- Bretherick's Handbook of Reactive Chemical Hazards [8]
- Incident reports
- Experiments
- Others

The next step is to look at the chemical and thermodynamic properties since they can provide information on potential hazards. Properties that can provide information are:
- Thermodynamic properties
- Chemical composition, structure and bonds

11.4.1.1 Thermodynamic properties and thermal runaway potential

The heat of reaction indicates the potential temperature increase in a mixture. The greater the heat of reaction the higher the potential for energetic, uncontrolled reactions.

Another important thermodynamic property is the heat of formation. Compounds having a high endothermic heat of formation (positive heat of formation) contain a lot of energy, and tend to be self-reactive. Correspondingly, they have a high exothermic heat of reaction.

Thermal runaway potential

In an exothermic reaction, if the heat produced surpasses the heat lost, the reaction rate can accelerate. This increased rate releases even more heat, creating a positive feedback loop that leads to an exponential rise in temperature. Eventually, the reactive mixture may undergo rapid vaporization and/or decomposition, a phenomenon known as "thermal runaway." A runaway reaction is defined as *"A thermally unstable reaction system which exhibits an uncontrolled accelerating rate of reaction leading to rapid increases in temperature and pressure."* [2]. Such a rapid increase in pressure and temperature can culminate in a thermal explosion. A common trigger for such explosions is the failure of a cooling system in a reactor containing a mixture capable of inducing a thermal runaway.

The severity of a thermal runaway can be assessed by calculating the maximum possible temperature rise. Using the heat of reaction and the heat capacity of the reaction mixture, one can calculate the adiabatic temperature rise, the maximum temperature rise that can theoretically occur without heat loss:

$$\Delta T_{ad} = \frac{-\Delta H_r}{C_p}$$

where ΔT_{ad} is the adiabatic temperature rise (K), ΔH_r is the heat of reaction (kJ/kg), and C_p is the heat capacity of reaction mixture (kJ/(kg·K)).

The sum of the maximum process temperature and adiabatic temperature rise can give important information about potential hazards. When looking at the result one should wonder if this temperature is below a temperature where
-   additional chemistry (such as decomposition),
-   phase transition (such as boiling or gas generation); or
-   overpressurization from increased vaporization can occur.

The adiabatic temperature rise indicates the potential maximum temperature of a reaction, and thus the severity, but it does not convey the dynamics such as how rapidly this temperature might be reached. The rate at which this temperature is attained is crucial when considering the controllability of a reaction. To gain insight into the speed of the runaway reaction, one can calculate the time to maximum rate of the reaction under adiabatic conditions (tmr_{ad}). For a zero-order reaction (where the reaction rate is independent of the concentration of the reactants) under adiabatic conditions, the tmr$_{ad}$ will be [46]:

$$\text{tmr}_{ad} = \frac{C_P R T_0^2}{q_0 E}$$

where tmr$_{ad}$ is the time to maximum rate of the reaction under adiabatic conditions (s), T_0 is the initial temperature from which the thermal explosion develops (K), R is the gas constant (kJ/mol·K), q_0 is the heat release rate of reaction at the initial conditions T_0 (kJ/kg·s), and E is the activation energy of the reaction (kJ/mol)

The tmr$_{ad}$ provides information on how quickly the runaway develops. It can be used to determine whether, after a cooling failure, there is sufficient time to implement emergency measures before the runaway accelerates to an uncontrollable rate [46]. Thus, it serves as a timescale to qualitatively assess the likelihood of a runaway. The tmr$_{ad}$ is a function of the reactor kinetics [46]. While this equation uses a zero-order assumption, it can be applied to other reaction kinetics as well. This is because the zero-order assumption provides a conservative estimate, ignoring the concentration depletion that would otherwise slow down the reaction [46].

When assessing the potential for thermal runaways, it is essential to understand the reactions that can contribute to these events. Broadly, these reactions can be categorized into "primary" and "secondary" reactions. Primary reactions are the intended reactions in a given process, typically optimized for maximum yield and efficiency. While this primary reaction might lead to a runaway, there can also be undesired reactions – termed "secondary" reactions – that occur either in parallel or subsequently at elevated temperatures. These secondary reactions can initiate new runaway reactions and will require to be evaluated in a similar way as the primary reaction. Secondary reactions are often the least understood and calorimetry experiments are required to evaluate the reactivity and thermal stability of these reactions. The information obtained can then be used to assess the severity and likelihood of the secondary reaction, just as with the primary reactions. For a more detailed discussion and guidance on the steps and methods for evaluating thermal hazards in chemical processes, readers are referred to the book *Thermal Safety of Chemical Processes* [46].

11.4.1.2 Chemical composition, structure, and bonds

Chemical structure and bonds also provide clues to potential chemical reactive hazards. Be suspicious of [9]:
- Compounds that have high bond strains
- Carbon–carbon double bonds not in benzene rings (e.g., ethylene, styrene)
- Carbon–carbon triple bonds (e.g., acetylene)
- Nitrogen-containing compounds (NO_2 groups, adjacent N atoms, etc.)
- Oxygen–oxygen bonds (peroxides, hydroperoxides, ozonides)

- Ring compounds with only three or four atoms (e.g., ethylene oxide)
- Metal- and halogen-containing complexes (metal fulminates; halites, halates; etc.)

Some chemicals have the tendency to produce strong oxidation reactions, which can be explosive in nature. The oxygen required for this oxidation can be stored in the compound itself or be supplied from outside during the reaction. One estimating parameter for the potential hazard and instability of substances or reaction mixtures containing oxygen is the oxygen balance (OB). The oxygen balance is the amount of oxygen, expressed as weight percent, liberated as a result of complete conversion of the material to CO_2, H_2O, SO_2, Al_2O_3, N_2, and other relatively simple oxidized molecules [10]. If the OB is positive, there is an excess of amount of oxygen available in the compound for a conversion to simple components. If the OB is zero, there is just enough oxygen available for the conversion to simple components. If the OB is negative, there is insufficient oxygen for a complete oxidation of the compound.

The following decomposition reaction forms the basis of the OB:

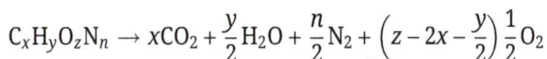

$$C_xH_yO_zN_n \rightarrow xCO_2 + \frac{y}{2}H_2O + \frac{n}{2}N_2 + \left(z - 2x - \frac{y}{2}\right)\frac{1}{2}O_2$$

In case of perfect oxidation, no excess oxygen is left after the reaction and the carbon and hydrogen atoms have fully oxidized to carbon dioxide and water. The heat of reaction tends to reach a maximum in this case. The parameter $z-2x-(y/2)$ is then equal to zero under such ideal conditions. The oxygen balance is defined as:

$$OB = \frac{1,600\left[z - 2x - \frac{y}{2}\right]}{MW}$$

where MW is the molecular weight of the compound. The closer the OB is to zero, the higher the hazards since the explosion sensitivity and severity will tend to reach a maximum at an OB of zero. The OB can thus be used to assess the degree of hazard [10]:

Hazard potential	Value of OB
Low	$-160 < OB < -240$
Medium	$-240 < OB < -120$
	$80 < OB < 160$
High	$-120 < OB < 80$

It should be stressed that the equation above is not applicable to molecules containing other atoms and that the OB only serves as an indicator. The way oxygen is bonded should be reviewed in relation to the OB. For an explosive such as nitroglycerine the oxygen balance is +3,5 which corresponds well to the high hazard potential of the table above. Acetic acid on the other hand has an OB of −106,7 but is not an explosion hazard since the oxygen present cannot be used efficiently for the oxidation of carbon.

When chemical and/or thermodynamic properties are not known, then the next step in the hazards assessment is to perform tests. Guidance for performing such tests is provided in [10, 11], and [12].

11.4.1.3 Chemical interactions' hazards

Chemicals may come into contact with other chemicals, materials, or steams in desired and undesired ways. The potential hazards of binary (two-component) interactions can be assessed using a chemical interaction matrix, which offers valuable insights for subsequent hazard evaluations. An example template of such a matrix is provided Fig. 11.2.

		1	2	3	4	5	6	7	8	9	10	11	12
1													
2													
3													
4													
5													
6													
7													
8													
9													
10													
11													
12													

Legend	
- :	Not Safety Relevant
E :	Explosion
F :	Fire
H :	Exothermal reaction
G :	Gas release
T :	Toxic
C :	Corrosion
N :	Ecotoxic
X :	Other hazard
? :	Unknown

Fig. 11.2: Chemical interaction matrix (adapted from [46]).

HAZID and ENVID

Plot plans preliminary layouts are developed early during the design before process and instrumentation diagrams have fully matured. A relatively simple method to identify hazards during this stage is by using qualitative techniques that use key words to think of causes, consequences, and safeguards in a design. One of the more common techniques that can be used during the feasibility up to detailed design is the HAZID study (HAZard IDentification). The HAZID is a study technique for early identification of potential hazards and threats.

The major benefit of a HAZID is that the early identification and assessment of the critical HSE hazards provides essential input to project development decisions. This will lead to safer and more cost-effective design options being adopted with a minimum cost of change penalty. The deliverable of the HAZID study is a set of detailed minutes, showing the summary of the discussion with action recommendations directed to the separate involved parties, which were present in the meeting.

The HAZID study has been developed specifically to reflect the importance of HSE issues on the fundamental decisions that are made at the inception of all development projects. The HAZID study is usually the first opportunity to assemble experienced operational, engineering, and HSE staff together to address the issues surrounding the project in a short time frame.

During a HAZID study review session, a guide word list is used and the session itself is facilitated by a chairman and supported by a scribe when required. An example guide word list and part of a HAZID table are given in Tabs. 11.2 and 11.3 for a HAZID performed for a conceptual design of a new offshore facility. The guide words should be adapted to reflect the specific circumstances or features of the design if possible.

Tab. 11.2: Example guide words for a HAZID for an offshore installation.

HAZID category	Risk element, system, or event
Process event	Unignited gas release
	Jet fire
	Flash fire
	Explosion
	Blowout
	Unignited liquid release
	Pool fire
	Chemical/toxic hazard
	Non-process fire
	Vent/exhaust/air intake
	Hazardous area classification
External event	Ship collision
	Dropped load
	Helicopter impact
	Environmental hazards
Safety system	Escape routes
	Muster station
	Lifeboat
	Life raft
	Active fire fighting system
	Firewalls/blast wall/shielding/PFP
	Fire and gas detection
	Area alarm systems
	ESD/EBD
	HVAC

Tab. 11.2 (continued)

HAZID category	Risk element, system, or event
Working environment, operations and maintenance	Access to maintainable equipment Limited workspace Interaction with rig LSA/radiation Noise Handling of chemicals/explosives Mechanical handling Control room operations Well intervention operations Helicopter operations Crane operations Supply vessel operations
Others	Drawing comments Environmental aspects (air, soil, landscape, noise, energy, radiation, water, waste, light)

Tab. 11.3: Example table for a HAZID.

Risk element/ system/ Event	Cause/ guide word	Consequences	Evaluation/ safeguards	No.	Action required	Action party	Due date
Process event Unignited HC release Jet fire Others							

For each guide word, expanders are provided to help the brainstorming session in identifying possible hazards. The HAZID team brainstorms to find all potential hazards and its causes.

Each hazard is examined to see what:
- consequences may arise from the hazard;
- controls are in place;
- and, if the team decides that the controls are not sufficient or can be improved, action recommendations are formulated and assigned to the responsible party.

An ENVID (ENVironmental IDentification) uses the same principle as the HAZID except that the purpose is to identify environmental hazards and thus the keywords are different. Below is an example list of keywords that can be used during an ENVID:

Tab. 11.4: Example guide words for an ENVID for an offshore installation.

ENVID category	Guide word
Impact –air quality	Greenhouse gas emissions
	VOCs and fugitive emissions
	NO_x and SO_x
	Ozone-depleting substances
	Particulate matter
	Toxic metals
	Odorous compounds
Impact –waste	Waste production
	Hazardous waste
	Domestic waste
Impact –resource use	Land use
	Land quality
	Loss of habitat
	Degradation of land quality
	Alteration of ecological/landscape processes
Impact –oil and chemical spills	Chemical use and storage
	Hydrocarbon use and storage
	Bunkering and equipment failure
	Human error and procedures
	Pipework, valves, and seals
	Bunding and drainage
Impact –chemical use and discharge	Chemical use and storage
	Hydrocarbon use and storage
	Bunkering and equipment failure
	Human error and procedures
	Pipework, valves, and seals
	Bunding and drainage
Impact –aqueous discharges	Drains
	Hydrocarbon and chemical contamination
	Sewage discharges
	Storm water
	Hydro test water
	Slurry
Impact –wildlife disturbance	Noise/vibration and light pollution
	Air pollution
	Water pollution
	Conservation of protected areas
	Conservation of protected species
	Ecosystem effects/biodiversity
	Displacement/loss of access

It is important that the actions are monitored after the session. A common pitfall is that action closeout is delayed due to demanding engineering schedules and not having assigned a person with the proper authority to follow up and expedite the actions with the engineers. Agreeing at the start of the project on whom this responsibility will fall on will help prevent costly rework during later stages of the design.

The HAZID and ENVID methodology is a tool that can be applied to process and product engineering assessments and is easy to understand.

HAZOP

H.G.Lawley of Imperial Chemical Industries (ICI) developed the HAZOP methodology in the 70s, in order to identify causes and consequences of disturbances in process installations. HAZOP stands for Hazard and Operability analysis and is a structured hazard identification tool, using a multidiscipline team. The methodology differs from the HAZID in the sense that HAZOP focusses on hazards that originate from the process itself (inside-out approach) while HAZID assesses a broader range of potential hazards from within but also from outside the process (outside-in approach). The HAZOP is accepted as the main technique for the identification of process hazards in the design and operation of a facility. A detailed description of the HAZOP methodology is available in the IEC code 61882 [13].

The HAZOP process is the systematic application of combinations of parameters (e.g., flow, pressure, and temperature) and guide words (e.g., no, more, or less) to produce deviations (no flow, less pressure) from the design intent or intended operational mode of the installation. The purpose of the HAZOP study is to:
–   review the design and consider whether any of the conditions, which may occur from operation, malfunction, or maloperation cause a hazard to people or cause damage to plant and equipment
–   review whether the precautions incorporated in the design are sufficient to either prevent the hazard occurring or reduce any consequence to an acceptable level
–   Consider whether the plant remains safely operable, goes to a safe hold state, partially shuts down, or shuts down under the circumstances in above given purposes

The deliverable of the HAZOP study is a set of detailed minutes presented on worksheets for each deviation. These show the summary of the discussion with action recommendations directed to the separate involved parties, which were present in the meeting.

A chairman will guide the meeting through the HAZOP and can be assisted by a scribe if required. Possible deviations/deficiencies from design will be addressed and consequences are identified during the session.

The installation being studied is divided into logical items (equipment and piping) that can be studied according the structured brainstorm method of the HAZOP study. These items are called nodes and are selected by the HAZOP Chairman. HAZOP nodes are usually divided by phase flow (liquid, gas) and isolatable sections (e.g., between emergency shut down valves), or installation parts. The node or installation part is reviewed by the application of:

– parameters (e.g., flow, pressure, temperature, or level), in combination with
– guide words (e.g., no, more, less, reverse, or other than).

The guide word/parameter combination is known as a deviation.

Example parameters and guide words are available in Tab. 11.5.

Tab. 11.5: Example parameters and guide words for a HAZOP.

Parameters		Guide words
Flow	Maintenance	No
Pressure	Utilities	More/high
Temperature	Start-up	Less/low
Level	Shut down	Reverse
Composition/phase	General	As well as
Operations		Other

The HAZOP team brainstorms to find all potential causes of the deviation. All potential causes are discussed. Each *cause* is *separately* examined to see what:

– consequences may arise with a certain deviation;
– safeguards are in place (i.e., design criteria, indications, alarms, procedures, etc.);
– and, if the team decides that the safeguards are not sufficient or can be improved, action recommendations are formulated and assigned to the responsible party.

An example of a HAZOP minute is provided in Tab. 11.6.

HAZOPs are used primarily in process risk assessment and are less commonly used in product hazard identifications. The Failure Mode Effect and Analysis (FMEA) is a technique that resembles the HAZOP in some aspects and is much more commonly used for product hazards assessments.

FMEA

Another well-known HAZID technique is the FMEA. This method reviews failure on a component level and assesses the effect on the system as a whole. The review results in recommendations to improve the product or system design with respect to reliability and safety. Originally developed by the US military after the Second World War, the first formal application of the FMEA was in the aerospace industry in the 1960s for the NASA Apollo missions. It is a technique that is nowadays used in the product manufacturing and services industries. Like the other HAZID tools discussed previously in this chapter, the FMEA is performed as a team session facilitated by a chairman and sometimes supported by a scribe.

Tab. 11.6: Example of a HAZOP minute.

Node	Deviation no.	Guideword	Cause	Consequence	Safeguard	Action no.	Action	Action responsible	Due date
2	50	High temperature	Failure of heater control causing too high temperature	Exothermic runaway, over-pressurization of reactor	Bursting disc on reactor prevents over-pressurization	12	Ensure venting of bursting disc is to safe location	Jane Doe (Process Engineer)	December 1, 20XX

The objectives of an FMEA include [7]:

1. identification of each failure mode, the sequence of events associated with it, and of its causes and effects
2. classification of each failure mode by relevant characteristics, including detectability, diagnosability, testability, item replaceability, compensating, and operating provisions
3. formulation of recommendation for improvement if required

One of the first steps during the preparation of a FMEA is to identify the systems' boundaries. An FMEA can be performed on a system level down to individual component level. For example, one could look at the failure of an ESD valve as a whole or break this system down to individual components (valve body, actuator, solenoid) and review failure of the individual components. The second step is choosing the appropriate failure or error modes. Failure modes for safety systems are provided in sources such as ISO 14224 [14]. During the team session, the failure causes, local and systems effects, detection, etc. will be identified by the team members.

The advantage of the FMEA is that it is a relatively easy method to apply and that it is suitable for hazard assessment at an equipment level. The FMEA approach also has a well-deserved reputation for efficiently analyzing the hazards associated with electronic and computer systems, whereas the HAZOP Analysis approach may not work well for these types of systems [15]. The disadvantage of the method is, depending on the system boundaries and level of review, the study can become a lot of work to perform. The methodology is also less suitable for systems where complex logic is required to describe system failure [7]. Tab. 11.7 shows an example FMEA table.

11.4.2 Dow index methods

The Dow Fire and Explosion Index (Dow F&E) [5] and Chemical Exposure Index (Dow CEI) [6] are internationally used ranking systems to evaluate the hazard category of a process plant and expected losses due to potential fires and explosions or chemical exposure. The Dow F&E uses the NFPA material factor, which is a measure of the hazards of the substance that falls between 0 and 40 in a matrix of instability/ reactivity and flammability [16]. The material factor is then modified by penalties and credit factors, which depend on the operating conditions. If the results of the assessment is an index value higher than 100, the risk is deemed to be unacceptable and risk-reducing measures need to be applied. Other outcomes of the assessment are an estimate of the potential area of damage, value of the equipment in the area, maximum probable property damage, and business interruption. The Dow F&I is a useful tool during early engineering stages since it helps assess multiple process options and identifies which process unit is the most hazardous. Modification or additional protection of mitigating measures can then be considered. The Dow CEI assesses the relative acute toxicity risks due to liquid spill, evaporation, and dispersion of a vapor cloud.

Tab. 11.7: Example list of FMEA table.

System	Component (sub-system)	Failure or error mode	Failure cause	Effect on other components	Effect on whole system	Detection method	Compensating provision	Actions	Action party	Remarks
Production export	Hydraulic actuated ESD Valve XXX	External leakage – process medium	Leakage of valve body	Hydrocarbon release, potential for fire and explosions	– Loss pf production– Inability to fully isolate system during ESD	– Large leakage detected by pressure drop PZTXXX– Gas detection at ESD deck	Shut down upstream production facilities to stop flow from ESD valve	None	None	None
		External leakage – utility medium	Leakage of valve body	Hydraulic spill on ESD deck, loss of hydraulic pressure, closure of valve	Loss of production	Hydraulic pressure detection PZTXXX	Platform will shutdown on loss of hydraulic pressure	None	None	None
		Failure to close on demand	Jammed gate valve	None identified	Failure to close export; in case of ESD, possible escalation of fires on platform	Limit switches, time-out alarms	– Shutdown upstream and downstream facilities– yearly functional tests	None	None	None

11.4.3 Bowtie assessments

The bowtie method is a risk assessment method that can be used to analyze and communicate how high risk scenarios develop. The essence of the bowtie consists of plausible risk scenarios around a certain hazard and ways in which the organization stops those scenarios from happening. The method takes its name from the shape of the diagram that you create, which looks like a man's bowtie [17]. The methodology in this paragraph is based for the most part on the work described in the CGE bowtie methodology manual [17] in which more detail is provided on creating bowties and how to facilitate bowtie sessions.

The bowtie method has several goals:
– Provide a structure to systematically analyze a hazard.
– Help make a decision whether the current level of control is sufficient (or, for those who are familiar with the concept, whether risks are As Low as Reasonably Practicable or ALARP).
– Help identify where and how investing resources would have the greatest impact.
– Increase risk communication and awareness.
– Provide a framework to plot/collate safety-related information from audits, incidents, maintenance backlogs, etc.

The advantage of using a bowtie is that it creates an easy picture to understand and communicate on multiple levels of the organization. A complete bowtie diagram, linked to the management system, is like a graphical table of contents – a map showing everything an organization does to controls its major risks. During engineering projects it can be used in various stages of design and is useful in helping identify safety critical systems. It contains operational hardware barriers, behavioral barriers and organizational management systems, which makes it a useful tool to holistically look at where investing resources would have the greatest impact.

Bowties are best created during a team brainstorm session using input from various engineers and operations personnel. Constructing a Bowtie can be done by the following steps:
1. Specify the hazard that will be assessed (anything that has the potential to cause harm, e.g., stored flammable hydrocarbons, toxic compounds, object at height);
2. Specify the top event, the event that occurs at the point in time when control over the hazard is lost (e.g., loss of containment, object falling down);
3. Identify threats, a possible cause that can release the hazard by producing the top event. (e.g., corrosion, overflow, breaking rope);
4. Assess consequences after the top event has occurred, which are unwanted events caused by the release of the hazard (e.g., fire, explosion toxic exposure to personnel, fatality from dropped object falling on personnel);

5. **Identify control barriers** against the threats, which can prevent or reduce the likelihood of the top event occurring (e.g., proper material selection, proper rope certification).
6. **Identify mitigation barriers** against the consequences (e.g., fire walls, evacuation procedures, dropped object protection frame);
7. **Identify escalating factors**; these are specific conditions or circumstances, which could cause a barrier to fail or be less effective;
8. **Identify escalating factor barriers**; these are barriers that can help to prevent or manage escalating factors.

A generic overview of an example bowtie is presented in Fig. 11.3 created using the CGE software Bowtie XP.

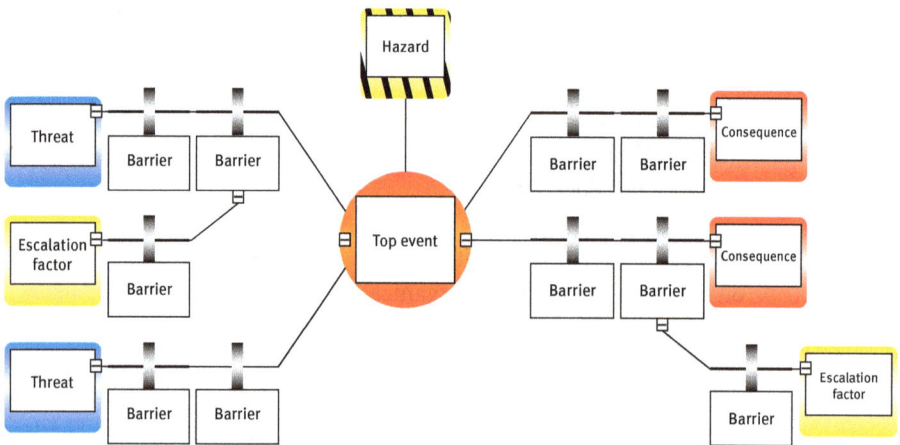

Fig. 11.3: Generic bowtie.

Unfortunately, barriers are seldom 100% effective. Escalation factors take this into account by stating a condition that reduces the effectiveness of a barrier. The following escalation factors categories are typically considered:

– Human factors: anything a person does to make a barrier less effective;
– Abnormal conditions: anything in the environment that causes a barrier to be put under strain;
– Loss of critical services: if a barrier relies on an outside service, losing that service might cause it to lose effectiveness.

When identifying escalation factors, the guideline is to keep in mind that they should not cause a top event or consequence, that they need to be credible, that lessons learned from other incidents are used, and to focus on the critical barriers. Questions that a team should ask themselves are: if there are any circumstances under which the barrier will not work, how the barrier can be "destroyed" and if the barrier has failed in the past.

Once the escalation factors have been established, the barriers for the escalation factors can be identified. Escalation factor barriers can be divided in the categories of human factors (e.g., training, supervision, etc.), abnormal conditions (e.g., maintenance, spares, etc.) and loss of critical services (backup systems, shutdown, etc.).

Bowties can be created for process-related hazards but also for product-related hazards. The challenge is to identify the proper hazard and top event. Some products are not hazardous by themselves and one should look more towards the intended use of the product to identify consequences.

11.5 Consequence assessment

Having identified the hazards, the next step in the risk management process is to assess the consequences. From a process or product safety perspective the consequences of toxic exposure, fire, and explosions will have to be reviewed. Depending on the engineering stage this can be a high level assessment for a risk screening purpose or a detailed one as input to a full QRA. The models discussed in this chapter provide the reader with relatively simple means to quantify consequences, which can be used as input for high level assessments. However, modelling performed as input for quantitative risk assessments or as input to engineering decisions such as safety distances or specifications should consult sources where consequence assessment is discussed in more detail such as Lees' Loss Prevention [7], Cameron and Raman [18], the Yellow Book [19], and Casal [20]. When performing consequence modelling, it is very important to know what the limitations and underlying assumptions in the models are. This is in order to prevent using models that are not suitable for the case assessed, leading to an underestimation or overestimation of the consequences and therefore the risk.

11.5.1 Fire models

Fire consequences assessments allow the calculation of heat radiation transfer to a target with heat radiation expressed in kW/m^2. Different models are used depending on the type of fire. The following industrial fires are commonly considered during process risk assessments:
-   Pool fires
-   Jet fires
-   Fireballs resulting from boiling expanding liquid vapor explosions (BLEVE)
-   Flash fires

Simplified models are presented in the following section for pool fires, jet fires, and fireballs. Flash fires are discussed only briefly. Computational fluid dynamic (CFD) models such as Fire Dynamic Simulator and KFX are used for more advanced assessments but are not discussed in this chapter.

To provide some feeling with what the effect of different levels of exposures are, some common tolerability levels used in the process industry are provided in Tab. 11.7.

Tab. 11.8: Heat radiation exposure effects [21].

Heat radiation	Effects
35 kW/m^2	Significant chance of fatality for people exposed instantaneously (can assume 100% lethality).
10 kW/m^2	1% lethality; structural failure of equipment can start to occur at these radiation levels and above.
6–9 kW/m^2	Pain within approximately 10 s, only rapid escape is possible.
3–5 kW/m^2	Will cause pain in 15–20 s and injury after 30 s exposure, special protected clothing of first responders provide just enough protection at 3 kW/m^2.
1 kW/m^2	Heat felt comparable to exposure to the sun during a hot summer day.

11.5.1.1 Pool fire

Pool fires result from the leakage and subsequent ignition of liquid substances. In an industrial facility they might be caused by mechanical impact, over-pressurization, leaking flanges or valves, small bore fittings, etc. Fire models can consider the flame as a point source, multiple point sources and solid flame, or modelled dynamically using computational fluid dynamics. The point source model is discussed in this section.

The point source radiation model models a flame as a single point radiator. The heat flux Q at a distance x from the flame center to the observer can be calculated by [18]

$$Q = \frac{M \Delta H_c f}{4x^2}$$

where Q is the heat flux (kW/m^2), M the rate of combustion (kg/s), ΔH_c the heat of combustion (kJ/kg), f the fraction of thermal energy radiated (–), x the distance from flame center to observer (m), and τ the atmospheric transmissivity.

The above expression assumes that the position of the observer towards the flame is such that the observer is at maximum exposure to the flame radiation.

The radiation fraction will be dependent on the type of fuel, pool diameter, flame temperature, etc. Larger pool diameters will lead to more soot formation in some type of fuels and thereby reduce the fraction of thermal energy radiated. The fraction is an important factor and, in practice, difficult to estimate. The following expression has been proposed to calculate the fraction of thermal energy radiated for hydrocarbon pools with a pool diameter D_p (m)[20]:

$$f = 0.35e^{-0.05D_p}$$

In practice, the value of f ranges between 0,15 and 0,35 for various hydrocarbon pool sizes [18].

The atmospheric transmissivity τ accounts for the absorption of thermal radiation by the atmosphere. An approximate relation is [18]:

$$\tau = \log_{10}\left[14.1(RH)^{-0.108}r^{-0.13}\right]$$

where RH is the relative humidity (%) and r the distance (m).

The maximum rate of combustion M for a liquid pool fire on land can be estimated by the burning rate m_b and the pool surface area A (m²) [22]:

$$M = m_b A$$

where M is the maximum rate of combustion (kg/s), m_b the burning rate (kg/(s·m²)), and A the pool surface area (m²).

The burning rate m_b can be calculated by

$$m_b = 110^{-3}\frac{\Delta H_c}{\Delta H^*}$$

where ΔH_c is the heat of combustion (kJ/kg) and ΔH^* the modified heat of vaporization at the boiling point of the liquid (kJ/kg).

The modified heat of vaporization ΔH^* can be calculated by

$$\Delta H^* = \Delta H_v + \int_{Ta}^{Tb} C_p dT$$

where ΔH_v is the heat of vaporization of the liquid at the ambient temperature (kJ/kg), C_p the heat capacity of the liquid (kJ/kg·K), T_a the ambient temperature (K), and T_b the boiling point temperature of the liquid (K).

A limitation of the point source model is that it overestimates the intensity of the thermal radiation near the fire due to the fact that it does not take into account the flame geometry [20]. It is recommended that the model is only used for far field prediction at distances greater than 5 pool diameters from the center of the fire, thus, using it to perform conservative calculations of danger to personnel as opposed to establishing distances between adjacent equipment [20].

11.5.1.2 Jet fire

Jet fires are high momentum ignited releases. Their size and flame intensity are a function of the release rate and fuel properties. Jet fires can occur in facilities where pressurized flammable gasses and/or liquids are stored. Unlike pool fires which can, in principle, be extinguished using deluge and foam systems, jet fires are difficult, and in most cases impossible to extinguish by active fire protection systems.

The best validated jet fire models are the Chamberlain model for vertical and angled jet fires [23] and the Johnson model for horizontal jet fires [24]. Calculation of thermal radiation effects on a target is an involved process and the reader should again refer to sources such as Lees' Loss prevention [7] and the Yellow Book [19] for details. The models treated here can best be used when an approximation of the flame dimensions is required or as input for a flame radiation model [25].

A simple correlation for the length L of a jet fire is [25]:

$$L = 18.5 m_r^{0.41}$$

where L is the flame length (m) and m_r the mass release rate (kg/s).

A generalized formula for different fuel types is [25]:

$$L = 0.00326 (m_r H_C)^{0.478}$$

where ΔH_c is the heat of combustion (J/kg).

11.5.1.3 Fire balls/BLEVEs

A BLEVE occurs when a container containing a superheated liquid or liquefied gas experiences a sudden loss of containment. If the medium is flammable, a fireball can occur. The most well-known disaster involving BLEVEs occurred in San Juanico, Mexico on 19 November, 1984, in which the town of San Juan Ixhuatepec housing 40,000 people located adjacent to a facility owned by the PEMEX State Oil Company was almost completely destroyed due to a series of BLEVEs. Approximately 500–600 people died and between 5,000 and 7,000 suffered injuries.

The models presented here assume a BLEVE of a flammable material. The fireball diameter, duration, and height can be calculated using the following correlations based on the TNO Yellow Book model [19].

The diameter of the fireball can be calculated using:

$$D_{fb} = 6.48 (m_f)^{0.325}$$

The fireball duration can be calculated using:

$$t = 0.852 (m_f)^{0.260}$$

And finally, the height the center of the fireball will reach is assumed to equal two times the radius of the fireball and thus equal to the diameter D_{fb}:

$$H = D_{fb}$$

where D_{fb} is the fireball diameter (m), m_f the mass of fuel (kg), t the fireball duration (s), and H the height of fireball center from ground (m). See Fig. 11.4 for the simplified model of the fireball.

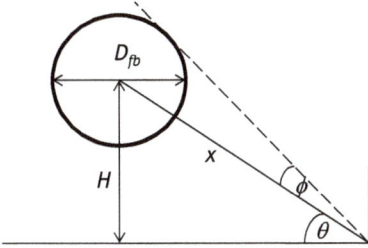

Fig. 11.4: Simplified model of fireball.

The radiation received by the receiver, assuming that the fireball is in full sight of the receiver, can be calculated using the emitted flux E of the fireball and the view factor F [18]:

$$I = \tau EF$$

where I is the received flux (kW/m^2), F the view factor = $(D_{fb})^2\cos\theta/4x^2$ for $\theta \leq 90° - \phi$, and E the emitted flux $\cong \varepsilon \cdot \sigma \cdot T_s^4$.

Here ε is the emissivity (~ 1.0), σ is the Stefan-Boltzmann constant (5.67 × 10^{-11} kW/ (m^{2}K^4), T_s the flame surface temperature (K), and τ the atmospheric transmissivity (–).

11.5.1.4 Flash fires

When a cloud of flammable gas forms over an area and finds an ignition source, the mass between the flammability limits will burn rapidly. People in the gas cloud will be exposed to very high heat fluxes for a short duration but this is enough to assume that anybody inside will suffer fatal injuries. Under some conditions, the flash fire can transform into an explosion. This is discussed in the following section.

11.5.2 Explosion models

The difference between a fire and an explosion is that the latter occurs in a premixed combination of the fuel and oxidant. The power of the explosion is dependent on the proximity of the fuel and the oxidant. Explosions common to the chemical industry can be divided into two types: deflagrations and detonations. Deflagrations are explosions in which the reaction front travels through the unreacted fuel in subsonic speeds and the propagation method is by heat transfer [26]. A detonation is an explosion with an extremely high, supersonic reaction speed. The propagation method is not heat transfer but an extremely rapid and sharp compression occurring in a shock wave. Under certain conditions, a deflagration could transition to a detonation. The deflagration type occurs most commonly in vapor cloud explosions and the detonation in dense phase explosions in cases where TNT or ammonium nitrates are involved. Thus, various models have been developed over the years to model the different types of explosion behaviors and the most recent ones can be divided into two classes: Blast curve models and CFD models. The most common models are provided below:

- Blast curve models
  - TNT equivalency
  - Multienergy (ME)
  - Baker Strehlow-Tang (BST)
- CFD models
  - FLACS

This section will discuss briefly the TNT model, which, in specific cases, is still used in the process industry and ME model, which is used for modelling vapor cloud explosions. The reader should refer to other sources such as Lees [7] for more information on the other types of explosion models. Dust explosions are another important type of explosion, which are not treated in this section. The reader should again refer to sources such as Lees [7] for further information. Damage levels from explosion effects are also discussed in Lees.

11.5.2.1 TNT equivalency

The TNT equivalent method is based on the relation between the flammable material and TNT factored by an explosion efficiency term. Since a lot of information is available with respect to TNT explosion characteristics and damage potential, this model is still widely used but the user should keep in mind that it is a poor model for vapor cloud explosions. TNT explosions are dense phase explosions producing a shockwave from the initial blast of the explosion. Vapor cloud explosion behaves differently. The initial flame front is laminar and its speed low but as it propagates into congested areas the flame front will become more turbulent and the peak pressure will rise. Thus, the explosion overpressure in a vapor cloud explosion is more dependent on the plant geometry. The blast shockwave has a different pressure-time profile than that of a TNT explosion. In the near field it is higher and in the far field lower compared to a vapor cloud explosion. Due to this, the TNT model underestimates overpressures in the far field of a vapor cloud explosion. When the results of this consequence assessment are used for building design against vapor cloud explosions, the TNT method is not a suitable model and the ME method and Baker-Strelow-Tang or CFD models are better suited for that purpose. The TNT method is better suited for condensed phase explosions such as with ammonium nitrate. Even then, it is important to know that there is at least an order of magnitude margin of error in the TNT model due to large scatter in the results of the underlying experiments [7].

There is experimental evidence that if a spherical explosive charge of diameter D_c produces a peak overpressure ΔP at a distance R from its center, as well as a positive-phase duration t^* and an impulse i (the integral of the pressure-time history), then a charge of diameter kD_c of a similar explosive in the same atmosphere will produce an overpressure wave with a similar form and the same peak overpressure p_0 (as well as a positive-phase duration kt^* and an impulse ki at a distance kR from the charge center. This is called the Hopkinson or "cube root" scaling law [20]. The blast curve characteristics of a shock wave showing the side-on peak overpressure, the positive phase duration, and positive impulse, is shown in Fig. 11.5

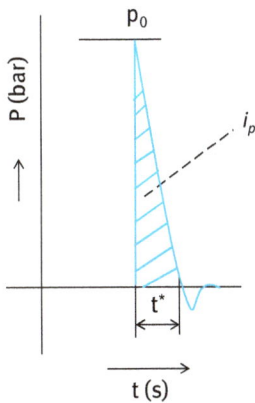

Fig. 11.5: Blast curve characteristics of a shock wave showing the side-on peak overpressure p_0, the positive phase duration t^*, and positive impulse i_p.

Typical blast curves for TNT have been obtained by military field experiments and these curves apply the cube root scaling law to provide blast characteristics versus scaled parameters. The first step is to calculate the equivalent TNT mass based on the heat of combustion, the explosion efficiency, and the mass of the fuel [22]:

$$m_{TNT} = \frac{\eta m_{fg} \Delta H_c}{\Delta H_{TNT}}$$

where m_{TNT} is the equivalent mass of TNT (kg), η the empirical explosion efficiency (–), m_{fg} the mass of hydrocarbon gas kg), ΔH_c the heat of combustion (kJ/kg), and ΔH_{TNT} the heat of combustion TNT (4,437–4,765 kJ/kg).

The explosion efficiency depends on the substance. Cameron [18] proposes the following classes based on various sources of information:
– Class I: $\eta = 0.05$ (propane, butane, flammable liquids)
– Class II: $\eta = 0.10$ (ethylene, ethers)

- Class III: $\eta = 0.15$ (acetylene)

With the TNT equivalent mass calculated, the next step is to calculate the scaled over-pressure properties [7]:

$$z = \frac{R}{(m_{TNT})^{\frac{1}{3}}}$$

where R is the real distance from the center of the explosion origin (m) and z is the scaled distance from the explosion origin (m/kg$^{(1/3)}$).

$$p_s = \frac{p_0}{P_a}$$

where p_0 is the real peak overpressure (Pa), p_s is the scaled peak overpressure (–), and P_a is the atmospheric pressure (Pa);

$$t_s = \frac{t^*}{(m_{TNT})^{\frac{1}{3}}}$$

where t^* is the positive phase duration (s) and t_s is the scaled positive phase duration (s/kg$^{(1/3)}$);

$$i_s = \frac{i_p}{(m_{TNT})^{\frac{1}{3}}}$$

where i_p is the real impulse (s) and i_s is the scaled impulse (Pa·s/ kg$^{(1/3)}$);

$$s = \frac{t_a}{(m_{TNT})^{\frac{1}{3}}}$$

where t_a is the real arrival time (s) and τ_s is the scaled arrival time (s/kg$^{(1/3)}$).

Kingery and Bulmash [27] created blast parameter diagrams where, using the scaled distance z, the scaled blast parameters can be read from the diagram and the real explosion parameters calculated from these using the above equations. The diagram for a surface burst (hemispherical symmetry) is provided in Fig. 11.6, which is the most representative type of geometry in case an explosion occurs at a process facility.

11.5.2.2 Vapor cloud explosions: multi-energy method

The most common type of explosion in the process industry is the vapor cloud explosion. A vapor cloud explosion occurs when a flammable gas between the concentration of the lower explosion limit (LEL) and upper explosion limit (UEL) ignites. The LEL is the minimum concentration of a flammable gas in air that is capable of igniting and causing an explosion. Below LEL the gas concentration is too lean for ignition.

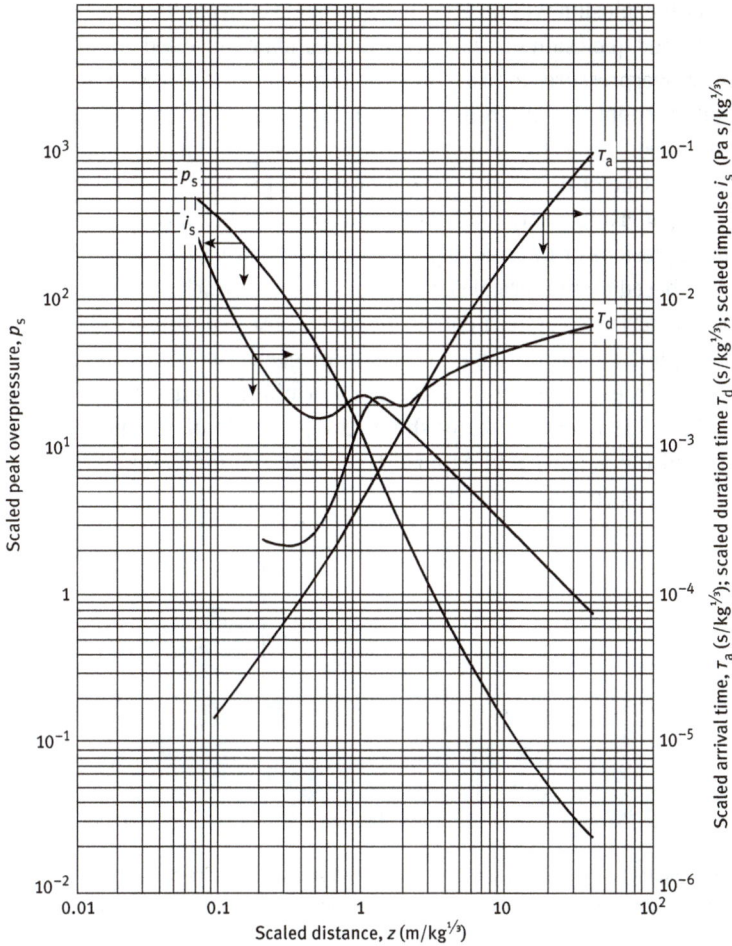

Fig. 11.6: Some side-on blast parameters for a TNT explosion for a surface burst, hemispherical symmetry explosion (*reused with permission from Elsevier*) [28].

The UEL is the maximum gas concentration in air at which the gas can be ignited. Above the UEL the gas is too rich to ignite. The flammable mass for a vapor cloud explosion is the gas that is within the LEL and UEL concentration range. The explosion overpressure from a vapor cloud explosion will be more severe the closer the mixture is to the stoichiometric combustion concentration in air. But analysis from vapor cloud explosion incidents have shown that the fuel mass of the vapor cloud alone is not the only parameter that defines the effect of the explosion. Depending on the confinement and congestion in different areas where the gas is located, the explosion characteristics can differ widely.

A vapor cloud explosion in an area of no congestion and confinement will have a low laminar burning speed, which is too low to cause any significant blast overpressures. The vapor cloud explosion will behave like a large flash fire in this case. On the other hand, a vapor cloud explosion in a heavy confined and/or congested environment can create high blast overpressures approaching the characteristics of dense phase explosions. To take into account this difference between dense phase explosion behaviors and vapor cloud explosions, specific models have been developed for the modeling of vapor cloud explosions. One of the most accepted models is the ME method developed by TNO [29]. The method is applicable to unconfined vapor cloud explosions and not to vented vapor cloud explosions or internal explosions in piping or vessel [30].

The ME model is based on numerical simulations of blasts, which produced scaled blast parameter characteristics versus a combustion energy-scaled distance (Fig. 11.8). The ME method is applied by identifying each blast source, characterizing the congestions and confinement in each source, and finally determining their contribution to blast generation. Figure 11.7 schematically shows the principle of the method: the blast characteristics of the vapor cloud explosion are assessed individually for each region where the gas cloud covers the facility.

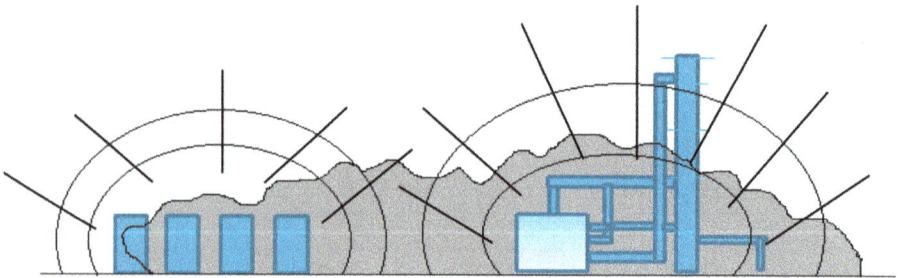

Fig. 11.7: Principle of the multi-energy method showing that the blast characteristics of a vapor cloud explosion are assessed individually for each region on the facility. Adapted from [19].

The following steps are followed to determine the explosion characteristics of a vapor cloud explosion using the ME method [19]:
1. Determine cloud size.
2. Identify the congested areas.
3. Determine the free volume of each congested area (unobstructed areas).
4. Estimate the explosion energy E of each area: volume of cloud times the explosion energy of the fuel ($3,5MJ/m^3$ combustion energy of most stoichiometric hydrocarbon mixtures in air assumed in the ME model).
5. Determine the free volume of each congested area.
6. Determine the source strength of each area (1 to 10).
7. Determine the location of the center of each area.
8. Calculate the combustion energy scaled distance from the center of each area.
9. Calculate the blast characteristics at the real distance by using the ME blast charts.

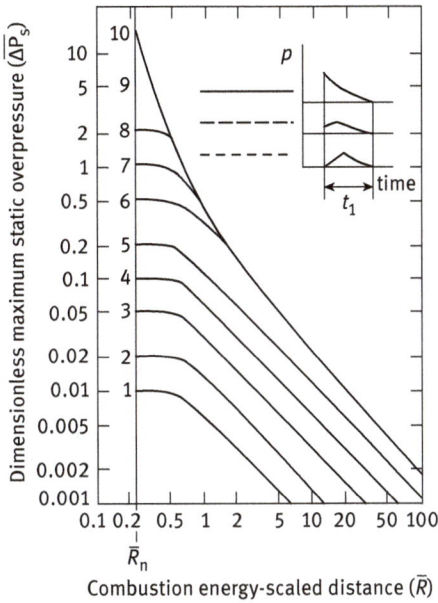

Fig. 11.8: Blast chart of ME method showing the scaled peak side-on overpressure and combustion energy-scaled distance for 10 different blast source strengths. Similar charts are available for blast characteristics such as impulse and dynamic overpressure. *(Reused with permission from Elsevier)* [28].

The combustion energy scaled distance r' is calculated as follows:

$$r' = \frac{r}{\left(\frac{E_{ex}}{P_a}\right)^{\frac{1}{3}}}$$

with:

$$E_{ex} = V_{cloud} E_v$$

where r is the real distance from center of area (m), r' the scaled distance (m$^{(2/3)}$), E_{ex} the explosion energy (J), V_{cloud} the cloud volume in area (m^3), E_v the combustion energy of hydrocarbon mixture $= 3{,}5 \cdot 10^6$ J/m^3, and P_a the atmospheric pressure (Pa).

The scaled side-on overpressure can then be calculated as follows:

$$P_S' = \frac{\Delta P_S}{P_a}$$

where ΔP_S is the real peak side-on overpressure (Pa) and P_S' is the scaled peak side-on overpressure (–).

Critical things to consider while using the ME model are: the validity limits of the model, potential blast interaction between blast sources, and the selection of blast source strength.

The ME is best suited for vapor cloud explosions where overpressure effects need to be considered at relatively large distances from the blast source. The ME model is not suitable for assessing overpressure effects close to the explosion source and in complex geometries such as offshore production facilities, drilling rigs, and floating production storage and offloading vessels (FPSOs). CFD models need to be used in such cases.

A point of consideration is up to what distance individual blast sources can still be considered to be independent. If two blast source areas have sufficient separation distance, the explosion will start to build up pressure in the area where ignition occurred, then reach the free field between the sources and lose flame speed and overpressure before reaching the second area. If the two sources are close to each other, the blast wave will lose some flame speed but then continue in the second area. In this case, the blast sources cannot be considered independent and actually behave as one large explosion source. Thus, the potential for interaction between blast sources also needs to be considered. The Yellow Book defines the critical separation distance as the parameter to help make this distinction. The project "RIGOS" [31] concluded that the guidance provided in the Yellow Book with respect to critical separation distance is not conservative and could lead to underestimating the explosion overpressures. The research showed that critical separation distance is dependent on the overpressure of the donor explosion. For further guidance, the reader should refer to the research report [31].

11.5.3 Toxic exposure

Another hazard that is important to assess is toxic substances. Toxic substances can enter a biological organism by various routes [26]:
– Inhalation: uptake via the lungs
– Dermal contact: penetration through the skin
– Ingestion (oral): uptake via gastrointestinal tract
– In utero: uptake by the fetus via the placenta

In process risk assessments, the inhalation and dermal contact routes are the most common. In product engineering, other routes can also become relevant. An example is in the assessment of pharmaceutical products. In pharmaceutical practice, additional routes of exposure are also possible such as intravenous, intraperitoneal, subcutaneous, and intramuscular [26].

From a consequence assessment perspective, the exposure time and concentration of the substance are the parameters to consider. Threshold values have been developed for different circumstances such as working environment concentration limits and emergency concentration limits. Some examples:
– TLV-TWA: Time-weighed average exposure over 8 h or 40 h;
– TLV-STEL: Short term exposure limit, not longer than 15 min;
– TLV-C: Threshold limit value, ceiling limit, which may not be exceeded at any time;
– IDHL: Immediately dangerous to life and health, exposure up to 30 min;
– ERPG: Emergency Response Planning Guidelines, exposure to 1 h assumed.

A threshold limit value is selected based on the goal of the assessment and, in some cases, regulatory requirements. Calculations then need to be performed to assess the concentration and exposure time and these are compared to the threshold limit value to estimate the effects.

To assess the consequences of toxic exposure to people, dose-response relationships are sometimes used. A "dose" is the amount of substance that is taken up by an organism. The dose is expressed as a concentration c raised to a chemical specific parameter n multiplied by the exposure time t. Since the amount taken up is time-dependent [20]:

$$D_{toxic} = c^n \cdot t$$

where D_{toxic} is the received dose, c the concentration of toxic substance exposure, and t the exposure time.

It should be noted that the toxic concentration in limit values or dispersion models are sometimes expressed in mg/m^3 or in ppmv. To transform the units from ppmv to mg/m^3 and vice versa, one can use the following equation assuming ideal gas behavior [20]:

$$c_{ppmv} = \frac{22.4}{M} \cdot \frac{T}{273} \cdot \frac{1,013}{P_{a\_bar}} \cdot c_{mg/m^3}$$

where M is the molecular weight (kg/kmol), T the ambient temperature (K), and P_{a_bar} the atmospheric pressure (bar).

Since the absorbed dose is time-dependent due to a varying concentration exposure during an emergency situation, the expression becomes:

$$D_{toxic} = \int_0^t [c(t)]^n dt$$

The dose is used in so-called probit relations, which will be explained briefly in this section. Probit stands for "probability unit" and probit functions are empirically derived mathematical expressions linking risk (effect, probability) to exposure (intensity, time, duration) [26]. It is assumed in the probit function that the log-exposure-risk relation follows a Gaussian normal distribution. By using the probit relation one can calculate the probability expressed as a percentage of an exposed population, which will exhibit a given adverse effect as related to a given exposure. Probit functions are expressed in the form:

$$Pr = a + b \cdot \ln(D_{toxic})$$

where Pr is the probit value, a and b are regression constants, and D_{toxic} is the toxic dose. The regression constants depend on the toxic effect that is assessed and are available for a limited number of substances. Transformation tables for probits to response fractions (%) have been developed by Finney [32] and adopted in the TNO

Green Book [33]. Analytical expressions are also available for calculations, which are useful in case computer models are used for the calculations [34].

The reader should refer to the Green Book [33] in which the subject is treated in more detail. Note that toxic exposure relations provided in 1992 might have changed, so be advised to mainly use the source to get familiar with the methodology and check more recent literature sources before using the dose response relations mentioned in the Green Book.

Note that the application of probit relations is not limited to toxic effects. There are probit relations available in literature that can be used to calculate the consequence for fire and explosions such as those provided in the Green Book [33] and Lees [7].

11.6 Codes, standards, and designing ALARP

Having identified the hazards and assessed the consequences, how does one choose the relevant engineering design cases? What is, and when is it "safe enough"? First, there are a number of information sources, approaches, and methods that can assist engineers making these choices, such as:

– regulations, codes, and standards;
– company-specific design engineering standards;
– as low as reasonably practical (ALARP) and risk analysis.

These will be treated in the following sections.

11.6.1 Regulations, codes, and engineer standards

Engineering regulations, codes, and standards all serve to ensure that products and processes are safe and of good quality. Each has its own characteristics and there are differences between them:

Technical regulations: Mandatory requirements set by a regulatory authority, defining the characteristics and/or performance requirements of a process, product, or service.

Engineering codes: Laws and regulations specifying minimum requirements to protect safety, health, and the environment. Codes will typically refer to standards for specific details on requirements that have not been specified in the code itself. Codes are enforced by one or more governmental entities and are important for establishing best engineering practices.

Engineering standards: Technical documents defining technical methods or characteristics of a product, process, or service. They are not laws and are thus not enforced by regulatory authorities but are considered as engineering practice.

In other words, codes specify what needs to be done and standards detail how to accomplish this.

Regulations, codes, and standards specify in some cases what the design cases should be. But in many instances this is not the case. And even when they do, there are cases during engineering projects where there are conflicting design requirements between standards or they provide too little information on how to deal with a specific design challenge. In such cases, further analysis or reviews are required to make the proper design choices. This is where the ALARP principle and risk analysis comes into the picture when reviewing design aspects that could have an impact on safety, health, or the environment.

11.6.2 ALARP principle

ALARP stands for "As Low as Reasonably Practical" and has its origin in the UK Health and Safety at Work Act, 1974 [35]. The principle's goal is to avoid making sacrifices grossly disproportionate to the benefits gained. An extreme example might be to spend $1 m to prevent five staff suffering bruised knees, which is obviously grossly disproportionate. But to spend $1 m to prevent a major explosion capable of killing 150 people is obviously proportionate [36].

The principle is outlined in Fig. 11.9:

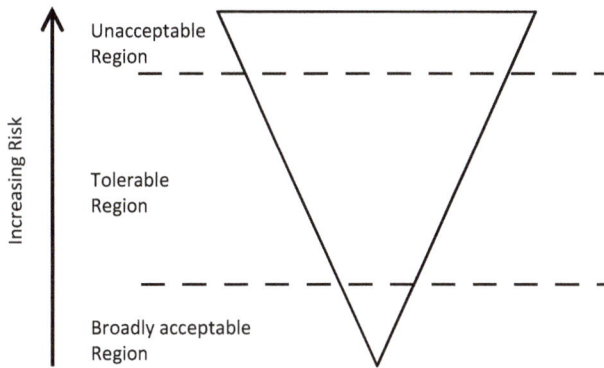

Fig. 11.9: ALARP triangle.

The risk is divided into three regions [37]:

Unacceptable region: Risks are too high and thus risk reduction measures are required, unless there are exceptional reasons for the practice or activity to be retained.

Tolerable region: Risk reduction should be applied as long as the sacrifices are not disproportional to the benefits gained.

Broadly acceptable region: Risks falling into this region are generally regarded as insignificant and adequately controlled. They are typical of the risk from activities that are inherently not very hazardous or from hazardous activities that can be, and are, readily controlled to produce very low risks.

The challenge of the ALARP principle lies in the practical application of the principle. What is considered "reasonable practical" can be highly subjective. The UK's Health and Safety Executive website provides some guidance to inspectors to assess if duty-holders have reduced risks to ALARP. Camaron et al. [18] state factors that come into consideration when assessing practicability under the ALARP principle:

1. Can the consequences be eliminated?
2. If elimination is not possible, can they be mitigated?
3. What are the possible mitigation measures?
4. Are the mitigation measures within engineering practicability?
5. If it is not practicable to design the mitigation measure due to engineering constraints, then can the likelihood of an occurrence resulting in the consequence be reduced?
6. What are the possible likelihood reduction measures?
7. Are the likelihood reduction measures within engineering practicability?
8. What are the benefits of the risk reduction measures? This may not be a reduction in public risk alone, but should also address risk to employees, benefits gained in minimizing interruptions to operations, etc.

After having established what is deemed "practical" it is required to assess if these measures are "reasonable." Cost-benefit analysis and risk assessment are some of the important tools to assess this. The UK Health and Safety Executive provides guidance on how to perform the cost-benefit analysis [36] but warns that it should not be the sole criterion to determine if the ALARP level has been reached. The book *Operational Safety Economics* [47] provides a detailed treatise on various approaches and theories for applying economic analysis in relation to process risk focused on the chemical and process industries.

11.7 Risk analysis

Risk analysis is a tool that can be used as part of an ALARP assessment and help in making design decisions. There are two types of assessments:
- Qualitative: where risk is expressed in a nonquantifiable way (e.g., low, medium, high);
- Quantitative: where the result of the assessment is a number and compared to a risk acceptance criteria.

The following sections provide a short overview of the methods and their advantages and disadvantages.

11.7.1 Qualitative assessments

A commonly used qualitative tool in the chemical and process industry is the risk matrix. The risk matrix is a graphical representation of the risk as a function of the consequences and frequency or probability of occurrence. Consequences categories can be defined for people, environment, assets, and reputation. An example is provided in Fig. 11.10. It has three risk categories (low, medium, high) but more could be allocated. The categorized regions on the matrix correspond with the regions of the ALARP triangle (Fig. 11.9). When the outcome of the risk assessment is in the "high" category the risk is unacceptable and risk reducing measures are required. If a risk is in the "low" category the risk must be managed for continual improvement. If a risk is "medium" risk reduction measures must be incorporated until ALARP.

The matrix in our example has a 5 × 6 dimension but other dimensions such as 3 × 3, 4 × 4, 5 × 5, 6 × 6, 4 × 6 are also possible with more or less allocated severity and frequency categories. Other examples are available in industry guidelines such as from the International Organization of Oil and Gas Producers [38] and ISO standards [39]. The same principles apply to all the matrixes, irrespective of their dimensions. Industrial companies have a corporate-defined risk matrix, which is used a basis for all the process safety-related risk assessment studies. The advantage of having a corporate-defined risk matrix is that it helps achieve an increase in consistency between risk assessments in a company.

Consequences			Frequency of occurrence					
Severity Level		People	1	2	3	4	5	6
			Remote	Improbable	Very Low	Low	Medium	High
			$<1*10^{-5}$ /yr	$1*10^{-5}$/yr -$1*10^{-4}$/yr	$1*10^{-4}$- $1*10^{-3}$ /yr	$1*10^{-3}$- $1*10^{-2}$ /yr	$1*10^{-2}$- $1*10^{-1}$ /yr	$>1*10^{-1}$ /yr
5		Multiple fatalities						
4		1 fatality, or lasting invalidity					**High**	
3		Severe injury or effect on health, partial invalidity			**Medium**			
2		Lost workday	**Low**					
1		First aid, no leave						

Fig. 11.10: Example risk acceptance matrix defining consequences for people.

Qualitative risk assessments, such as the risk matrix, have the advantage that they are relatively easy and quick to use and can help prioritize actions. Despite its apparent ease of use, there are in fact a number of areas where the risk could be incorrectly addressed leading to a wrong prioritization of actions and incorrect screening of high risks. An important point to remember when using the risk matrix is its limited suitability for assessing complex and high risks. In those cases, decisions should be based on more quantified risk assessment methods to decide if the risks have been reduced to ALARP.

11.7.2 Quantitative assessments

In case of complex systems with the potential for major hazards, quantified risk assessment can be employed to assist in making engineering decisions. This section will briefly discuss some common techniques and their advantages and disadvantages.

Quantitative risk analysis requires failure frequency of equipment and an assessment of the consequences as input for the assessment. Consequence assessment models have been discussed previously in this chapter. Frequency data can be obtained from online and offline information sources. Several databases containing failure frequencies are available free of charge from sources such as from the International Association of Oil and Gas Producers [40] and the UK Health and Safety Executive [41]. The quantitative assessments that will be discussed in the following sections are the Fault tree analysis, Event tree analysis, and the Quantitative risk analysis.

11.7.2.1 Fault tree analysis

A fault tree analysis (FTA) investigates how a top event (fault), such as an exothermic runaway, loss of containment, etc. originates from individual causes. It is a deductive technique meaning that one starts at the final outcome. The goal of the FTA is to calculate the probability or frequency of the top event. It also provides insight into how a top event scenario develops from its individual causes. The name is derived from the fact that the top event is the apex that branches to individual causes thus resembling a tree. A fault tree considers that the system components are in one of two states: a correct state or a fault state. Note that there is a difference between a failure and a fault in a component. A fault is an incorrect state, which may be due to a failure of that component or may be induced by some outside influence. Thus, fault is a wider concept than failure. All failures are faults, but not all faults are failures [7]. A fault tree is constructed by using "gates." Gates are points in the diagram where branches meet. Two common gates are the "AND" and "OR" gates (See Fig. 11.11). At the "AND" gate all the components meeting at this gate must fail in order to progress further towards the top event. At the "OR" gate one of the components meeting at this point must fail in order to further progress towards the top event.

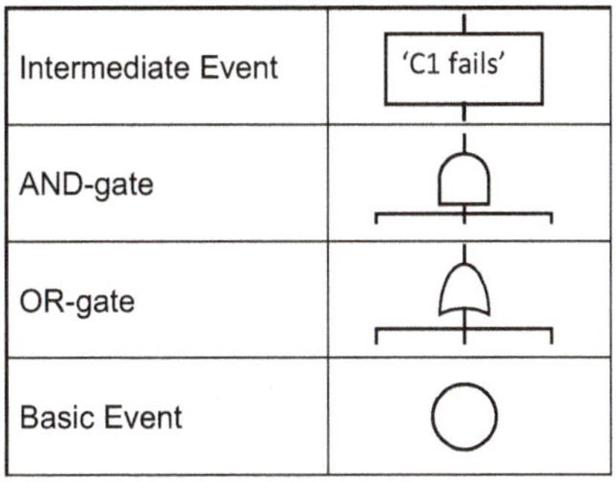

Intermediate Event	'C1 fails'
AND-gate	
OR-gate	
Basic Event	

Fig. 11.11: Fault tree symbols.

To effectively use the fault tree the following needs to be prepared and decided before the analysis:
– The goal of the analysis such as screening of scenarios, verification of codes, comparing design alternatives, etc.
– Scope of the system to be investigated
– Frequency and probability data

A fault tree diagram can then be constructed by the following steps:
1. Define the top event.
2. Decide where the analysis will start on the system diagrams such as P&ID and PFDs. For safety related assessments, this will typically be the point where the safeguarding system starts to act.
3. Move through the system against the flow of liquid, gas, energy, current, etc. and take into account the faults of all relevant components.
4. Continue modelling up to the point where the goal of the analysis has been met.

The FTA and has the advantage that risk contribution to the top event from individual components can be quantified. The limitation of an FTA is that only Boolean states are assessed (fail/correct) and thus no account is given to partial functioning, temporal effects, and spare parts [16]. There are various references for the construction of fault tree diagrams such as Lees' Loss Prevention in the chemical industries [7].

The FTA also forms the basis of the Layers of Protection Analysis (LOPA). The LOPA is becoming more common in the process industry as a tool to determine how reliable independent barriers will have to be to prevent the realization of a single process hazard. The steps of an LOPA consist of:
– Calculation of initial event frequency without taking barriers into consideration;
– comparison of result to a quantitative criteria and determination of the "gap";
– Calculation if current barriers reduce the risk sufficiently;
– If not, redesigning, increasing reliability of barriers and/or introducing new independent barriers to ensure criteria are met.

The LOPA technique is sometimes used during Safety Integrity Level (SIL) assessment. During an SIL assessment the level of required protection of an instrumented system is based on the IEC61508 and IEC61511 standards. For more information on the LOPA methodology the reader should refer to the book "Layer of Protection Analysis: Simplified Process Risk Assessment" [42].

Example 11.1
A reactor requires constant addition of chemicals to keep its pressure under control (see Fig. 11.12).

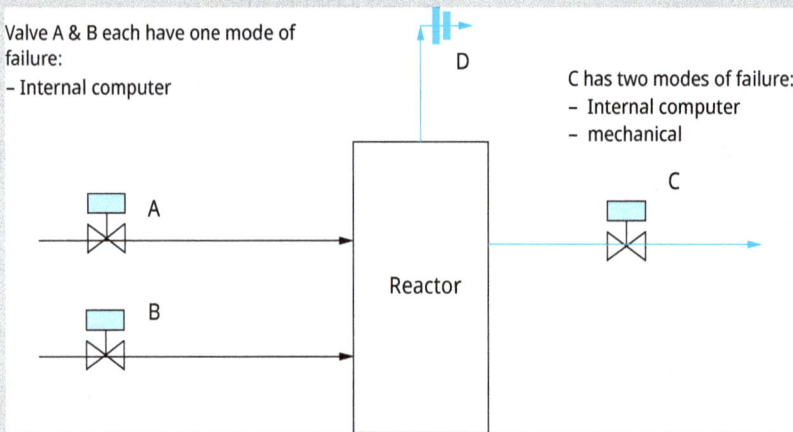

Valve A & B each have one mode of failure:
– Internal computer

C has two modes of failure:
– Internal computer
– mechanical

Fig. 11.12: Reactor kept under control by addition of chemicals via two inlet streams control valves A and B, and one outlet control valve C. A pressure relief device D will activate in case the pressure becomes too high in the reactor.

Draw the fault tree diagram for the reactor rupture top event. The following events are relevant:
- If one of the valves A, B, or C closes due to a failure in the local valve computer, pressure would build up;
- In addition, valve C can also fail if a mechanical component would spontaneously break in that valve, thus having two modes of failure;
- A mechanical release device D would open in case A, B, and/or C closes, preventing rupture of the reactor by pressure buildup;
- Should D fail to open and one of the valves A, B, or C be closed, the reactor would overpressurerize and rupture.

Answer: The answer of this example is provided in Figure 11.13

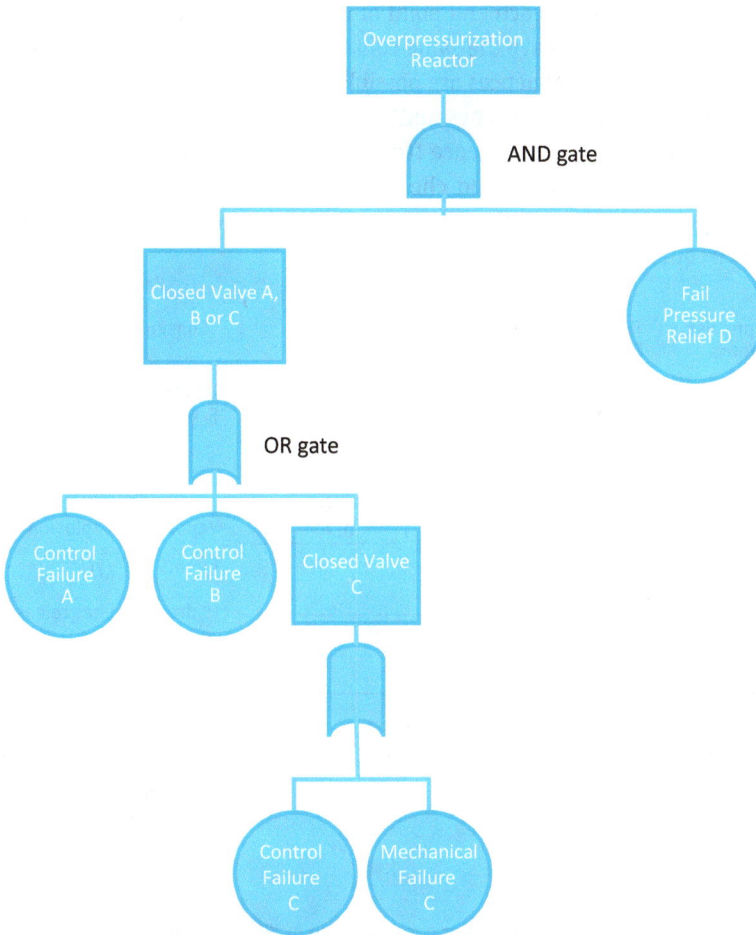

Fig. 11.13: Fault tree answer of Example 11.1.

11.7.2.2 Event tree analysis

The event tree analysis (ETA) is a complementary method to the FTA, employing an inductive method to assess branching consequences from an initial cause. Probabilities or frequencies can be assigned to each of the branches. An example for a hydrocarbon gas scenario is provided in Fig. 11.14.

Construction of an event tree consists of the following steps:
1. Define the initiating event.
2. List the sequence of protections of the system.
3. Start from the first layer and branch for a yes/no for the success or failure of the layer.
4. Continue branching for each layer until the sequence has been completed.

5. Describe the consequences at each end point of the event tree. Note that it is possible that multiple branches have the same consequences thus meaning that multiple pathways to these consequences are possible.
6. Assign numeric values to the branches and calculate the frequency of the consequences. Calculation of the consequence frequencies is performed by multiplication along the branches. One should check that each probability node on the event tree sums up to unity and that the consequence frequencies sum up to the initial event frequency.

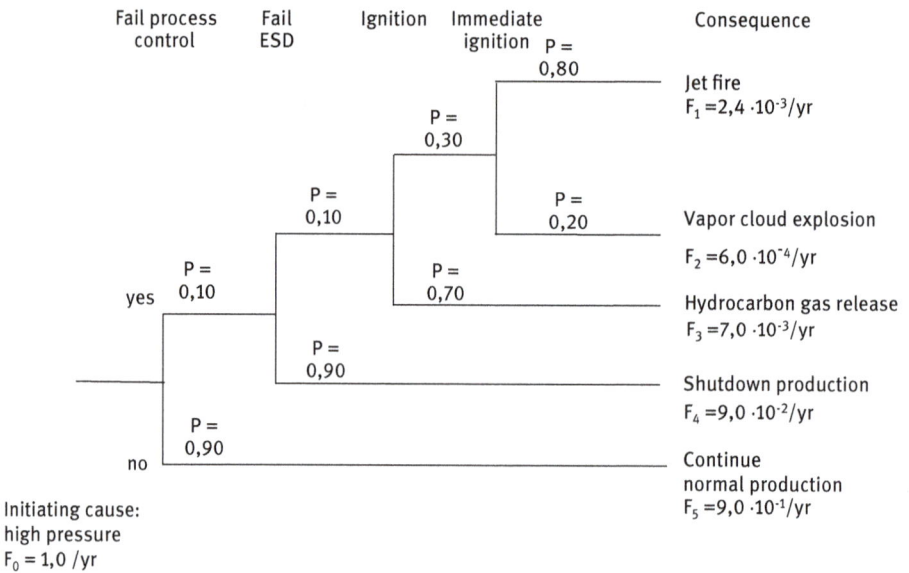

Fig. 11.14: Example event tree for a pressure peak in a gas production facility. Frequencies and probabilities are only for illustrative purposes.

The advantage of the event tree is that it is a good method to show many possible consequences from a single scenario. The event tree is also used as a precursor for quantitative risk analysis (QRA).

11.7.2.3 Individual and group risk

The tools mentioned in this chapter can be used as input to a full QRA of a system. A QRA will assess the frequencies and consequences of a variety of scenarios and is therefore a more involved study compared to other assessments. The goal of the QRA can vary from land use planning to assessing if the risk to operators on the industry site is acceptable. In the process industry, it is common to investigate the risk of a fatality since this is a severe consequence that is more easily definable compared to the risk of injury. General assumptions in QRA assessments are that contribution of all incidents on the facility are additive and that the risk receiver is exposed to the presence of the risk source over a certain period of time (usually 1 year). There are two categories of risks that can be investigated:

- Individual risk (IR): The frequency at which an individual may be expected to sustain a given level of harm from the realization of specified hazards [43];
- Group risk (GR): Expresses the relationship between the frequency and the number of people suffering from a specified level of harm [18].

If consequence severity values are calculated for a meshed area around a risk source, points of equal risk can be connected to produce individual risk contours at various levels, for example, 10^{-4}, 10^{-5}, 10^{-6}, or 10^{-7} [16]. If directional effects are absent, the contours are circular; if a dominating wind direction exists or directional blast or fragment projection occurs, contours become ellipsoidal or even irregular [16].

GR (also called societal risk) differs from IR in the fact that it expresses the relationship between the frequency and the number of people suffering from a specified level of harm as opposed to the individual approach [16].

11.8 Exercises

1. What is inherently safer, driving by car or flying a plane from Amsterdam to Madrid? Explain why.
2. A new reactor is being designed for a hydrogenation with a novel liquid organic hydrogen carrier. The heat of reaction is: – 375.6 kJ/mol and the heat capacity of the mixture is 610 J/mol·K.
   a. Calculate the maximum temperature that can be theoretically achieved by this primary reaction if the temperature of the process is initially maintained at 150 °C when a cooling failure occurs.
   b. The tmr$_{ad}$ was measured at 140 °C as 24 h. What would you expect the tmr$_{ad}$ to be at the process temperature of 150 °C? Hint: Consider the van't Hoff rule, whereby the reaction rate doubles for a temperature increase of 10 K.
   c. Discuss the potential challenges in controlling the process at higher operating temperatures, especially considering the reduced tmr$_{ad}$. How might a higher operating temperature and faster reaction rate impact the stability and safety of the process?
3. In December 2005, the Buncefield oil storage facility in the UK was the site of one of the largest peacetime explosions and fires in European history. The explosion was caused by the overflow of a large petrol filled tank, leading to the formation of a vapor cloud that ignited explosively. In the Buncefield case, overfilling was the cause of loss of containment but more causes for loss of containment and associated consequences can be defined. Create a bowtie diagram of a petrol filled storage tank that suffers from a loss of containment using the following steps:

    a. Define the hazard.

    b. Define the top event.

    c. Define threats and consequences.

    d. Define barriers for the threats and consequences.

4. Continuing from Example 11.1 in the fault tree exercise, it has been decided to install an additional, independent computer system (E) as a backup. Refer to Fig. 11.15 for details. In the event of a control computer failure in A or B or C, this backup system (E) would prevent the valves from closing. Therefore, each valve can now only close unexpectedly if both the control computer and the new protection computer (E) fail. However, this does not apply to valve C, which still has an additional mechanical mode of failure (C2) that the backup system (E) cannot prevent.

Draw the new fault tree diagram.

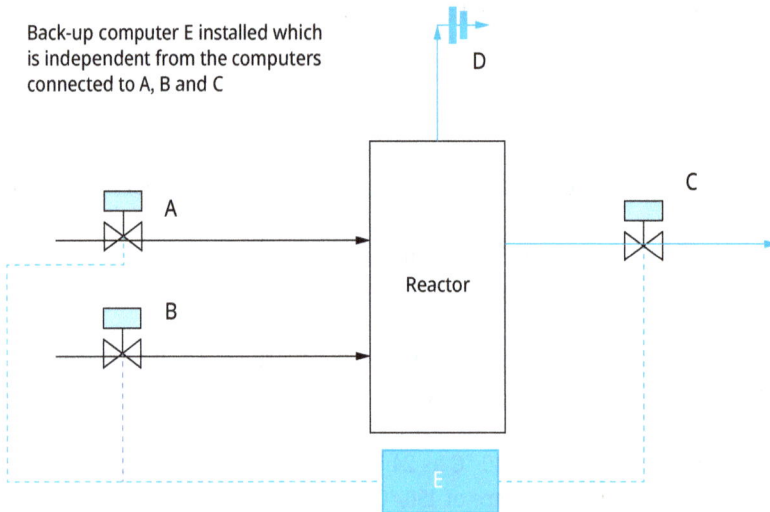

Fig. 11.15: A backup computer E is installed to keep valves A, B, and C open should a control failure cause them to spontaneously close.

References

[1]   Pat-El IE et.al. Major accidents and their impact – The EU directive for offshore safety. Society of Petroleum Engineers conference paper 172094, 2014.

[2]   CCPS Glossary of Terms. https://www.aiche.org/ccps/resources/glossary, (accessed August 2023).

[3]   CCPS. Chemical Process Quantitative Risk Analysis, 1999.

[4]   Kletz TA. Inherently safer design – the growth of an idea, Process Safety Progress, 15, 1996, 5–8.

[5]   AIChE. Dow's Fire & Explosion Index Hazard Classification Guide, 7th edn, New York, N.Y.: American Institute of Chemical Engineers, June 1994.

[6]   AIChE. Dow's Chemical Exposure Index Guide, 1st edn, New York, N.Y.: American Institute of Chemical Engineers, 1994.

[7]   Mannan S, Lees' Loss Prevention in the Process Industries, (ed). Volume 1, Chapter 8 3rd edn, Burlington: Butterworth-Heinemann, 2005.

[8]   Urben PG (ed). Bretherick's Handbook of Reactive Chemical Hazards, 7th edn, Oxford: Academic Press, 2007.

[9]   CCPS. Essential Practices for Managing Chemical Reactivity Hazards, 1st edn, New York, N.Y.: American Institute of Chemical Engineers, 2003.

[10]   CCPS. Chemical reactivity evaluation and applications to process design, 1995.

[11]   CCPS. Guidelines for safe storage and handling of reactive materials, 1995.

[12]   Leggett DJ. Chemical Reaction Hazard Identification and Evaluation: Taking the First Steps, New Orleans: AIChE Loss Prevention Symposium, 2002.

[13]   IEC 61882. Hazard and Operability Studies –, Application guide, 2001.

[14]   ISO 14224. Petroleum, petrochemical and natural gas industries – Collection and exchange of reliability and maintenance data for equipment, 2016

[15]   CCPS. Guidelines for Hazard Evaluation Procedures, 2nd edn, New York, N.Y.: American Institute of Chemical Engineers, 1992.

[16]   Pasman. Risk Analysis and Control for Industrial Processes – Gas, Oil and Chemicals, 1st edn, Waltham USA: Elsevier, 2015.

[17]   CGE. Bowtie XP methodology Manual, Rev.15, March 2015.

[18]   Camaron, et.al. Process System Risk Management, 1st edn, Sydney: Australia, Vol. 6 Elsevier, 2005.

[19]   VROM. Methods for the Calculation of Physical Effects – 'Yellow Book' Publication Series on Dangerous Substances PGS 2, 3rd edn, SZW: The Hague, 2005.

[20]   Casal. Evaluation of the effects and consequences of major accidents, Industrial Safety Series, 8, 2007.

[21]   HSE. Methods of approximation and determination of human vulnerability for offshore major accident hazard assessment, 2010.

[22]   CCPS. Guidelines for Consequence Analysis of Chemical Releases, 2nd edn, New York, N.Y.: American Institute of Chemical Engineers, 1999.

[23]   Chamberlain GA. Developments in design methods for predicting thermal radiation from flares, Chemical Engineering Research and Design, 65, July 1987, 65:4, 299–309.

[24]   Johnson AD, et.al. A model for predicting the thermal radiation hazards from large scale horizontally released natural gas jet fires, Transactions of the Institution of Chemical Engineers, 72, Part B 1994.

[25]   OGP. Consequence modelling, Report No. 434 – 07, March 2010.

[26]   Lemkowitz SM, Pasman HJ. Chemical risk management. In: Faculty of Applied Sciences, Chemical Engineering, Delft University of Technology, Delft: The Netherlands, 2002.

[27]   Kingery CN, Bulmash G. Air blast parameters from TNT spherical air burst and hemispherical surface burst, Tech. Rep. ARBRL-TR 02555, US Army, Ballistic Res. Lab., Aberdeen Proving Ground, MD, 1985

[28]   Mannan S, Lees' Loss Prevention in the Process Industries, 4th edn, Burlington: Butterworth-Heinemann, page 1496, figure 17.59b & and page 1535 figure 17.79a, 2012.

[29]   Van den Berg AC. The multi-energy method – A framework for vapor cloud explosion blast prediction, Journal of Hazardous Materials, 12(1), 1–10, 1985.

[30]   TNO. Application of correlations to quantify the source strength of vapor cloud explosions in realistic situations, final report for the project "Games", Report PML 1998-C53, October 1998.

[31]   HSE Research to improve guidance on separation distance for the multi-energy method (RIGOS, Research Report, 369, 41, 2005.

[32]   Finney DJ. Probit Analysis, 3rd edn, Cambridge University Press, New York: USA, 1971.
[33]   Green Book TNO. Methods for Determining Possible Damage, 1st edn, SZW: The Hague, 1992.
[34]   Vilchez JA, Montiel H, Casal J, Amaldos J. Journal of Loss Prevention in the Process, 14, 193–197, 2001.
[35]   HSE. guidance, http://www.hse.gov.uk/risk/theory/alarpglance.htm, accessed on 18-12-2016.
[36]   HSE. cost benefit, http://www.hse.gov.uk/risk/theory/alarpcheck.htm, 2016, accessed on 18-12-2016.
[37]   HSE. Reducing Risks, Protecting People – HSE's Decision Making Process, London: HMSO, 2001.
[38]   OGP, 2010. HSE management – Guidelines for working in a contract environment, report 423, June 2010.
[39]   ISO 31000. Risk Management, 2009.
[40]   OGP. Risk assessment data directory, Report 434, March 2010.
[41]   HSE. Failure Rate and Event Data for use within Risk Assessments, available online at http://www.hse.gov.uk/landuseplanning/failure-rates.pdf, (accessed January 2017).
[42]   CCPS. Layer of Protection Analysis: Simplified Process Risk Assessment, 1st edn, New York, N.Y.: American Institute of Chemical Engineers, 2001.
[43]   Jones DA. Nomenclature for Hazard and Risk Assessment in the Process Industries, 2nd edn, Rugby, England: Institution of Chemical Engineers, 1992.
[44]   CCPS. Inherently Safer Chemical Processes: A Life Cycle Approach, 3rd edn, 2020.
[45]   CCPS. Inherently Safer Chemical Processes checklist, https://www.aiche.org/sites/default/files/book-downloads/p277iscp3rded-appendixachecklist.pdf, accessed on 21-08-2023.
[46]   Stoessel F. Thermal Safety of Chemical Processes, 2nd edn, Weinheim: Germany, Wiley-VCH, 2020.
[47]   Reniers GLL, Noël Van Erp HR. Operational Safety Economics: A Practical Approach Focused on the Chemical and Process Industries, Wiley, West Sussex: UK, 2016.

Further reading

The following references are recommended for further reading:
[9]   CCPS. Essential Practises for Managing Chemical Reactivity Hazards, 1st edn, 2003.
[10]   CCPS. Chemical reactivity evaluation and applications to process design, 1995.
[11]   CCPS. Guidelines for safe storage and handling of reactive materials, 1995.
[15]   CCPS. Guidelines for Hazard Evaluation Procedures, 2nd edn, 1992.
[16]   Pasman H. Risk Analysis and Control for Industrial Processes – Gas, Oil and Chemicals, 1st edn, 2015.
[18]   Camaron, et.al. Process System Risk Management, 1st edn, Vol. 6 Elsevier, 2005.
[19]   VROM. Methods for the Calculation of Physical Effects – 'Yellow Book' Publication Series on Dangerous Substances PGS 2, 3rd edn, 2005.
[20]   Casal. Evaluation of the effects and consequences of major accidents, Industrial Safety Series, 8, 2007.
[22]   CCPS. Guidelines for Consequence Analysis of Chemical Releases, 2nd edn, 1999.
[28]   Lees FP. Lees' loss prevention in the process industries. In: Mannan S (ed). 4th edn, Burlington: Butterworth-Heinemann, 2012.
[33]   TNO. Green Book, Methods for Determining Possible Damage, 1st edn, 1992.
[42]   CCPS. Layer of Protection Analysis: Simplified Process Risk Assessment, 1st edn, 2001.
[44]   CCPS. Inherently Safer Chemical Processes: A Life Cycle Approach, 3rd edn, 2020.
[46]   Stoessel F. Thermal Safety of Chemical Processes, 2nd edn, Wiley-VCH, 2020.
[47]   Genserik LL, Reniers HR, Van Erp N. Operational Safety Economics: A Practical Approach Focused on the Chemical and Process Industries, Wiley, 2016.

12 Evaluating for sustainable development, environmental impact, social acceptance

12.1 Introduction

This chapter treats evaluation methods for new product and process designs. The methods start with the UN Sustainable Development Goals (SDGs), for two reasons. The first reason is that these SDGs are accepted both by governments and companies worldwide. The second reason is that these SDGs cover society (people), environmental (planet), and economy (prosperity), and hence cover most context dimensions. For environment impact assessment, life cycle assessment (LCA) methods are provided. For social acceptance determination two simple methods are provided.

The methods are very suitable for concept designs but can also be used for evaluation in later innovation stages. For students the methods will be useful for evaluating their design and for reporting their design. The results of the evaluations will also be useful for reporting to a stage-gate evaluation panel in industry.

12.2 Contributions to the UN sustainable development goals

Sustainable development started with a report of the World Commission on Environment and Development of the United Nations, titled "Our Common Future" in 1987 [1]. Brundtland, chairing the commission, provided the definition of sustainable development [1]:

> Sustainable development is not a fixed state of harmony but a process of change in which the exploitation of resources, the direction of investments, the orientation of technological development, and institutional change are made consistent with future as well as present needs. Sustainable development is development that meets the needs of the present generation without compromising the ability of future generations to meet their own needs.

From then on, in subsequent UN world conferences the content of sustainable development was further defined. Financial and business parts of society also took on the concept of sustainable development and defined three dimensions: People (social dimension), Planet (environmental dimension), and Profit (economic dimension. Elkington modified this in business terms and called it the triple bottom line [2]. The World Business Council for Sustainable Development was set up [3] and this council became an official partner to the UN in formulating Sustainable Development Goals in 2015 [4].

The last UN general assembly, September 25, 2015, set up the Sustainable Development agenda for 2030 with the aim to end poverty, protect the planet, and ensure prosperity for all, as part of a new sustainable development agenda, setting up the SDGs for 2030. These SDGs should be achieved over the next 15 years. They are sum-

https://doi.org/10.1515/9783110782127-012

marized in the Tab. 12.1 now with five dimensions: People, Planet, Prosperity, Peace, and Partnership [4]. For companies People, Planet, and Prosperity goals are most relevant.

Tab. 12.1: United Nations Sustainable Development Goals for 2030 [4].

People (social) goals	
Goal 1.	End poverty in all its forms everywhere.
Goal 2.	End hunger, achieve food security and improved nutrition, and promote sustainable agriculture.
Goal 3.	Ensure healthy lives and promote well-being for all at all ages.
Goal 4.	Ensure inclusive and equitable quality education and promote lifelong learning opportunities for all.
Goal 5.	Achieve gender equality and empower all women and girls.
Prosperity goals	
Goal 6.	Ensure availability and sustainable management of water and sanitation for all.
Goal 7.	Ensure access to affordable, reliable, sustainable, and modern energy for all.
Goal 8.	Promote sustained, inclusive, and sustainable economic growth, full and productive employment, and decent work for all.
Goal 9.	Build resilient infrastructure, promote inclusive and sustainable industrialization. and foster innovation.
Goal 10.	Reduce inequality within and among countries.
Environmental (planet) goals	
Goal 11.	Make cities and human settlements inclusive, safe, resilient, and sustainable.
Goal 12.	Ensure sustainable consumption and production patterns.
Goal 13.	Take urgent action to combat climate change and its impacts.
Goal 14.	Conserve and sustainably use the oceans, seas, and marine resources for sustainable development.
Goal 15.	Protect, restore, and promote sustainable use of terrestrial ecosystems, sustainably manage forests, combat desertification, and halt and reverse land degradation. and halt biodiversity loss.
Peace goal	
Goal 16.	Promote peaceful and inclusive societies for sustainable development, provide access to justice for all and build effective, accountable, and inclusive institutions at all levels.
Partnership goal	
Goal 17.	Strengthen the means of implementation and revitalize the Global Partnership for Sustainable Development.

For a design evaluation, these 17 goals can be used for all innovation stages. The goals are further specified in target sections and indicator subsections. These are even more suitable for determining how and how much a design contributes to the SDGs.

For the very first innovation stage, ideation, very little information about the design subject is available, so the evaluation should be very simple. Here is a simple guideline for assessing a design, product, process, or service, in the ideation stage on sustainability by answering the following questions:

- Does the novel product, process, or service provide in a need?
- Does it not harm the environment over the life cycle?
- Does it create prosperity over the value chain?
- Does it benefit societies over the value chain?

If an answer to any of these questions is negative, then technology does not contribute to sustainable development. If some answers are positive, it is likely that the design, when further worked out, can contribute to the SDGs. Or better stated: The design can be modified to contribute to some of the development goals.

For the concept stage when more information is available and also some time can be spend on finding information, a rapid SDG evaluation can be made, by searching for relevant SDGs in each of the sections: People, Planet, Prosperity. Subsequently the most relevant targets within the SDGs should be searched for to define specified and quantified contributions to these targets.

For the development stage and for the detailed engineering stage more specific and quantified contributions to the SDGs can be made. For these stages a check can be made for all SDGs and targets to see what contributions by the design will be made, when commercialized.

Apart from this consensus on the definition of the UN SDGs there are numerous other descriptions of sustainability. These descriptions are of little use for companies and for education as they lack global consensus and have therefore little authority. In addition to this it should be remarked that in industrialized countries, the meaning of the word sustainability is often reduced to living in harmony with the environment, so the society and prosperity dimensions are not considered.

To summarize: The only powerful term and definition of sustainable development reached by global consensus is the one reached within the United Nations, and which has been accepted by the World Business Counsel of Sustainable Development (WBCSD) [4].

12.2.1 Example use of SDG in process concept design

University of Delft students, following the postgraduate EngD program, made a process concept design as part of the course sustainable design. The design is a plastic-to-methanol process in which flexible packaging plastic waste is first converted into syngas through pyrolysis and catalytic steam reforming. The syngas produced is then converted to methanol. The design's annual plant capacity is 100 kton/year and is foreseen to be located at Rotterdam Botlek industrial complex in the Netherlands.

The competing established reference case is a waste-to-energy (W2E) process, which involves burning plastics to produce thermal energy, which is then converted to electricity. This process is carried out by ReEnergy, a subsidiary of SUEZ group. This plant has an annual capacity of 291,000ton plastics/year. It is located at Roosendaal, Netherlands.

The feedstock for both processes is a mixed flexible packaging plastics waste stream, which is not separated for mechanical recycling. W2E is chosen as the reference case because it is currently used by the industry for plastics, which are not mechanically recycled.

The following three sustainable design goals (SDGs) and specific targets were chosen by the students for this project. Their choices and reasoning for these choices are given below:

SDG 12: "Ensure sustainable consumption and production patterns." This goal was chosen because the design process corresponds to plastic recycling, which is a method to reduce waste disposed. This process aids the goal of sustainable production patterns.

The target chosen was 12.5: "By 2030, substantially reduce waste generation through prevention, reduction, recycling, and reuse. Keeping in mind this target, the aim is to improve the design to process increased amounts of waste plastic."

This target 12.5 was chosen in view of reducing waste generated through prevention, reduction, recycling, and reuse. The indicator for this SDG is the amount of material recycled in a year. This could be measured by estimating the amount of plastic that is recycled or brought back into the supply chain. In the W2E reference case, plastic feed is burned to generate electricity, thereby recycling no plastic. In the design plastic is converted into methanol and is brought back into the supply chain. This type of recycling is called tertiary recycling. The design input is 100,000 ton/year of plastic feed and converts it to methanol at a conversion of 80%. This is equivalent to 40% of the plastic waste stream that goes through the W2E reference process.

SDG 9: "Build resilient infrastructure, promote inclusive and sustainable industrialization, and foster innovation."

The chosen target is 9.4: "By 2030, upgrade infrastructure and retrofit industries to make them sustainable, with increased resource-use efficiency and greater adoption of clean and environmentally sound technologies and industrial processes, with all countries taking action in accordance with their respective capabilities."

This target was chosen to ensure optimum use of resources, along with a reduction in carbon footprint. The indicator is CO_2 emission per converted unit feed. The reference process has 1.2 kg of CO_2/ kg feed, whereas the current design has 0.7 kg of CO_2/ kg feed.

Although the design already has a lower emission value compared to the W2E reference case, a target to reduce the net emissions to 0 kg CO_2/ kg feed is set to make the process design even more sustainable.

SDG 7: "Ensure access to affordable, reliable, sustainable, and modern energy for all." (The 17 Goals, n.d.) The target chosen for this goal is 7.2: "By 2030, increase substantially the share of renewable energy in the global energy mix."

This SDG 7 and its target 7.2 were chosen because they directly relate to the electricity source to be used. The design goal is to use as much electricity produced from renewable energy sources, such as solar and wind, as is feasible. Electricity is required in the process for pumps, refrigeration, and steam production. The target is that 100% of this electricity is taken from renewable sources [5].

12.3 Environmental impact evaluation by life cycle assessment

12.3.1 Introduction to environmental impact evaluation by life cycle assessment

Environmental impact assessment has increased enormously in the last 60 years. In the 60s of the last century, environmental legislation was set by individual countries and was very limited. In the Netherlands for instance, laws were implemented to reduce low level smog, by limiting low level acid gas emissions. To that end very tall factory chimneys, 200 meters and higher, were then constructed, to make sure that acid flue gases no longer caused local low layer smog in the Netherlands.

In the 70s and 80s, these tall chimney emissions then caused acid rain and thereby caused forests to die in other countries, notably in Sweden. This in turn caused stringent flue gas emission legislation in several countries in Europe, which were later adopted by the European Union. Other industrialized regions such as North America, East Asia, and South America followed suit.

Due to global acceptance of the SDGs described in detail in Section 12.2, environmental concern has now spread worldwide both for governments and companies. The commonly accepted method of determining the environmental impact of products and processes is now LCA.

LCA has been defined by Klöpferr [6] as: LCA studies the environmental aspects and potential impacts throughout a product's life (i.e., cradle-to-grave) from raw material acquisition through production, use, and disposal. The general categories of environmental impacts needing consideration include resource use, human health, and ecological consequences.

LCA is now a well-established method. A good textbook about the practice of LCA is provided by Klöpferr [6]. For reliable and accurate results ISO methods 14040 and 14044 should be applied, based on reliable databases and executed by experienced LCA experts. In general, these reliable LCAs are made for new products and are executed in the development stage. Guinee provides a good introduction to these ISO methods [7].

To establish quickly whether a new product or process is significantly better than an established product or process for the same function, these elaborate methods however are not practical for the concept stage. A full LCA may take a half year to complete and require large amounts of data, often not available at the early innovation state. A rapid LCA method is therefore needed for the ideation and concept stages to compare various concept design options on environmental impact and determine whether the new concept is significantly better than the established commercial product or process. Such a method is described below.

12.3.2 Rapid LCA method for discovery and concept innovation stages

In the discovery and concept stage very little environmental information is available or can be generated on the novel product and/or process concept idea. On the other hand, only a little information is needed as the only question to be answered is: Is the idea better than the commercial product and process with which the idea competes in view of environmental impact?

To this end, a rapid LCA method has been developed by Jonker [8]. It follows all steps of the generic LCA method but requires far less information. That it is far less accurate is not a big concern as the concept idea should be far better than the established product or process.

In my course on sustainable product and process design for the EngD program this method has been taught and applied by the students for over 10 years. It appears that groups of four students can make a rapid environmental impact assessment in less than 20 h for their new design and for the reference case. The reference case is the commercial scale established product or process. The new process or product should have a factor 2–10 lower environmental impact than the reference case. The rapid LCA is in most cases accurate enough to determine whether this is achieved with the new design. This rapid LCA method of Jonker [8] is described below.

Rapid LCA steps:

1: Define functional unit.
2: Define life cycle study goal.
3: Define life cycle scope.
4: Select one or two key environmental impact types.
5: Identify all major emission streams for each life cycle step.
6: Quantify the environmental impact.
7: Compare novel design with the reference case.
8: Conclude whether the novel design has a lower environmental impact or not.

Step 1: Define functional unit (FU).
Defining the functional unit (FU)is officially part of step 3 in LCA methods. However, for product and process designs it is very useful to first define the functional unit. As in design, one of the first steps is setting the design goal. Defining the FU is part of setting the design goal, so for this design book it is logical to put defining the FU as step 1.

The word function in design defines what should be achieved, not how it should be achieved. The FU in LCA describes what performance is needed for the product. This definition facilitates the product designer to design different products with the same FU and compare the alternative designs on environmental impact and chose the best alternative. It also facilitates a comparison with an established commercial product or process.

The FU of the new design should be the same as the reference case. The reference case is the established commercial product with the same FU. Only by having the same FU definition for both the new design and the reference case, a fair comparison on environmental impact of the new design and the reference case can be made.

Defining an FU can be simple. If for instance a company wants to change their throwaway plastic coffee cups for a design with a lower environmental impact over its life cycle, then the FU is: a throwaway coffee cup, which holds 100 ml of warm coffee. Various alternative materials can then be evaluated for that functional unit. However, alternatives where the coffee cup is reused are then left out. If those reuse options are included then the FU may be defined as a reusable coffee cup holding 100 ml of warm coffee and cleanable by using a cold water rinse. Thus, the definition of the FU needs consideration.

Defining the FU, however, can also be complicated especially for consumer products, where the performance of the product matters and not the absolute amount. Take for instance, shampoo. One needs less of an active shampoo for a hair wash than a weak shampoo. Also, for a hair wash, the size of the hair matters. Consumer product manufacturers therefore often have their own standardized methods for measuring the amount needed for a hair wash.

If the novel product design replaces several existing products, then the FU should be defined in such a way that it is clear how much of each product of the reference case makes up the FU, so that a fair comparison is made.

The FU for a process is simply the product. If, however, the process produces two products then the FU should be defined for the two products. If for instance the two products are styrene and propylene oxide, the FU could be defined as 1 mole styrene and 1 mole propylene oxide.

If it is desired to know the individual environmental impact of each of the two products, then the emissions have to be allocated in some way to each of the two products. This is the so-called allocation problem. There is an endless discussion going on for centuries amongst economists starting with Stuart Mill (1848) on how to make a fair allocation [6]. For engineers, my advice is to simply use the mass ratio of the two products and allocate the emissions based on this ratio. If you use a different rationale, then explain the rationale in detail. However, spend little time in defending the rationale, as then you probably end up in the discussion started in 1848 and still going on.

Step 2: Define life cycle assessment goal
For any LCA, the goal of the LCA has to be set. For concept design of a product or process, the goal of the LCA will be in most cases an environmental impact assessment of the design (or alternative designs) in comparison with the established commercial product or process in such a way that it can result in a conclusion that the new design has a higher or lower environmental impact compared to the established commercial product or process.

Step 3: Define life cycle scope
The scope definition contains the main life cycle boundaries. These boundaries can be chosen with the help of Fig. 12.1.

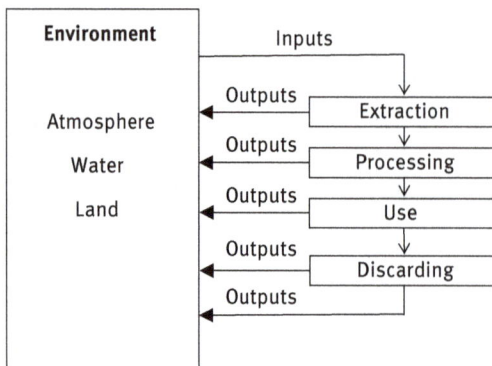

Fig. 12.1: Life cycle assessment system elements.

In general, the first life cycle step is to be extracting material from the environment (cradle), which after transformations enters the product. The last step is often the disposal of the product after its use (grave), or its upgrade step from waste to a reused product.

Many more life cycle steps can be considered; for instance, the steps leading to the process equipment for mining and further processing. For product and process concept design these steps can be neglected, as for most cases these contributions to the environmental emissions are small compared to the emissions in the main life cycle string.

Transportations in between some main life cycle steps can be included in the life cycle. This is relevant if carbon dioxide emissions of transportations are significant compared to other carbon dioxide emissions. For bulk products transported by ship this contribution to the overall carbon dioxide will be small and can be neglected in the concept stage of the design.

Life cycles are best represented by block flow diagrams in which each block represents a process step, and each arrow represents a mass flow. Standard methods for mass balance checks can be applied to ensure the quality of the life cycle.

Pitfall in defining the scope

Often designers chose the product factory gate as the limit to the LCA scope. They call this a cradle-to-gate LCA. They often do this because accurate data of a new product for its use and disposal is not available in databases. This means first of all that by definition no LCA is executed. But secondly and more importantly, the results of such an LCA can lead to totally wrong conclusions regarding the environmental impact of the new design compared to the established reference product.

An example of a wrong conclusion based on would-be cradle-to-gate LCA is an aircraft manufacturer who develops a novel construction material for aircrafts. The new construction material has a far lower global warming environmental impact in its materials manufacturing step. The aircraft manufacturer concludes from a cradle-to-gate "LCA" that the aircraft with the new material will have a lower environmental impact over its life cycle than the present construction material.

However, the new construction material applied for the aircraft is heavier than the existing material. This means that the in the use step of the aircraft, flying, the aircraft needs more fuel per mile and the emitted carbon dioxide causes a much higher global warming gas emission over its life cycle and outweighs the reduction obtained in the materials manufacturing by far. This can be easily understood based on the statistics that aircrafts last at least 30 years. They spend 75% of their time in the air. In a single Transatlantic flight about 100 tons of fuel is burned, which is about the same mass as the airplane construction material. This means that in its lifetime it emits a factor $30 \times 1{,}000$ of carbon in the form of carbon dioxide more than its own mass. Typical carbon dioxide emissions during material manufacturing are a factor 5 more than the construction material mass. In conclusion it is far better to be inaccurately right than to be accurately very wrong.

In many cases, global warming gas emissions occur during the use of a product, compared to all other life cycle steps. Cradle-to-gate analysis for global warming gas emissions are therefore of little use and conclusions can be very wrong. A reasonable estimate using conservative reasoning with some statistical information to get a cradle-to-grave environmental impact assessment will in most cases be of use.

Step 4: Select one or two key environmental impact types

Table. 12.2 shows all environmental impact types of an LCA. For a concept design it is not necessary and also not desired to do a full LCA for all impact types. A literature search on the reference product or process for environmental impact can be executed to find the impact types of most concern. If no information is available about the ref-

Tab. 12.2: Environmental impact types for life cycle assessment [6].

Atmosphere	Water	Biotic	Soil
Global warming	Eco-toxicity	Eco-toxicity	Eco-toxicity
Ozone layer depletion	Nitrification	Habitat reduction	Physical degradation
Photochemical pollutants	Heat	Biotic resource reduction	Dehydration
Acidification		Physical ecosystem reduction	Resource depletion
Radiation			
Noise			
Smell			
Visual landscape			

erence product or process, then global warming may be chosen as the main impact type, as global warming is a concern for most products and processes.

Step 5: Identify all major emission streams for each life cycle step
The best way to identify all major inputs from and emissions to the environment is to execute a literature search on the life cycle of reference product. If that is not available, then search for an LCA of a product made from the reference product. That LCA will then also give the major emissions of the reference product.

For the new product no LCA data may be available. However, for the input material to make the new product LCA information may be available. LCA software packages can also be used to produce the major emission streams.

Step 6: Quantify the emission streams and the resulting environmental impacts
The emissions to the environment should be quantified. The emissions should also be related to the function unit dimension chosen. If the function is for instance expressed as 1 kg, then the emissions should be quantified relative to the functional unit. Often the emissions can then be expressed as a dimensionless unit, for instance, in the case of carbon dioxide emissions as kg CO_2/kg product. As soon as the production capacity of the product is defined, the total emissions for each emission type can also be defined.

Quality check on emission calculations
The quality of the emissions calculated can be quickly evaluated by a simple atom balance over the process input and output streams. Any atom that enters the process will leave the process. Often when a crude input such as biomass is used as input to the process the concept designer neglects or forgets that the biomass contains atoms such as potassium and sulfur will leave the process in one of the outlet streams. A simple atom balance analysis then reveals that the output streams are not sufficiently defined and that the envisaged output destination of some output streams has become infeasible.

Pitfall in emission dimensions

A pitfall in emissions is the wrong dimension obtained from some publications. Economist and also policymakers often express emission amounts in absolute mass dimension (ton), while they mean the amount per year (ton/year). Hence, check what is meant and also do not follow this habit of expressing the mass flow in mass. Sooner or later, errors will be made further down the road while using the report.

Step 7: Compare novel design with the reference case

If more than one impact type is used in the comparison between the new design and the reference case, and for one impact type the new design has a lower impact and for the other impact type it has a higher impact, then a conclusion on the new design cannot be simply drawn.

In LCA literature this problem is called valuation [6]. The problem is solved in textbooks on LCA by using weight factors for each impact type, so that the results of the two impact types can be added up to a single number.

However, the weight factors are subjective and depend on the world view of the person or group of persons that make the weight factors. Also, within a company, the discussion on whether the new design is better or not on environmental impact will often be endless.

It is far better to reconsider the new design and modify it in such a way that it will have lower impacts for both impact types. Such a design is then robust to different opinions within and outside the company and to the opinions of various stakeholders on the relative importance of impact types. Also, for marketing purposes the novel product will then be conceived as far better.

A helpful method for a rapid evaluation in the concept stage is to fill in template Tab. 12.3 for all environmental impacts for which data are available. For impacts with no data "no data yet" is filled in. For some of those impacts it may be possible to reason that the impact for the modern design will be lower than the reference case. Then "lower" can be filled in in the 4th column.

Tab 12.3: Template environmental impact ratio: innovative design versus reference case.

Impact type	Reference impact	Innovative design impact	Impact Ratio new/reference

The numbers of the last column are used for the evaluation. If all ratios are less than 1 then the environmental evaluation has a positive outcome. If for some impacts the ratio is less than 1, but for at least one impact the ratio is higher, then the outcome is negative. It has resemblance to a Pareto improvement.

Step 8: Conclude whether the novel design has a lower environmental impact or not

Finally, a conclusion should be drawn on the environmental impacts of the new design compared to the reference case. As the rapid LCA method is very inaccurate, the calculated emissions can be 50% lower or 100% higher. Firm conclusions can only be drawn if the design has at least a factor 4 lower emission to the environment.

12.3.3 Environmental evaluation feasibility and development stages

In the feasibility and development stages a full LCA is the best method to fully assess environmental impacts associated with the new product. Now a reliable and accurate LCA will be needed. It is beyond the scope of this book to treat this subject in detail. The reader is referred to

LCA methods as part of the ISO 14000 environmental management standards [7].

A complete LCA of the novel product should then be determined using the certified well-defined LCA method ISO 14040 and 14044 [7]. These are the leading standards for LCA. In addition to applying the LCA method properly, reliable LCA databases are needed to get a reliable LCA of the novel product and the reference case. These databases may be purchased from institutes specialized in LCA data gathering and processing.

The database can be based on measured data for the case at hand (called foreground data) or based on averaged data, holding for a certain country or region. Then, choosing the relevant country is very important.

For a detailed description and application, the reader is referred to LCA handbooks such as by Klöpferr [6]. A nice overview of all LCA methods and ranking, the reader is referred to a booklet generated by a European commission [9].

In addition to the full LCA, also environmental laws and regulations of the countries where the product will be marketed, where the process will be located, and the countries where the feedstock will be obtained, need to be considered. Often a public environmental impact report must be made to facilitate local civilians' objections to certain emissions or risks of emissions.

12.3.3.1 Fugitive emissions evaluation process
EPA provides the amount of fugitive emissions in kg/h for most unit operations and process equipment in combination with a simple classification of streams as gas, light liquid, or heavy liquid directly [10]. An estimate of fugitive emissions from flanges for a process design can also be made with an EPA method [10].

In a full LCA, the emissions are converted into environmental impacts. Global warming emissions are converted to global warming potential impact using a specified characterization factor. For carbon dioxide the factor is set at 1 and, for instance, for methane emissions it is set at 25 [6]. For concept design it may be decided to limit the assessment to the quantified emissions and avoid the conversion to the environmental impact. It facilitates reporting and communications about the LCA results.

12.3.3.2 A remark on scope 1, 2, and 3 environmental emissions

Apart from LCA methods to evaluate environmental impact analysis for products and process over the entire life cycle, companies also are also analyzing their whole portfolio of activities on environmental impact. However, in most cases the analysis is limited to their own process outlet (cradle to-gate), or even limited to emissions from their own process (gate-to-gate).

For greenhouse gases (GHG) a systematic method is available for these limited scopes. The method uses three scopes. Scope 1 relates to direct GHG emissions from the company processes. Scope 2 relates in addition to scope 1 to emissions by imported utilities. Scope 3 relates in addition to all imported goods and services. These organizational GHG impact analyses are not LCAs of specific products or processes and are therefore not part of this book.

12.3.3.3 Example case LCA: new infant milk formula product and package

The example case is an extract from a student report [11].

Product case design

Infant milk formula is a product, which has been in the market for decades now. Mothers all over the world, who cannot feed their babies themselves, rely on infant formula milk. However, this infant formula, when stored improperly, is susceptible to biological contamination. Ingestion of such a contaminated mixture can lead to diseases, raise malnourished children, or can prove fatal in some cases. In this context, the design of a novel product (infant formula powder plus its packaging) that is safer for babies to consume in India, was undertaken.

The design was divided into two items, formula design and packaging design. The overall design had five key design elements:
1) addition of garlic derivatives to infant powder
2) addition of FDA approved acids to infant powder
3) implementation of additional sterilization technique: Pulsed Electric field (PEF) pasteurization in the production line
4) tableting of the product formula
5) multilayer packaging container.

These design elements focus on killing and mitigating the growth of bacteria during the shelf life of the product.

Rapid LCA new design compared to reference cases

The rapid LCA method of Jonker as described above was applied.

Two main impact types were selected:
–   global warming (carbon dioxide emissions)
–   freshwater depletion

The LCA was undertaken for the two design sections, infant powder design and packaging design, separately.

For the liquid infant formula design the FU was defined as one serving of liquid infant formula. As reference case for infant powder a process patented by Nestlé was taken.

For both the design and the reference case the FU (one serving of liquid infant formula) amounted to 30 g of infant powder.

For the multilayer packaging design one package was taken as the functional unit.

Tetra Pak packaging was taken as reference case for packaging, also one package as the FU.

Subsequently, it was calculated that 2.4 g of 2 Tetra Pak packaging was required for the FU, compared to 3 g packaging required by the new design. The new design also required an additional 5.4 g of material for the tableting of the powder, which was absent in the reference case.

The LCA scope boundaries for the calculation were defined and comprised of five steps namely: mining, transportation, processing, use, and end-of-life.

For the global warming impact (reduced to carbon dioxide emissions) it was calculated that the designed product has a higher carbon footprint impact (+ 201 g CO2 eq per FU) than the reference case.

For the water depletion impact, the infant powder had a lower impact (−27 kg H_2O per FU) compared to the Nestlé powder reference case, while the new packaging design had a higher impact (+137 kg H_2O per FU) than the Tetra Pak reference case. This high environmental impact for the design was ascribed to the presence of an extra tableting processing step in the packaging line, which was absent in the reference case. For the overall water depletion impact, the product design has higher water depletion impact than the reference case.

12.4 Evaluating for social acceptance

12.4.1 Social acceptance concept stage

Social acceptance of novel products and processes is enormously important. It can be a showstopper. Think for instance about nuclear power plants in Germany, which are no longer accepted by the government, and will be phased out.

For a new process, a license to operate for decades is therefore essential. That license is not only governed by law but also by social acceptance. For consumer products in a free market with competition, social acceptance is also the first criterion for success. Social acceptance is related to the beliefs and culture of societies in which the product and process will be located. When, for instance, animal material is used in a product then sales in certain societies will not be successful.

In the concept stage, an estimate of social acceptance of the new product or process can be made with the help of checklists. Such a checklist is provided in Tab. 12.4. The first column, modal aspects (of artefacts) is obtained from Verkerk [12].). The second column has been filled in using information from Potter [13] and de Soto [14], and authors' own observations. A brief description for each modal aspect related to novel product acceptance follows thereafter.

Tab. 12.4: Checklist for social and market acceptance of novel product and process (derived from Verkerk [12], Potter [13], and de Soto [14]).

Modal aspect	Essence social and market acceptance
Beliefs	The innovation is acceptable to believers.
Ethical	The innovation agrees with moral values.
Juridical	The innovation agrees with laws.
Aesthetic	The innovation is appealing.
Economic	The innovation price fits its functions (value for money).
Social	The innovation fits in the social context.
Linguistic	Symbols and language provide the correct perception of the innovation.
Historic	The history of the innovation and company is attractive to customers.
Logical	The logic of the innovation is clear.
Sensitive	Senses (smell, taste, hearing, sight, and feeling) and emotions associated to the product and process are appealing.

Beliefs of a society can have a strong influence on social acceptance of the novel product and or process. This does hold for societies with a named governing belief, such as, Hindu (no animal-derived feeds), Muslim (no ethanol), and humanism (no human rights violations).

Ethical and moral aspects of a product over its value chain are often very important for social acceptance. Think for instance of child labor associated with production of an intermediate in the value chain.

Juridical aspects such as laws and regulations can make the product unacceptable, for instance, because the product (or its manufacturing process) violates laws related to safety and health. Also, the laws on property and on land ownership regulations are very important [14].

Aesthetic aspects such as the shape of the product and its packaging) can make or break a consumer product.

The economics of a novel product are often very important for acceptance. The product price should fit the product use and expectation.

Society aspects of the novel product should be such that it fits with the local society and enhances its welfare. Contributing to social SDGs will help in this respect. Also, the market type (centrally governed, free market, or regulated market, informal market, presence of workers unions) are relevant for introducing a novel product or process. Also, the government structure, i.e., democracy or dictatorship, is of importance for introducing novel products and processes.

Linguistic aspects are particularly important for new consumer products. The product is often not only seen as an artefact fulfilling a function, but also as providing meaning to the person and his/her social relations. For example, a novel nail polish may send a message to the person who sees it on the nails of the lady.

Historic aspects can be very important. When the new product is intended to replace an existing product, while the history related to the reference product has such strong ties in society, then the new product is not accepted.

Logical aspects of the new product and process should be applied. If the novelty makes no sense, the new idea will not survive.

Sensitivity is an enormously important aspect of consumer products. The impression it has on the human senses will make the product attractive or unacceptable.

12.4.2 Social acceptance at the end of the development stage

Social acceptance has to be determined by consulting the people concerned with the new product or process. For new products, companies have their own methods of determining market and society acceptance. It is beyond the scope of this book to describe these methods.

For the new process acceptance, Gary Scholes provides a practical method called red flag. In the local community where the new process is envisaged, civilians reflecting social and environmental interests form a panel and are invited to determine the acceptance of the new process. In a session the new process is explained in layman terms, part by part. If a panel member has a concern about a certain part, he raises a red flag and expresses his concern. The concern may be removed by further explaining the process part or by changing the design [15].

Case example application social acceptance ethylene glycol plant

The red flag method was applied by Shell Chemicals Canada for a new ethylene glycol plant. Apart from the products, the plant design also releases a stream of carbon dioxide to the atmosphere, which is a global warming gas. A civilians' panel did not accept this emission. One civilian knew farmers who burned natural gas to provide carbon dioxide to their greenhouses to grow tomatoes. Carbon dioxide availability was ingredient limitation for tomato growth. The civilian proposed that carbon dioxide should be provided to the farmers. After investigations it appeared that traces of ethylene in the carbon dioxide stream had to be removed before the stream could be used for tomato growth; ethylene is a plant hormone that enhances ripening. This idea was implemented in the process design so that burning natural gas and emission of carbon dioxide was avoided [15].

12.5 Exercises

Given: chewing gum that does not foul streets

Chewing gum cleans teeth and thereby can contribute to human health. However, a lot of spent chewing gum ends up after its use on streets and pavements, which is an ugly sight. Most gums finally end up in rivers and oceans. Since the gum base is polypropylene it does hardly breaks down and contributes to ocean fouling.

A concept design of a new chewing gum that does not foul streets is made by students. The gum base is Polyvinyl alcohol (PVA). It furthermore contains menthol for taste. The PVA depolymerizes into water-soluble molecules within days or weeks and then it is further broken down by microorganisms into water and carbon dioxide. Thus, on open pavements it is washed away by rain. In closed rooms it can be easily removed by water [16].

Task 1

Determine the contributions of the new chewing gum to the SDGs for People, Planet, and Prosperity by identifying relevant specific goals and targets.

Task 2

Perform a rapid LCA for the new chewing gum and for the reference case, the present chewing gum, and conclude on the environmental impact of the new chewing gum relative to the reference case.

Task 3
Determine the social acceptance of the new chewing gum.

References and further reading

[1]   World Commission on Environment and Development, Our Common Future, (Commonly Referred to as the Brundtland Report), Oxford University press 1987.
[2]   Elkington J, Rowlands IH. Cannibals with forks: The triple bottom line of 21st century business. Alternatives Journal, 25(4), 42 1999.
[3]   Grayson D, Nelson J. Corporate Responsibility Coalitions: The Past, Present, and Future of Alliances for Sustainable Capitalism. Stanford University Press, 2013.
[4]   UN. Transforming our world: The 2030 Agenda for Sustainable Development, Resolution adopted by the General Assembly on 25 September 2015. Resourced from: http://www.un.org/sustainabledevelopment/sustainable-development-goals/.
[5]   Bhatraju C, Jenkins JJ, Foglia J, Padmanabhan P. Waste Plastics to Methanol, Process Concept Design Report for Course Sustainable Design, TU Delft, 2022.
[6]   Klöpffer W, Grahl B. Life Cycle Assessment (LCA): A Guide to Best Practice. John Wiley & Sons, 2014. Also contains a quote from Stuart Mill in 1848.
[7]   Guinee JB et.al. Handbook on Life Cycle Assessment. Operational Guide to the ISO Standards. Dordrecht: Kluwer Academic Publishers, 2002.
[8]   Jonker G, Harmsen J. Engineering for Sustainability: A Practical Guide for Sustainable Design, Amsterdam: Elsevier, 2012.
[9]   General Guide to LCA. 2010. Resources from:http://publications.jrc.ec.europa.eu/repository/bitstream/JRC48157/ilcd_handbook-general_guide_for_lca-detailed_guidance_12march2010_isbn_fin.pdf.
[10] Protocol for Equipment Leak. Emission Estimates. Emission Standards, EPA-453/R-95-017. 1995. Resourced from: https://www3.epa.gov/ttnchie1/efdocs/equiplks.pdf as sourced 25-1-2017.
[11] Diaz CC, Joglekar C, Liu FH, Nagra H, Xiao Y. Sustainability assessment of infant milk formula product, Sustainable Design, student report 7th of July 2020, TU Delft EngD course.
[12] Verkerk MJ et.al. Philosophy of Technology, New York: Routledge, 2017.
[13] Potter D. Dimensions of Society, Sevenoaks: Hodder and Stoughton, 2nd edn, 1978.
[14] De Soto H. The Mystery of Capital, London: Bantam Press, 2000.
[15] Scholes G. Integrating sustainable development into the shell chemicals business. In: Weijnen MPC, Herder PM (eds). Environmental Performance Indicators in Process Design and Operation, Oral Presentation, and Reported in, EFCE Event 616, Delft University Press, 1999, 69–82
[16] Harmsen GJ. Chewing Gum that Does Not Foul the Streets, Oral Presentation, and Abstract Proceedings ECCE 2007, Copenhagen, September 2007, Conf CD ISBN 978-87-91435-57-9.

13 Communicating

13.1 Communicating: project team and stakeholders

A design project comprises many activities carried out by the team members and in interaction with stakeholders. This was made clear in Chapter 4 and also becomes clear from all other chapters. In this chapter, we want to focus not only on the reporting of the (final) design results, but want to also emphasize that "communicating" is far more than completing the final project documentation.

To support the communication during the project with the project team members and with the project's stakeholders, a number of communication tools and tips will be discussed. From experience, it is clear that in order to arrive at the best design solution within the resources (time, tools, and designers) made available, communicating about the design goal, approach, intermediate results, evaluation and decision-making, and argumentation with the team and with the stakeholders are crucial activities. Only by sharing this information in an effective and efficient way, reflecting, and making improvements based on team members' and stakeholders' feedback, design improvement is possible. Based on our experience, project teams may spend up to 40% in communication activities!

In this chapter, we will link a number of key communication tools to the Delft Design Map (DDM) (see Section 4.3).

13.2 Communicating using the Delft Design Map (DDM)

In the DDM (see Section 4.3), a "reporting" step is included in all design levels (see Table 13.1). This is done for a special purpose, and links to this Chapter's introduction.

By introducing the "reporting" step, we want to emphasize the need for capturing the goal, approach, results, decisions, and next steps in a series of design-level reports. It invites the project team to collect and report their findings at least at each design level.

These reports also form the basis for project review meetings, for, e.g., Stage-Gate^{TM}, and decisions about the project's next steps.

As explained in Section 8.1, "designing" is "taking decisions" about many different aspects and at many different levels to arrive at the optimal design. For reporting these design results, design models (schematic, mathematical, verbal, and sometimes even physical/iconic models) may be used – the design decisions that were made and the argumentation why these were made is crucial.

Many design projects will not make it to full implementation and will be stopped in the Stage-Gate^{TM} process because of various business, technical, environmental reasons. However, the knowledge gained from these projects is extremely valuable for

https://doi.org/10.1515/9783110782127-013

Tab. 13.1: Delft Design Map (DDM) design levels and design steps, including the "reporting" step.

Stage-Gate™:		Concept			Feasibility			Development			
AR: activity report(s) DLR: design-level report	Framing	Supply chain imbedding			Process technology			Process engineering			Final
Design levels ⇒ Design steps/ activities ⇓	(PF)	(CW–PC)	(PC–PF)	(I-O)	(SP)	(TN)	(UN)	(PI)	(ED)	(OI)	(FO)	(Final)
1 Scope	AR	AR	AR	AR	AR	AR	AR	AR	AR	AR	AR	AR
2 Knowledge	AR	AR	AR	AR	AR	AR	AR	AR	AR	AR	AR	AR
3 Synthesize	AR	AR	AR	AR	AR	AR	AR	AR	AR	AR	AR	AR
4 Analyze	AR	AR	AR	AR	AR	AR	AR	AR	AR	AR	AR	AR
5 Evaluate and select	AR	AR	AR	AR	AR	AR	AR	AR	AR	AR	AR	AR
6 Report	DLR	DLR	DLR	DLR	DLR	DLR	DLR	DLR	DLR	DLR	DLR	DLR

AR, activity report(s); DLR, design-level report.

the company's decisions at that time and in the future. A well-documented design project that includes what decisions were made, why they were made, learnings, and future outlooks is highly valuable. Business conditions can change and so these design results and learnings can be brought to life again later. With many team members having moved on to other parts in the company (or to other companies), it is only the documentation that remains. With this documentation, a new design team can have a head start in tackling this project, or in the absence of this historical documentation, can be outperformed by competitors.

It should be stressed that reporting the design activities only at the end of each design level is far from sufficient in the communication between the team members and with the stakeholders. These reports are a collection of information and decisions from many more activities. In fact, in each of the design steps, design activities that take place at each level should be communicated.

A good working habit for a design team is to use a mix of reporting and communicating tools. These can be used by each team member and by the project team in order to support communication.

In the next paragraphs, the communication tools that are in use by MSc students and EngD trainees in the TU Delft Design courses will be discussed:
– activity reports
– meetings: agenda and minutes of meeting (MOM)
– models (including brain writing and brain drawings).
– presentations
– reports

13.3 Activity reports

Well organized and written project documentation is not only valuable at the end of each design level but also at the completion of the entire project. *During* the project, reports and other documentation are needed even more.

After project planning meetings (see Section 13.4), team members start executing their (individual or team) design activities. Reporting the design activities in "activity reports" assists the designer(s) to first formulate the objective and the deliverables of the/their activity. Then, he/she/they can choose and report the approach and the resources needed. Finally, results, discussions, and learnings from the activity can be reported. Typically, activities that take a couple of hours to a day will be concluded with a concise activity report. A format for such a design activity report is shown in Appendix A13.1.

The "activity reports" greatly assist the designer to focus on the objectives of the activities, in relation to the overall design. Reporting the results forces the designer to think over these and gives time for alternative options or argumentation to surface. It also avoids executing too many and possibly unnecessary activities, and thereby losing valuable time. A proven guideline for designers to consider is: *"if a design activity is not worth reporting, it is not worth executing."*. The creation and use of "schematic models" like figures, drawings, and tables is highly recommended. The proper use of these models cannot be stressed enough. Schematics and drawings take time to prepare, but are far more effective in communicating design results and learnings than text. It is very tempting to create schematic models that are very close/similar to information found in other sources. It is much more productive to construct these schematic models to what is required for *your design*, always focusing on your target(s)! Some examples can be mentioned here:

1. House of quality (HoQ) diagrams (see Sections 4.3.3 and 8.3).
2. Stream specifications Table (streams crossing battery limit, see Table A13.1).
3. Concepts/criteria decision matrix or Pugh matrix [1] (see Table A13.2).
    a. *Do* list your requirements first and indicate per design option or technology, how each option fits in meeting these requirements.
    b. *Do not* make a long list of available technologies or of design options and compare their strengths/weaknesses (in general). Link it directly to your project's needs.
4. Pure component properties table (see Table A13.3):
    a. *Do* make a list of components properties that are essential/useful for your design and fill out what is easily retrievable and what is not, or not yet.
    b. *Do not* make a list of pure components with only properties that are easily retrievable.

Activity reports are not only valuable for the team member executing the activity; they are equally valuable for other team members, as they can keep themselves abreast with other activities than their own and they can do this at their convenience. Reading is much faster than listening to a team member's presentations in team meetings. Team members can come well-prepared for team meetings, and these meetings can be used for taking decisions on the design direction and future planning. This greatly improves information exchange and knowledge sharing.

The storage of the activity reports can be very well organized using the DDM's design-level/design-step coding: reports can be coded from PC–PF/scope, ..., all the way to Final/report. At each design level, the "report" design step can be used to consolidate the various activity reports, e.g., the design-level report: "PC–PF/report" (see Table 4.3).

Reporting/documentation when done with the "target setting, knowledge delivering approach," as explained above, communicates a clear result-oriented mindset in the project. In the DDM, the target-setting approach becomes already very apparent in the first design-level "project framing." Before starting the actual design work, the overall project goals, objectives, deliverables, etc. are determined and "framed."

Reporting and documenting the results of this design level gives an "intermediate result" that can always be consulted by the team members. From here on all team members will adopt the approach of reporting while designing, so that design activities results become available to the team in a continuous workflow.

Last but not least, it is a very good practice to decide very early on in the project on the reporting tools/software to use. This ensures that all stakeholders (also the ones outside the organization!) are able to open the files, read the report(s), and give (digital) feedback. Of course, this is equally important for mathematical modeling and simulation software, spreadsheets, and presentation files.

13.4 Meetings: agenda and minutes of meeting (MOM)

13.4.1 Standard agenda, MOM structure (internal and external)

During a design project, multitude of meetings are organized. The design team will schedule (internal) progress meetings for the team, and (external) meetings with stakeholders. Reports of these meetings (MOM) are special versions of activities reports. In order to set the objective(s) and duration of the meeting, time planning is essential, as is the preparation of an agenda. Also, meeting rooms and other facilities (audio/video equipment for presentations and access for remote audiences) need to be arranged. The devil is in the detail as each hiccup in the availability or proper functioning of these items will interfere with the meeting's effectiveness. Meticulous care should be put in these preparations, especially for the stakeholder meetings, and nothing should be taken for granted.

The preparation and recording can be enhanced by using standard templates for agenda and MOM (see Section A13.5).

13.5 Models

Models play a very important role in design, as explained in Section 8.1. The verbal, schematic, mathematical, and physical models form the backbone of the design results and decisions. Preparation of the schematic, mathematical, and physical models and communication thereof is an important task. These models should be prepared and explained in such a way that they can be understood by stakeholders without any additional extensive text. Designers need to give this aspect sufficient thought and attention, as misunderstanding about insufficient clear models will lead to confusion, mistakes, delays, and frustration. Well-prepared models all together represent the final design: a number of mutually consistent representations of the designed product and/or process: the objective of the design project after all.

Many models have been discussed in previous chapters. Some further examples are:

- mind maps,
- brainwriting or braindrawing maps,
- fishbone or Ishikawa diagrams,
- process block scheme (PBS) (Figure 8.1),
- process flow sheet (PFS) (Figure 8.2), and
- process stream summary (PSS) (Section A13.8) [2]..

13.6 Presentations

During the course of the design project, the team not only prepares and delivers reports, but also presentations. These presentations will be held at different phases of the project, and for various audiences/stakeholders. The design team should be aware that their presentations should always be prepared with their audiences in mind. Also, the limitations of presentations and their effectiveness in conveying information, knowledge, and message to the audience should be known to the design team members.

The goal of the presentation to the audience should be clear. During design projects, presentations at key project milestone meetings are to inform the audience about the project's progress, design decisions made, proposals on how to continue the project, and to seek stakeholders' approval and feedback.

The main "story line" can be supported (but not overshadowed) by supportive information. The relationship between design results, decisions, and the overall project context and objectives should always stand out. Supportive information, mostly on design options analysis and evaluation, comes second: it is needed but not at the expense of omitting the main story line.

Discussions and feedback of the stakeholders should be promoted, as this information is crucial for the next phases. This implies that the information presented and discussed should be easily understood by the audience, and thus the exchange of ideas is realized. Recording the discussions and feedback in MOM (Sections 13.4 and A13.4) is the main deliverable of most presentations/meetings. The next section provides further guidelines for preparing and delivering presentations.

13.6.1 Quality checks: FOOFI list for presentations

At regular intervals, formal presentations are given to the design project's stakeholders. The preparation and delivery of these presentation should not be underestimated. The structure, content, and layout of the slides need particular attention. Audiences have a limited concentration span. The presentation's content should be very clear, and the audience should be focused on the main "story line" and message. Stakeholders need to be taken from a helicopter view of the project's objective to the results, conclusions, and recommendations. This main story line should be clear. Detailed information is only supportive.

Over the years, a quality checklist for design project presentations has been composed by PLJ Swinkels, who supervises and coaches MSc and EngD trainee design project teams. This list has been named FOOFI, which translates to "frequently occurring opportunities for improvements." "FOOFI" is the phonetic equivalent of (the Amsterdam) slang word for the Dutch *"foefje,"* meaning "knack or agility."

This FOOFI list for presentations is presented in Section A13.6. A similar list is also available for reporting quality checks (see Sections 13.7.3 and A13.8).

13.7 Reporting in stage-gate reviews

In the following sections, the contents of the consolidated reports for the "concept and feasibility stages" are presented.

13.7.1 Concept stage

For the "concept" stage gate, a consolidated report is prepared for the concept "stage-gate" review. This consolidated report will be composed of selected contents from the consolidated reports at each of the design levels:

- framing,
- consumer wants/product concept,
- product concept/property function, and
- input/output and sub processes.

Often, this document is also called the "concept stage report," or "basis of design (BOD)" report. It is important that, at this stage, all key data for the design, as agreed by the design team and client/stakeholders, are tabulated and provided with background information.

As can been seen from the previous chapters, the first design levels concern the consumer needs, requirements and specifications, possible product compositions and structures, and possible processing technologies using various feed streams, etc.

Not only the specifications of the product, but also of (saleable) byproducts, commercially available feed streams and utilities, and last but not least, any environmental limits on gaseous, liquid and solid waste streams should be clearly specified (for an example, see Table A13.1). At TU Delft, for MSc and EngD level design projects, a report structure for this report is suggested, and standardized tables and formats are proposed (see Tab. 13.2 for product-focused design and Tab. 13.3 for a process-focused design. Further information on this topic can also be found in [3].

Tab. 13.2: Suggested concept stage report structure and contents: product design.

Section (DDM levels: Tab. 4.3)	Contents	Topics
1 PF	Description of the design; project incentive (opportunity assessment)	Project brief/charter: background; framing (BOSCARD – background, objectives, scope, constraints, assumptions, risks, and deliverables); stakeholder needs; design type and driver (market pull, technology push, innovation map) Market assessment (customers, markets, competitors, "SHEETS" (safety, health, environment, economic, technically feasible, and social criteria)-based opportunity definition); market segmentation; value chain analysis
2 PF	Design approach	Design challenges, design methodology, creativity methods, and group methods and tools (SWOT 0 strengths, weaknesses, opportunities, and threats); project and time planning
3 CW–PC	Consumer needs and requirements; project opportunity assessment	Stakeholder/consumer needs (voice of customer); product quality requirements, in relation to product application and entire life cycle
4 PC–PF	Product requirements	House of quality (HoQ) (composed by quality function deployment (QFD), relating consumer needs, product quality performance, product property function and manufacturing requirements

Tab. 13.2 (continued)

Section (DDM levels: Tab. 4.3)	Contents	Topics
5 PF; CW–PC; PC–PF	Database creation	Literature and patent search; product application process conditions and knowledge; chemical components; materials; pure component properties (physical, chemical, SHE (safety health, and environmental criteria), price); formulation and physical/chemical models and data; thermodynamic models and data; stoichiometry
6 CW–PC; PC–PF	Preliminary product concepts synthesis	Based on the product application process' understanding and functional requirements, generate product concepts – both top-down idea generation, and bottom-up, based on molecular identity, product (nano) and microstructures, combined into a product/application process combination
7 CW–PC; PC–PF; I-O; SP; TN; UN	Product concepts and preliminary process synthesis	Product concept/criteria evaluation/selection; chosen superior product concept (components, structure, physical/chemical product properties, in use performance). Based on product structuring/manufacturing technology choices, stoichiometry, yields and split factors, determine: plant capacity and location; battery limit; available utilities; other design determining assumptions; *quantitative* input/output diagrams; (sub)process block diagrams; stream summaries
8 I-O; SP	Preliminary business case (product and process design) "SHEETS" analysis	Product: performance versus requirements in use; product and process: performance vs. "SHEETS": safety ("Bowtie"), health, environment, economy (gross margin), technology, social analysis using concept stage knowledge
9 PF	Planning review and update	Review project and time planning, and use lessons learned to improve future project and time plan; work allocation between team members
10 PF	Creativity and group process tools	Review the team's creativity and group performance. Plan any improvements for the remainder of the project.
11 PF to UN	Conclusions and recommendations	At completion of the concept phase and feed forward to the subsequent phases
	List of symbols	
	Literature	
	Appendices	

Tab. 13.3: Suggested concept stage report structure and contents: process design.

Section (DDM levels: Tab. 4.3)	Contents	Topics
1 PF	Description of the design; project incentive (opportunity assessment)	Project brief/charter: project background; framing (BOSCARD); stakeholder needs; customer requirements; technical requirements Market assessment (customers, markets, competitors; "SHEETS"-based opportunity definition
2 PF	Design approach	Design challenges, design methodology, creativity methods, group methods and tools (SWOT); project and time planning
3 PF; CW–PC; PC–PF; I-O; SP	Database creation	Literature and patent search; chemical components; materials; pure component properties (physical, chemical, SHE, price) thermodynamics; stoichiometry
4 I-O; SP; TN; UN	Preliminary process concepts synthesis	Based on technology choice, stoichiometry, reaction yields and separation split factors, determine: Plant capacity and location; battery limit; available utilities; other design determining assumptions; *quantitative* input/output diagrams; (sub)process block diagrams; stream summaries
5 I-O; SP; TN; UN	Preliminary business case "SHEETS" analysis	Business case analysis based on "SHEETS": Safety ("Bowtie"), health, environment, economy (gross margin), technology, social analysis using concept stage knowledge
6 PF	Planning review and update	Review project and time planning, and use lessons learned to improve future project and time plan; work allocation between team members
7 PF	Creativity and group process tools	Review the team's creativity and group performance. Plan any improvements for the remainder of the project.
8 PF to UN	Conclusions and recommendations	At completion of the concept phase and feed forward to the subsequent phases
	List of symbols	
	Literature	
	Appendices	

13.7.2 Feasibility and development stages

A suggested report structure and contents for the feasibility and development stage reports can be found in Tables 13.4 (product focus) and 13.5 (process focus). The concept stage report is further extended. Parts of the concept stage report become part of the feasibility and development stage report.

Tab. 13.4: Suggested feasibility and development stages report structure and contents: product design focus.

Section (DDM Levels: Tab. 4.3)	Contents	Topics
	Summary	In not more than one page, the highlights of the project (objective, approach, results, including quantification of the key design results, and performance criteria, conclusions and recommendations) should be mentioned.
	Table of contents	Use descriptive chapter and section titles.
1 PF	Description of the design; project incentive (opportunity assessment)	Project brief/charter: background; framing (BOSCARD); stakeholder needs; design type and driver (market pull, technology push, innovation map) Market assessment (customers, markets, competitors, "SHEETS"-based opportunity definition); market segmentation; value chain analysis
2 PF	Design approach	Design challenges, design methodology, creativity methods, group methods and tools (SWOT); project and time planning
3 CW–PC	Consumer needs and requirements; project opportunity assessment	Stakeholder/consumer needs (voice of customer); product quality requirements, in relation to product application and entire life cycle
4 PC–PF	Product requirements	House of quality (HoQ) (composed by QFD), relating consumer needs, product quality performance, product property function and manufacturing requirements
5 PF; CW–PC; PC–PF	Database creation	Literature and patent search; product application process conditions and knowledge; chemical components; materials; pure component properties (physical, chemical, SHE, price); formulation and physical/chemical models and data; thermodynamic models and data; stoichiometry
6 CW–PC; PC-PF	Preliminary product concepts synthesis	Based on product application requirements and understanding of application process, generate product concepts, both top-down idea generation and bottom-up, based on molecular identity, product (nano) and microstructures, combined into a product/application process combination.

Tab. 13.4 (continued)

Section (DDM Levels: Tab. 4.3)	Contents	Topics
7 CW–PC; PC–PF; I-O; SP; TN; UN	Product concepts and preliminary process synthesis	Product concept/criteria evaluation/selection; chosen superior product concept (components, structure, physical/chemical product properties, in use performance) Based on product structuring/manufacturing technology choices, stoichiometry, yields and split factors, determine: plant capacity and location; battery limit; available utilities; other design determining assumptions; *quantitative* input/output diagrams; (sub)process block diagrams; stream summaries
8 I-O; SP	Preliminary business case (product and process design) "SHEETS" analysis	Product: performance versus requirements in use; product and process: performance versus "SHEETS": safety ("Bowtie"), health, environment, economy (gross margin), technology, social analysis using concept stage knowledge
9 PF	Planning review	Review project and time planning; use lessons learned to improve future project and time plan; specify work allocation between team members.
10 PF	Creativity and group process tools	Review the team's creativity and group performance. Plan any improvements for the remainder of the project.
11 CW–PC; PC–PF; I-O; SP; TN; UN	Detailed product design and validation; product/process interaction	Superior product concepts (concept stage) are further designed in detail. Product performance validation (by product modeling, product testing, product prototyping) is carried out. The design of the manufacturing process, the processing equipment and their interactions with the product physical properties is further modeled, tested, and validated.
12	Detailed business case evaluation	Product: performance versus requirements in use; Product and process: performance versus "SHEETS": safety ("Bowtie"), health, environment, economy (gross margin), technology, social analysis using concept stage knowledge
13 PF to final	Conclusions and recommendations	Based on the design knowledge gathered at this phase, present the strengths and weaknesses of the product/process design (in relation to the needs and requirements). Depict the major risks (technical, financial, etc.) and their consequences in failing to achieve the project objectives. Suggestions are given on how to mitigate/eliminate these risks, and what should be the next steps.
	List of symbols	
	Literature	
	Appendices	

Tab. 13.5: Suggested feasibility and development stage report structure and contents: process design focus.

Section (DDM levels: Tab. 4.3)	Contents	Topics
	Summary	In not more than one page, the highlights of the project (objective, approach, results, including quantification of the key design results, and performance criteria, conclusions and recommendations) should be mentioned.
	Table of contents	Use descriptive chapter and section titles.
1 PF	Description of the design; project incentive (opportunity assessment)	Project brief/charter: project background; framing (BOSCARD); stakeholder needs; customer requirements; technical requirements Market assessment (customers, markets, competitors; "SHEETS"-based opportunity definition)
2 PF	Design approach	Design challenges, design methodology, creativity methods, group methods and tools (SWOT); project and time planning
3 PF; CW–PC; PC–PF; I-O; SP	Database creation	Literature and patent search; chemical components; materials; pure component properties (physical, chemical, SHE, price) thermodynamics; stoichiometry
4 I-O; SP; TN; UN	Preliminary process concepts synthesis	Based on technology choice, stoichiometry, reaction yields and separation split factors, determine: Plant capacity and location; battery limit; available utilities; other design determining assumptions; *quantitative* input/output diagrams; (sub)process block diagrams; stream summaries.
5 I-O; SP; TN; UN	Preliminary business case "SHEETS" analysis	Business case analysis, based on "SHEETS": Safety ("Bowtie"), health, environment, economy (gross margin), technology, social analysis using concept stage knowledge
6 PF	Planning review	Review project and time planning; use lessons learned to improve future project and time plan; specify work allocation between team members.
7 PF	Creativity and group process tools	Review the team's creativity and group performance. Plan any improvements for the remainder of the project.

Tab. 13.5 (continued)

Section (DDM levels: Tab. 4.3)	Contents	Topics
8 I-O; SP; TN; UN; PI	Detailed and integrated process synthesis	Superior process concepts from the preliminary process concepts evaluation are designed, developed in further detail, and analyzed and evaluated (SHEETS) in the task network and unit network levels, leading to (a) superior process flow scheme(s) (PFS), process stream summary (PSS), and mass and energy balances. Mass and heat integration are applied in the process integration level. Utility requirements and product yields are discussed. SHEETS performances are re-assessed based on the more detailed analysis. SHEETS performances are reported in subsequent sections. The superior PFS is described in detail using the PFS and PSS. Design choices and possible alternatives are described in relation to the SHEETS performance.
9 ED	Equipment summary lists and equipment unit design specification sheets	Based on the functional specification of the equipment in the PFS, the equipment design variables are determined and sized to fulfill these specifications. It is especially important to identify the key equipment design variables (sizes, internals, material of construction, etc.), and how values are assigned to these variables for the best performance. See Section A13.7 for equipment summary lists and equipment specification sheets.
10 I-O; SP; TN; UN; PI;	SHE and S	Safety, health, environment and sustainable development detailed evaluation (HAZOP, DOW's Fire and Explosion Index (F&EI)
11	Economic evaluation	Detailed economic analysis: in addition to gross margin, at this stage, equipment costs, total capital investment and operating costs can be calculated and economic performance criteria – rate of return (ROR), pay out time (POT), discounted cash flow rate of return (DCFROR), net present value (NPV) – can be determined (see Chapter 10).
12 PF to Final	Conclusions and recommendations	Based on the design knowledge gathered at this phase, present the strengths and weaknesses of the design (in relation to the needs and requirements). Depict the major risks (technical, financial, etc.) and their consequences in failing to achieve the project objectives. Suggestions are give on how to mitigate/ eliminate these risks, and what should be the next steps.
	List of symbols	
	Literature	
	Appendices	

13.7.3 Quality checks: FOOFI list for reports

At major design project Stage-Gate™ milestones, design reports are prepared for the stakeholders by the design team. In the previous section, the contents for the concept, feasibility and development milestones were presented. In this section, guidelines are provided on ways to improve the quality of how the information is presented, considering the different roles of the readers of the report. What many (junior) designers do not appreciate is that very few readers will read the complete report (except university lecturers in design courses). Different stakeholders are interested in different parts of the report; they need to be able to navigate efficiently through the report and absorb and understand the information they are after with minimal effort. This means that, in order of priority, the following parts of the report need to be self-explanatory and understandable as a stand-alone text:
– Summary
– Table of contents
    – Should contain descriptive text about content
– Project objectives
– Conclusions and recommendations
– Figures and tables, especially of the design requirements, and design results:
    – Stakeholder needs, project and product and process (performances) requirements
    – Process stream summary (PSS)
    – Process block scheme (PBS) and process flow scheme (PFS)
    – And so on

Attention to figures and tables is compulsory, as (scanning) readers tend to focus on these report components first. Figure and tables titles, together with the content, should be sufficient for the reader to understand without reading the main text. The text surrounding the figures and tables also receive attention by the scanning reader, so it makes sense to put important information in these positions. Readers should be able to quickly retrieve the main storyline first, and then focus on the domain knowledge that is important for their role(s), e.g., related to the SHEETS performance criteria. These detailed chapters should be clear and make sense when read in isolation. One should include key information from other chapters that is crucial to understanding the individual chapters.

As already mentioned in Section 13.6.1, a quality checklist for design project reports has been composed (see Section A13.8).

13.8 Exercises

Exercise 13.1: Communication tools

(1) State whether the following communication tools can be used for design projects:
    (a) Activity reports
    (b) Slack
    (c) Reports
    (d) Models (brain writing, brain drawing)
    (e) Instagram
    (f) WhatsApp
    (g) M.O.M.

(2) State whether the following statements are true or false:
    (a) A design activity not worth reporting is not worth executing.
    (b) Reading is much faster than listening to team member presentations.
    (c) Schematics should never be made too specific to individual project targets.
    (d) Meetings are needed in projects for taking decisions and moving forward.

(3) The following are good ways of presenting ideas as models (state true or false):
    (a) Mind map
    (b) Block diagrams
    (c) Artistic illustrations
    (d) Process stream summary
    (e) Ishikawa diagrams

(4) Which among the following are considered FOOFI for presentations (indicate "Yes" or "No" and explain)?
    (a) Page numbers
    (b) Information in sentence form
    (c) Reading sentence text
    (d) Number of slides/minute
    (e) References to literature

(5) Which of the following are part of the Health and Safety data of a pure component properties table (indicate "Yes" or "No")?:
    (a) LC_{50}
    (b) Chemical reactivity
    (c) MAC value
    (d) Smell
    (e) LEL and UEL
    (f) All of the above

(6) Indicate which among the following are good practices while reporting (indicate "Yes" or "No")?:
    (a) The conclusions represent a summary of the results.
    (b) Clear definitions of battery limit and considerations of storage and waste treatment.

(c) Qualitative descriptions without references.

(d) Lumping of operating costs and investment costs.

(e) New formats created for final report.

Abbreviations

BOD	Basis of design
BOSCARD	Background, objectives, scope, constraints, assumptions, risk, deliverables
CW–PC	Consumer wants – product concept (DDM)
DCFROR	Discounted cash flow rate of return
DDM	Delft design map
DOW's F&EI	DOW's Fire & Explosion Index
ED	Equipment design (DDM)
FO	Flow sheet sensitivity and Optimization (DDM)
FOOFI	Frequently occurring opportunities for improvement
HAZOP	Hazard and operability study
HoQ	House of quality
I-O	Input-output structure (DDM)
MOM	Minutes of meeting
NPV	Net present value
OI	Operability integration (OI)
PBS	Process block scheme
PF	Project framing (DDM)
PFS	Process flow scheme
PI	Process integration (DDM)
P&ID	Piping and instrumentation diagram
PC–PF	Product concept – (product) property function (DDM)
POT	Payout time
QFD	Quality function deployment
ROR	Rate of return
SHE	Safety health environment
SHEETS	Safety health environment economy technology social
SP	Subprocesses (DDM)
SWOT	Strengths weaknesses opportunities threats
TN	Task network (DDM)
TOC	Table of contents
UN	Unit network (DDM)

References and further reading

[1]   Pugh S. Creating Innovative Products Using Total Design, New York, NY, USA: Addison-Wesley-Longman, 1996.

[2]   De Haan AB, Swinkels PLJ, de Koning PJ. Instruction Manual Conceptual Design Project CH3843 – From Idea to Design. Delft, Netherlands: Department of Chemical Engineering, Delft University of Technology, 2016.

[3]   Seider WD, Seader JD, Lewin DR, Widagdo S. Process and Process Design Principles: Synthesis, Analysis and Evaluation, 3rd edn. Hoboken, NJ, USA: John Wiley & Sons Inc., 2010, 681–692.

.

Part D: **Education**

14 Education

Engineering education and research are increasingly emphasizing systems engineering (SE) and design. Universities and industries recognize that addressing societal challenges such as climate change, energy transition, materials circularity, food, health, and global housing demands requires systematic solutions.

In the discipline of process systems engineering (PSE), Dutch technical universities (Delft, Eindhoven, and Twente) have observed a notable expansion of research activities, accompanied by the appointment of full professors in the PSE field at each of these universities.

At TU Delft, research groups focused on design and engineering across all faculties have proactively initiated the formulation of research programs centered on systems engineering approaches and tools. Additionally, efforts are underway to align education in the systems engineering domain, which has thus far developed independently within various expertise areas. This alignment aims to produce engineering designers proficient not only in designing and engineering systems within their respective disciplines but also capable of seamlessly collaborating across interfaces on larger systems.

In this chapter, the authors give an overview of the BSc, and truly international MSc, PhD, and EngD (engineering doctorate) programs at TU Delft in the field of (bio) chemical engineering, and with a focus on the interaction between education, research, and design. The cooperation with industry and public/private partnerships and the valorization of knowledge in support of innovation is presented from a "systems point of view." We hope this forms an inspiration for other universities and countries to adopt a similar approach.

14.1 (Bio)chemical design education: a long history

The authors of this book have built their diverse product and process design and innovation skills through their education at the Universities of Technologies in Delft, Eindhoven, and Twente, and through their industrial careers at various companies (Corbion, Cosun, DSM, ICI, Shell, and Unilever). The authors have over two decades of experience in teaching product and process design at the Delft University of Technology (TU Delft), and also at the Technical University Eindhoven (TU/e), University of Twente (UT) and Rijksuniversiteit Groningen (RUG). Inspired by their predecessors, and having worked closely with many of them in the last decades, they would also like to mention them in this chapter on education. At TU Delft, the process design courses – including a large open-ended design problem – combined with the setup of a computer simulation course and facilities, have been in existence since the mid 1980s. As a special recognition for their drive to setup and build a world class design course, including projects, the authors would like to mention the predecessors and

https://doi.org/10.1515/9783110782127-014

current co-lecturers in (bio)chemical product and process design at BSc, MSc, and engineering doctorate (EngD, previously known as PDEng – professional doctorate in engineering curricula (see Tab. 14.1).

In this chapter, the developments in product and process design education in the (bio) chemical engineering discipline, and the fit in the TU Delft BSc, MSc, and EngD programs are described.

14.2 Education programs

Since the year 2000, the higher education at the Technical Universities in The Netherlands has quickly converted to the Bologna first/second and third cycles of education (Fig. 14.1):
- First cycle: BSc: 3 years (180 European Credit Transfer system credits (ECTS credits)
- Second cycle: MSc: 2 years (120 ECTS credits)
- Third cycle: PhD: 4 years, research orientation
- Third cycle: EngD: 2 years (120 ECTS credits), design orientation

Dutch university education: 3 Bologna cycles

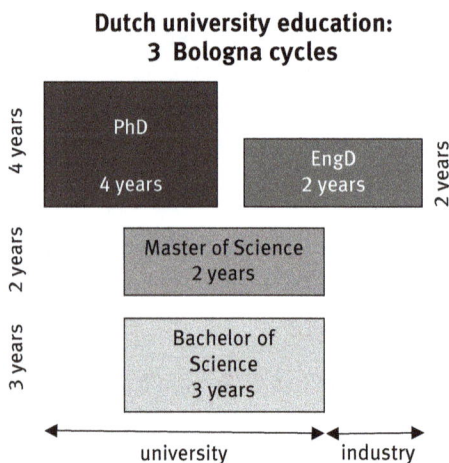

Fig. 14.1: Three cycles of higher education at the Dutch universities of technology (1st (BSc), 2nd (MSc) and third (PhD or EngD: Engineering Doctorate cycle), based on Bologna declaration.

The three cycles, based on the Bologna declaration, provide the opportunity to students to complete different parts of their studies at different universities at different countries. In addition to the European Erasmus programme (European Region Action Scheme for Mobility of University Students), this is stimulating the international mobility of students between European countries. It has been named the largest success of the European Union.

Tab. 14.1: TU Delft lecturers in (bio)chemical product and process design since mid-1980s – predecessors and current lecturers.

Predecessors in lecturing (bio)chemical product and process design at TU Delft	Current (co)lecturers in (bio)chemical product and process design
Em.Prof.ir. A.G. Montfoort	Dr.ir. G.M.H. Meesters
Drs. F.A. Meijer	Dr. A.J. Houtepen
Em.Prof.dr.ir. J. de Graauw	Dr.ir. J.L.B. van Reisen EngD
Em.Prof.ir. J. Grievink	Dr.ir. P.J. Daudey
Dr.ir. P.J. Verheijen	Dr. A.J. Groenendijk
Prof.ir. K.Ch.A.M. Luyben	Prof.dr.ir. L.A.M. van der Wielen
Dr.ir. C. van Leeuwen	Dr.ir. A.J.J. Straathof
Dr.ir. J.C. Goebel	Prof.dr.ir. M. Ottens EngD
Ing. M. Krijnen	Prof.dr.ir. H.J. Noorman
Ing. M. Valentijn	Prof.dr.ir. M.C.M. van Loosdrecht
Em.Prof.dr. G. Frens	Dr.ir. C. Haringa
Dr.ing. G.J.M. Koper	Prof.dr.ir. A.A. Kiss
Em.Prof.dr.ir. P.W. Appel	Dr. A. Somoza-Tornos
Prof.dr.-ing. J. Gross	Dr. J.A. Posada Duque
Prof.dr.ir. P.J. Jansens	Dr. A.M. Schweidtmann
Ir. C.P. Luteijn	Dr. J.P.C. Carvalho Pereira
Ir. J. Dijk	Dr.ir. F. Mousazadeh
Em.Prof.dr.ir. K. van 't Riet	Prof.dr.ir. C.A. Ramirez Ramirez
Dr.ir. J.F. Jacobs	Prof.ir. G.J. Harmsen
Prof.dr.ir. J.J. Heijnen	Prof.dr.ir. A.B. de Haan
Ir. M.W. Lambrichts	Ir. P.L.J. Swinkels
Ir.drs. G. Bierman EngD	
Dr.ir. H.W. Nugteren	
Dr. P.J. Hamersma	
Prof.dr.ir. G.J. Witkamp	
Dr.ir. M.C. Cuellar Soares EngD	
A.M. Echavarria Alvarez EngD	
Dr.ir. L.D.M. van den Brekel	
Dr.ir. E.H. Wolff	

The connectivity and curricula of the BSc, MSc, and the third-cycle programs (PhD or PDEng) will be discussed with a special focus on the design and systems engineering content.

14.2.1 BSc programs: TU Delft

The TU Delft BSc programs, "molecular science and technology (MST)" and "life science and technology (LST)," are offered in cooperation with the Leiden University. Students attend courses and take exams at both universities and, at graduation, earn a joint degree from TU Delft and Leiden University. This cooperation was started over fifteen years ago to attract more students to these education fields. It allowed high school graduates to delay their choice to the second year for the more science-related chemistry or life science direction in Leiden, or the more technology-oriented direction at TU Delft. This has been a very large success. Student numbers rose quickly to larger values than the separate bachelor programs before. TU Delft even saw a substantial increase after this cooperation. Being exposed to technology-oriented lectures and research group research activities, made more students choose to continue their BSc education at TU Delft. The same holds for the biotechnology-focused BSc program, life science and technology.

The MST and LST BSc curricula have a study load of 180 ECTS credits, are full time, and nominally take 3 years to complete. The program is largely held in Dutch, textbooks are mostly in English, and international lecturers may teach in English. The MST program combines chemistry and (chemical) technology. The LST program combines life sciences (cell biology and chemical processes) with biotechnology. Students are introduced to the specializations within chemistry and chemical technology (MST) and life sciences and biotechnology (LST) listed below. Students are made to also gain ample laboratory experience in the very well-equipped laboratories in Delft and Leiden. The exposure to research work and to the staff in the research sections starts in the first year. Experience shows that this boosts the students' understanding of, and interest in the field. The application of knowledge and research questions inspires the students. It greatly helps in their decision process whether to continue in the chemistry (MST), life sciences (LST) field at Leiden University or in the technology directions at TU Delft.

The BSc-MST chemistry track focuses in more depth in disciplines like inorganic, theoretical physical, organic, and molecular chemistry. The BSc-MST technology track covers a wider range of mathematics, numerical techniques, thermodynamics, physical transport phenomena, chemical reaction engineering, separation technology, and a (chemical) product design project [1]. As mentioned before, building competencies in experimentation skills and participating in scientific research projects are important in both tracks. In the third BSc program year, a more in-depth or a broad minor can be selected and the BSc final research project completes the curriculum.

The BSc-LST curriculum [2] is also run at Leiden and Delft, but all students follow the same program in the first two years. The more fundamental topics and laboratory projects (cell biology, biochemistry, molecular cell biology and immunology, microbial physiology, and molecular genetics) are covered in Leiden, and the more biotechnology topics (mathematics, statistics, thermodynamics, transport phenomena, bio-informatics, and the design project on sustainable biotechnology processes) are taught in Delft. The choice of the third-year direction (minor and BSc final research project) depends on the MSc choice. MSc-LST in Delft requires the third-year BSc to be completed in more technology-related topics at TU Delft. The MSc-LST in Leiden requires the third-year BSc to be completed at the Leiden University.

14.2.2 MSc programs: TU Delft

The MSc programs, chemical engineering (CE) and life science and technology (LST), are truly international programs and are fully taught in English. The fraction of international students joining this program is currently about 30%. This number is rising, as is the total number of MSc students. The very high rankings of the TU Delft (bio) chemical engineering field in various indices like QS World University Rankings and Shanghai Ranking, supported by the excellent research and teaching facilities, have a large attraction on international talent. Both MSc programs cover 120 ECTS credits and take 2 years to complete.

The first year of the MSc chemical engineering comprises compulsory core courses, elective courses, and a design module. Students can select any combination of electives, but combinations are suggested to align with the four science and engineering orientations: circularity, energy, health, and nuclear. The energy and circularity engineering orientations attract most students, but the other two tracks are steadily growing. The design module focuses on both process and product design, ranging from new construction materials, materials for energy conversion and storage to food ingredients and pharmaceuticals. The nuclear orientation focuses on nuclear engineering and its applications that make use of radioisotopes. The TU Delft nuclear reactor is a key facility for these radioisotopes, for example, medical applications field is developing rapidly into other health and industrial applications.

The MSc life science and technology has three tracks: biocatalysis, biochemical engineering, and cell factory. These two MSc programs comprise deepening domain knowledge, executing a large design project, an internship placement, and an individual research project.

Top-talent students in these MSc programs can apply for the Bologna third cycle of education: PhD (Section 14.2.3) or EngD positions (Section 14.2.4). MSc students with exceptional academic performance are invited after the first two quarters of the MSc program to participate in the "honors" program. During this program, the students select an (additional) 20 ECTS credits. This program should be completed within the 2 years – MSc program. "Honors" specializations can have an "academic research focus" (preparing for a PhD position), a "design focus" (preparing for an EngD position) or an "industry focus," with a research project with an industrial partner.

14.2.3 PhD programs: TU Delft

Research in the area of (bio)chemical engineering at TU Delft is clustered in the "TU Delft Process & Product Technology Institute" (Pro2Tech, former DPTI) [3]. Pro2Tech is the research incubator for process and product technologies at TU Delft. The Pro2Tech researchers work at about 20 research groups from the departments biotechnology, chemical engineering, process and energy, energy technology, and water processing

(Fig. 14.2). Pro2Tech's research, education, and design activities are clustered in four main areas:

- Food and pharma
- Energy
- Water processing
- Processing of advanced materials
- Solid and fluid mechanics

Pro2Tech aims for aligning the activities in the research groups, and intensifying collaboration with other (inter)national universities and industries. Pro2Tech offers industry and knowledge institutions a wide variety of partnership opportunities. It ranges from BSc or MSc internships to participation in large international consortia for PhD and EngD (engineering doctorate) projects.

Pro2Tech also has a strong impact on the visibility and impact of its research and education. Over the last few years, the TU Delft in the chemical engineering subject always reached top fifteen ranks in the QS World University Ranking [4].

In general, PhD research tracks are 4-year projects, funded by national, European, and worldwide initiatives. PhD tracks are mostly part of larger projects with many partners in public/private -partnerships (PPP). Top-performing MSc graduates are hired as PhD candidates after a careful selection process. The applicants' educational performance, research and personal competences, and motivation are tested in interviews. Once hired, the PhD candidate will soon setup a personal development plan in addition to their research project plan. PhD candidates select several personal development courses organized by the TU Delft's Graduate School. After 1 year, a "go/no-go" presentation and assessment is held. After the "go ahead" decision, PhD candidates are expected to complete their research project with minimum delay. The PhD candidates' progress in executing their personal development plan, research project, and realizing publications is monitored by their scientific supervisors. The Graduate School organization aids in recording the said plans, execution progress, and in advising PhD candidates and supervisors during this process.

14.2.4 EngD programs: TU Delft

During more than 35 years, post-Master design-oriented Engineering Doctorate (EngD, previously called "PDEng": Professional Doctorate in Engineering) programs are a very successful cooperation between the Dutch technical universities and the industry. Because of its unique setup, and the fact that these programs hardly exist in other countries, a rather detailed overview is presented here.

The text below is largely cited from Swinkels [5], who has given a concise description on the history, nature, and current situation of the over 35 years old PDEng (now called "EngD") traineeships at the Dutch technical universities. In some places, the

TU Delft Process & Product Technology Institute (Pro2Tech)
5 clusters of multidisciplinary research collaboration

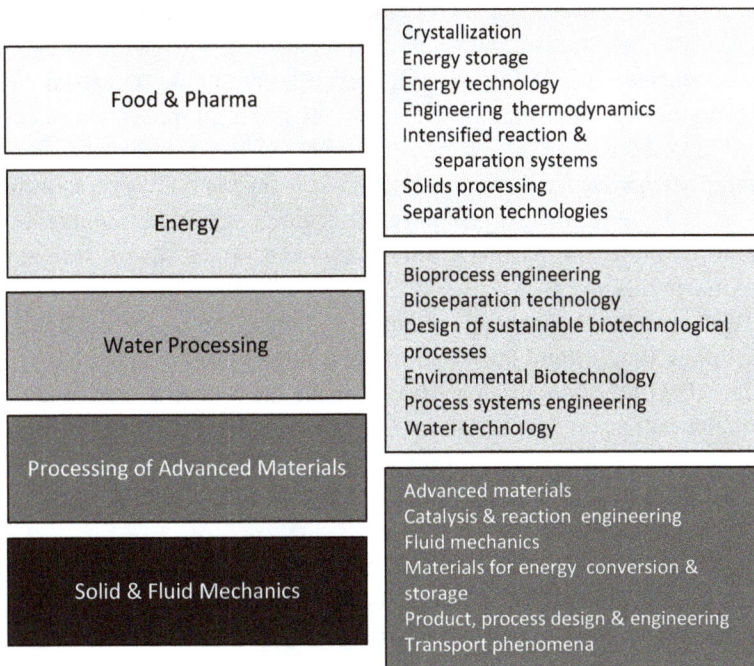

Food & Pharma	Crystallization Energy storage Energy technology Engineering thermodynamics Intensified reaction & separation systems Solids processing Separation technologies
Energy	
Water Processing	Bioprocess engineering Bioseparation technology Design of sustainable biotechnological processes Environmental Biotechnology Process systems engineering Water technology
Processing of Advanced Materials	Advanced materials Catalysis & reaction engineering Fluid mechanics Materials for energy conversion & storage Product, process design & engineering Transport phenomena
Solid & Fluid Mechanics	

Fig. 14.2: Overview of the five key research clusters of the TU Delft Process & Product Technology Institute (Pro2Tech), based on [3].

text has been adapted to the current (2023) situation. At the end of this section, the contents of the TU Delft EngD traineeship programs in the (bio)chemical product/process design fields are presented:

The three universities in The Netherlands – Delft University of Technology, Eindhoven University of Technology and University of Twente – offer two-year post-MSc technological design programmes (traineeships) to selected top level MSc graduates recruited from all over the globe. The Dutch Government and the Dutch industries jointly initiated these traineeships in 1986, with the aim to train top talent in design competencies, and in business and personal skills, to prepare them as key "knowledge workers" who strongly contribute to the design of new products and processes and can boost innovation. The PDEng (EngD) traineeships are designed to train and educate top MSc graduates to become qualified excellent designers capable of designing "fit for purpose" and "first of its kind: products, processes and systems. The PDEng (EngD) trainees are encouraged to actively look beyond the perimeters of their own disciplines and recognize the challenges and restrictions imposed by product chain management, time and money.

Over a period of thirty-five years many of these programs are still going strong – supported by long lasting and multinational industrial partners who actively and continuously pursuing (open) innovation and cooperation with the technical universities. Other programmes were terminated, and yet new ones matching with emerging innovation areas were started. This lifelike dynamic very closely follows industrial demand and opportunities for innovation, the demand for technical innovation talent and the universities' research, design and education capabilities and strengths. Contribution from, and participation by these stakeholders in innovation, is a prerequisite for every successful PDEng (EngD) program. Regular contacts and active cooperation projects between the industrial partners, universities and PDEng (EngD) trainees prove to be essential in this process.

In the late 1990's the Dutch Universities started to internationalize their MSc curricula, and the PDEng (EngD) (and PhD) recruitment followed suit. Currently over 90% of the PDEng (EngD) trainees originate from outside the Netherlands, and both genders are equally represented. The PDEng (EngD) traineeships have turned into truly global and gender-balanced doctoral training centers. This has increased the attractiveness of the traineeships for the industry, university and last but not least, for the PDEng (EngD) trainees themselves.

The position of the PDEng (EngD) programs in the Dutch technical university education system is shown in Fig. 14.3. It builds on the first (BSc) and second (MSc) cycles of the Bologna agreement, and forms together with the research-oriented PhD track, the third tier of academic education. The design oriented PDEng (EngD) education reaches out into the industry domain, as after one year of competence building and knowledge acquisition, the full second year is spent on an individual design project initiated by and executed (mostly) at the industrial partner's site.

On successfully completing the PDEng (EngD) program, the graduates receive a certified diploma and are entitled to use the academic degree "Professional Doctorate in Engineering (PDEng, now: Engineering Doctorate (EngD)". All diplomas are registered in a central register kept by the Royal Dutch Society of Engineers (KIVI: "Koninlijk Instituut Van Ingenieurs").

Since 2006 the PDEng (EngD) designer programmes have been coordinated under the flag of the 3TU.School for Technological Design – Stan Ackermans Institute (3TU.SAI), a cooperation between the three technical universities in the Netherlands (Delft, Eindhoven, Twente)". Currently also *Wageningen University & Research (WUR)* has joined the now called 4TU.SAI cooperation.

Nowadays between 200–250 trainees join – after a stringent recruitment and selection process – one of the 20 PDEng (EngD) programs in the various disciplines every year. They work as a salaried employee (PDEng (EngD) trainee) for one of the three technical universities on strengthening their skills through workshops and design cases, and on solving a real design case for (and at) an industrial partner during the full second year of their employment.

Engineering Doctorate
Post-MSc designer programmes

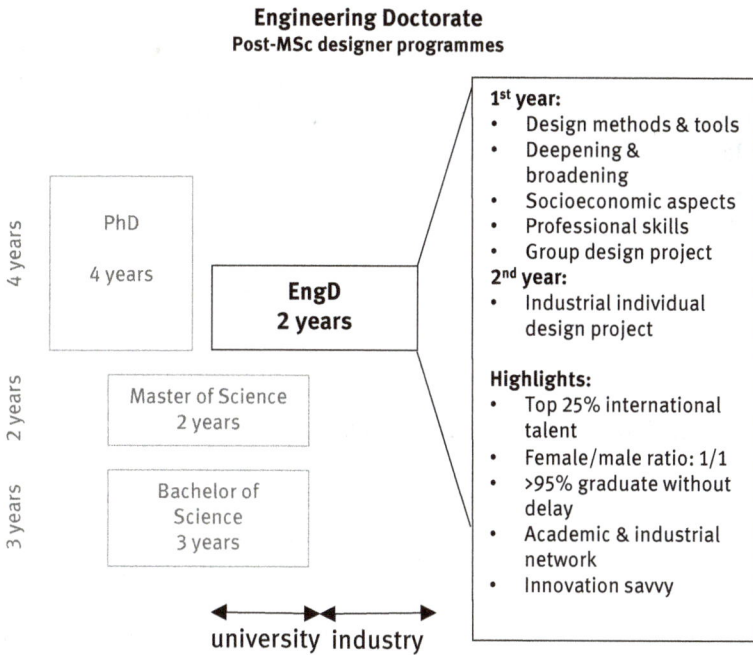

Fig. 14.3: Position of EngD traineeships at TU Delft (bio)chemical engineering fields and specifics regarding graduation efficiency and gender balance in the Dutch technical universities' education system.

On October 1st 2014 TU Eindhoven Rector Magnificus Prof.dr.ir. C.J. van Duijn awarded the 3,500th PDEng (Professional Doctorate in Engineering)." On October 22nd 2020 the 4500th PDEng graduate was celebrated.

It should be noted that over 80% of the 3,500 PDEng graduates originate from countries outside the Netherlands, and that the female/male ratio has steadily increased to over 50%. *More than 80%* of the PDEng (EngD) graduates find *technological challenging* and permanent jobs in the Netherlands. From a recent alumni overview from the (then) 23 year-old programme Process & Equipment Design (TU Delft), it was also evident that the large majority of these alumni progress their industrial careers in the technology or technology management career track. This is a very positive observation, as traditionally many (MSc) graduates from Dutch universities move into operation management, financial of general management roles rather soon in their careers, and stop contributing directly to technological innovations. As their business and financial counterparts don't make the reverse career switch, this loss of technical talent could easily become a threat. PDEng (EngD) graduates seem to very naturally and effectively occupy and these positions, and keep pursuing a technical career track for a very significant part of their careers. Their technological design talent and ambitions leads and keeps them in this high-in-demand career track.

Training PDEng (EngD) designers parallel to educating PhD researchers is an approach that has catalyzed and delivered many innovation benefits to the industries based in the Netherlands. With vast worldwide challenges ahead of us in the areas of energy, food, water, scarce resources, circular chemistry, and health, it is an example that could be adopted worldwide by more countries. The start-up and implementation of such PDEng (EngD) programs is not an easy task. To initiate these programs it requires joint initiative and conversion on programme content, but also financial and intellectual property aspects between universities, companies and government. To sustain these programmes a continuous dialogue between these stakeholders is absolutely necessary. These are conditions that are not quickly and easily met. Dutch industry took the initiative in 1986 and universities and government converged quickly. Over twenty-eight years (now already 35 years) these partners managed to keep aligned in the interest of the Dutch innovation agenda and of the many (inter)national technological design talents that found their career for life (and are living it).

The PDEng (EngD) programs thrive on the industry's interest in cooperating, and supporting (both in supervising the PDEng (EngD) trainee, and in financial contribution) the one year individual design assignments as topic for the PDEng (EngD) design thesis. Many of DPTI's (now: Pro2Tech's) multinational industrial partners in large multinational consortia, also regularly provide design problems for the PDEng (EngD) traineeships. Also, small and medium enterprises and start-up companies are able to find their way via the TU Delft Valorization Centre to this cooperation with the PDEng (EngD) traineeships. All industrial partners (small or big) not only benefit from a novel design solution for their design assignment (by the inputs of fresh minds, a sound design methodology, project management approach, and the latest academic research results), but also build and maintain a strong and lasting network with the university and with top level designer trainees who mostly start their industrial career in the Netherlands.

In order to prepare the PDEng (EngD) trainee for the challenging one year individual design project (IDP) for an industrial partner, over the years a technical training programme has been developed to strengthen selected knowledge and competence areas. The structure of the programs as described by Swinkels [6] hasn't undergone major changes over the last fifteen years. Strengthening the design methodology tools, and fine-tuning and extending the advanced domain knowledge into the quickly developing research and education strengths are the major developments. The PDEng (EngD) trainee's knowledge and competencies continue to be trained and coached during the full year Individual Design Project executed at the industrial partner.

The learning objectives of the EngD program (as laid out in the TU Delft Regulation on Engineering Doctorate are [7]:
– an independent attitude that is conducive to achieving innovation,
– a critical approach to the information and opinions provided, with regard for the societal context,
– the ability to translate creativity into concrete results,

- the ability to achieve usable innovations,
- to direct innovation projects with regard to time, budget, and quality, and
- soft skills that support optimum functioning in professional teams.

The contents of the key PDEng (EngD) design courses are presented in paragraphs 14.3.3 and 14.3.4.

14.3 Design-oriented courses at TU Delft's BSc, MSc, PhD, and EngD levels

14.3.1 BSc molecular science and technology (MST) and BSc life science and technology (LST)

14.3.1.1 Chemical product design (BSc-MST, 4052TLEON3, 6 ECTS)

The course provides insight into the aspects that play a role in developing chemical products. Teams of 5–6 students need to effectively use their creativity and acquired chemical technology knowledge to develop an innovative, but realistic new product. Course topics comprise: performing as a team, product design and development, relation between product properties and physical-chemical composition, developing physical-chemical models to describe product properties, identifying customer needs, and idea generation and selection. The course setting mimics the company situation where a development team of engineers (=students) has been assigned a product development task by their manager (= a staff-member of the Department of Chemical Engineering). The teams are graded on a written report about the functioning of an existing product (20%) and their report (60%) and presentation (20%) of their own product design. All students will fill in a peer-review form on their fellow teammates. This peer review is used to modulate the average group-grade and to determine the grade of the individual students.

14.3.1.2 Design of sustainable biotechnological processes (BSc-LST, LB2611, 5 ECTS)

The course provides insight into the aspects that play a role in integrating the various scientific and technological approaches and their application for the design of sustainable biotechnological processes at industrial scale. Different microorganisms, products, and processes are covered by TU Delft and industrial guest lecturers. Teams of students will apply the learned knowledge to redesign a biotechnological process, aiming to improve its economics, safety, and sustainability aspects. The students are graded on their report of the analysis of the standard process for the case study (35%), an executive summary and presentation of the redesigned/improved process (45%), and a presentation on the reflection on the industrial guest lectures (20%).

14.3.2 MSc chemical engineering

14.3.2.1 Product and process design (CH3803, 6 ECTS)

This course is an obligatory preparatory course for the Conceptual Design Project (CDP, CH3843), and therefore full participation by proven presence, taking assignments and tests, and sitting for the exam are required. Topics covered include:
- product design and material supply chains;
- relation between product performance and composition;
- process design methodology;
- process integration;
- process flow sheet modeling (Aspen Plus®);
- energy and mass integration;
- process and product evaluation and optimization;
- health, safety, and environmental aspects of design
    - economic performance estimation and evaluation,
    - creativity methods and team performance.

Lectures are combined with multiple graded product and process design assignments, in which teams of students directly apply the obtained knowledge. The students are graded through a combination of individual tests (20% product, 20% flow sheet modeling, and 20% equipment design) and a team assignment (40%). An individual mark for the team assignments is established by conducting a peer review in which all students evaluate the contribution of their fellow teammates.

14.3.2.2 Conceptual design project (CH3843, 12 ECTS)

The design project includes the generation of a conceptual design for an integrated manufacturing process, and/or a chemical product/device. To offer students a realistic experience, all assignments are of high practical relevance and mostly delivered and cosupervised by experts in the industry acting as "clients." The projects are carried out in teams (between five and seven students from mixed educational backgrounds and nationalities), with a stage gate-based intermediate reporting and presentations to train in teamwork and communication skills. Each project is supervised by a university team coach. Furthermore, the team is supported by multiple university scientific staff members, who are selected based on the project content and act as technical advisors. Creativity is stimulated through a creativity workshop and active application of the various creativity tools during the project execution. Final grading is based on the concept design report, the final design report, the presentations (kickoff, concept, and final) and project defense during final presentation and for supervisors, divided over:

- Design theory and concepts (quality of design) (40%)
- Implementation (design project organization, planning and execution, creativity, and innovativeness) (30%)
- Communication (progress meetings, reporting, and presentations) (30%).

With input collected from all supervisors and advisors, individual marks for the team members are established by conducting a peer review as part of the final assessment.

14.3.3 Process and product design EngD courses

14.3.3.1 Advanced principles in product and process design (ST6064, 6 ECTS)

The "Advanced Principles in Product and Process Design" (ST6064) course is divided into three parts:
- Part 1: online individual product design assignment
- Part 2: the core of the course on product design and process design methodologies
- Part 3: co-mentor and critically review and evaluate an MSc-level conceptual product or process design (draft final report)

The first part provides the EngD trainees with the basics of design work and is executed as an individual self study and administration of assignments. This first part of the ST6064 course starts immediately after starting the EngD traineeship.

The core of this course consists of four parts. Course participants:
- Read the course materials on product & process design principles and methodologies discussed in this book and based on the Delft Design Map (DDM) for product and process design, and other referenced textbooks. They engage in online discussions on the material. This preparation is followed up in seven half-day workshops. During these workshops, the material is summarized by the lecturer, examples are provided, and a question and answer session is held.
- Apply the design methodology, project planning, team working skills, and communication tools in an open-ended team project. Teams consist of three to five EngD trainees, who are coached by the lecturer(s). Each team member spends 80 h on this project. In three workshop meetings, the teams present and discuss progress and receive feedback on the design approach, results, and team cooperation. The final results are reported, and formally presented and discussed with an EngD trainee peer group.
- Attend and contribute to a one-day national conference or workshop organized by, for example, Netherlands Network Process Systems Engineering – NL (PSE-NL) [8], Process Intensification Network (PIN-NL) [9], or others).

14.3.3.2 Process simulation laboratory (ASPEN Plus®) (ST6063A, 2 ECTS)

This course's objective is to train the trainees to build a flow sheet in *Aspen Plus®* using a few example cases, understand an existing flow sheet simulation, simulate the process with Aspen Plus®, and solve problems related to the modeling and conversion of these simulations. Also, practical skills in modeling and analysis of complex systems is acquired, and insight into the simulation of several processes is gained.

The course covers mass and energy balances, sequential modular flow sheeting, application of various thermodynamic models, use of simulation convergence methods, and finally a two-week process design and simulation project on the simulation of a complete (bio)chemical process.

The course is completed with presenting and discussing the project results at the Fluor Haarlem office for a professional audience, and with the assessment of the project report.

14.3.3.3 Advanced process energy analysis and optimization (ST7101, 3 ECTS)

Industrial processes often have large utility requirements for heating and cooling process streams to be able to convert, separate, and transport raw materials to products. Efficient use of utilities reduces operating costs and CO_2 emission per unit product. Since the 1970s, analysis and optimization techniques have been developed to assess heat requirements and integration opportunities within process systems and industrial sites. The basic principles are generally available in chemical engineering textbooks. This course gives an advanced guide to applying basic and advanced energy analysis techniques to common industrial cases. It uses these techniques to optimize industrial heat exchanger networks, both in grassroots and retrofit designs. Additionally, attention is given to the design of common utility systems for process heating and cooling. Cases will be elaborated using *SuperTarget™ Software*. The course covers both individual and team assignments. The team assignments are presented at the McDermott The Hague office in front of an audience of experienced designers.

14.3.3.4 Technology management, economical evaluation in the process industry (ST6612, 6 ECTS)

The present course offers a tailor-made package of knowledge and practical skills in the fields of strategy and marketing as well as economics and finance. Particular attention is paid to industry analysis, formulation and analysis of options, capital and operating cost estimating, investment appraisal, and evaluation of corporate financial performance. The course teaches product & process developers and engineers to put a price tag on their ideas and so offer the best solutions from an integral technical and economic point of view. Finally, the course offers EngD trainees a broad orientation on the nontechnical (business administration) aspects of their discipline and, hence, the opportunity to put their affinity with such issues into a better perspective.

The course starts with interactive lectures, combined with workshops. The workshops comprise exercises and case studies. The course is concluded with an assignment in which the economic feasibility of an actual project is assessed.

Course outline:
- Part 1
  - Introduction of industry/economy general
  - Process (chemical) industry business
  - Case study
  - Project organization
  - Economic evaluation/investment appraisal
- Part 2
  - Capital cost estimating
  - Case studies
  - Manufacturing cost estimating
  - Evaluation of profitability/discounted cash flow (DCF) analysis
- Part 3
  - Cost of capital, risk, and uncertainties
  - Finance and accounting fundamentals
  - Case study
  - Management accounting
  - Case study/project

14.3.3.5 (Personal and) project management (ST6111, 2 ECTS)

The (Personal and) Project Management training is set up to learn how to combine personal, project, and process management theory with daily (project) work. In addition, it demonstrates how to apply these methods into practice during the various projects in the EngD traineeship. "Hard" project management skills like project structuring, project planning, and scheduling, are covered. However, an analysis of personal competencies and preparing and working on a personal improvement plan are also key parts of this course. The course builds on multicultural differences, and peer sharing and learning. Individual coaching sessions are held with all participants in the first year and the personal improvement plan is further monitored during the design projects, GDP and IDP.

14.3.3.6 Sustainable design of processes, products, and systems EngD course (ST7111, 3 ECTS)

Educating design, in general, and educating sustainable design, in particular, to students (and EngD trainees is paved with difficulties. On the one hand, the students should be free to use their creativity to come up with breakthrough novel designs that address a need and are a factor of four to ten better in environmental, economic, and social impacts than conventional solutions. On the other hand, because they are still inexperienced in design, they can become stifled due to this difficult task and their little experience in designing in general. Moreover, students learn designing in various ways. Dorst [10] distinguishes six different forms of intelligence relevant for design: linguistic, logical, spatial, musical, bodily, and personal. He also observed different characters such as pragmatic problem solvers, autonomous designers, visionary artists, etc. It is therefore important to have the teaching and coaching adapted to these differences in personal make-up. To teach the EngD trainees, a method has been developed that facilitates different learning methods and that provides an atmosphere of trust between the trainees and the teacher. The course is set up in five steps, spread over three half-days with two to three weeks in between for assignments. Each step consists of:

- A plenary teaching lecture of a course element group design work, starting in the classroom with the teacher available for coaching, on request
- Assignment task work: a course book by Jonker and Harmsen [11] is used that allows students to look up information
- Presentation of intermediate results
- Feedback by the teacher on these results

In this way, the design effort is divided in smaller steps and allows for a large amount of coaching, specifically geared to each group.

The purpose of the course is that EngD trainees learn how to make a breakthrough sustainable concept product or process design, or a combination of both, using the UN Sustainable Development Goals and supporting methods on environmental, economic, and social dimensions, and on robustness to future uncertainties. With that capability, EngD engineers can provide companies a breakthrough novel concept design of a product and or a process that contributes to the Sustainable Development Goals of the United Nations.

Course program schedule:

- Step 1: Select and apply UN Sustainable Development Goals (and targets) belonging to People, Planet, and Prosperity dimensions to a design case.
- Step 2: Improve designs for environmental impacts with a life cycle assessment method.
- Step 3: Make a design fitting to an economically viable circular economy concept, with no waste streams.
- Step 4: Use a scenario set to analyze the design for future social uncertainties and make it robust.
- Step 5: Report the sustainable design, containing the learning outcomes 1–4 in writing and an oral presentation to the EngD peer groups and lecturer.

14.3.3.7 Group design project (ST6802, 21 ECTS, ST6814, 17 ECTS)

The objective of the group design project (GDP) for both EngD programs (Product, Process & Equipment Design – ST6802, and Designer in Bioprocess Engineering (ST6814) is to enable the participants to develop their design skills in a real-life design case in teams consisting of three to five members. The group, as a whole, is responsible for the final outcome of the project. In turn, each member of the group will be assigned the group leadership. The participants are forced to use their deepened skills of product and process fundamentals and acquired design skills in an integrated way. They are made aware of other engineering disciplines in the areas between conceptual product and process design and an operational plant. The theory and learnings from the prerequisite courses, Advanced principles of product and process design (ST6064), and (Personal and) Project management (ST6111), are applied.

- Each participant spends 590 (ST6802) or 476 (ST6814) hours on the assignment.
- The total amount of (elapsed) time for the assignment is limited to half-a-year, so each participant spends at least 24 h a week (60%) on the assignment, on average.
- The hours not spent on the project (400 h) are used for other EngD program subjects.
- The implementation of the assignment is the responsibility of the team as a whole.
- Generally, three reports are written during the project, and two intermediate and one final presentation(s) are held at the following milestones meetings:
  - Scope and basis of design
  - Results of conceptual design (including product composition and structure, process flow diagrams and balances)
  - Final review meeting: evaluation of the design, final conclusions and recommendations
- These meetings take place at TU Delft or at the location of the (industrial) principal and for a wider audience of the principal's employees.
- The team members manage their own planning and project execution. During the project, the supervisors act solely as consultants. After project completion, the supervisors review and assess the results and the team performance.

Progress and consultation meetings are scheduled on a regular basis and are attended by the participants and supervisors (at least every three weeks, often, more frequently). The project is carried out under supervision of TU Delft staff members who are in an advisory role of an experienced process engineer/designer from the industry. The industrial supervisors join in the progress meetings (at TU Delft or at their location). In between the meetings, communication takes place by online meetings, telephone, or e-mail. The supervision by (an) experienced designer(s) from industry is a directive of the Netherlands Committee for Certification of (post-MSc) Technological Designer Programmes (CCTO).

14.3.3.8 Loss prevention in process design (ST6042, 5 ECTS)

The main objective of the course, loss prevention in process design, is the design of safe chemical processes. Important subjects are: Inherent safe process design, equipment design, selection of material, piping systems, heat-transfer systems, thermal isolation, process monitoring and process control, High pressure systems, electrical systems, control of dangerous emissions (for instances flares) and documentation.

The goal of the course is to familiarize EngD trainees with advanced knowledge of risks of fire, explosion, and toxicity of chemical products and processes. Students should be able to design a safe and environmentally friendly chemical product and plant using modern loss prevention design methods upon completion of the course. The course consists of the following parts:

– lectures/exercises on chemical risks (fire, explosion, and toxicity).
– student presentation on chosen safety topic.
– preparation of written conceptual design for hazardous chemical process, with detailed analysis of and solution to safety, health and environmental risks, making use of various approaches: layer of protection analysis (LOPA), inherently safer design (ISD), green chemistry, and green chemical engineering.
– presentation of project assignment results. Students will work in groups of two to four.

The objective of the assignment is to create a conceptual design of a chemical process for safe and environmentally friendly production of a hazardous substance (where "hazardous" = toxic, flammable, and explosive), using a modern design philosophy. The scope of the assignment comprises: synthesis route and reaction stoichiometry, fire/explosion/toxicity data on all substances involved (reactants, products, by-products), basic thermodynamics, basic kinetics, process flow sheet including most important unit operations, and description of process including process conditions, with special emphasis on design of most hazardous part of process.

14.3.3.9 Individual design project (ST6902, or ST6903, 60 ECTS)

All EngD trainees carry out an Individual Design Project (IDP) during the full second year of their EngD traineeship. The necessary skills needed for this project are provided during their prior studies and during the first year of the traineeship.

The IDP aims for the (re)design of a (bio)chemical product (material, formulation, configured product device), process or method, including attention for the imbedding in the supply chain.

In each case, it should comprise a design problem that requires a novel design for a client, rather than applying already existing designs. Each individual design assignment generally contains design as well as research and development aspects. In all cases, the research and development must be beneficial to the design: for example, acquiring required knowledge or experimental data.

The design part should amount to at least 60%, leaving at most 40% for the research and development part.

The design task can be part of a larger project, for example, a product and process development/design project. The designer will always interact with other project participants.

14.3.3.9.1 Confidentiality and working location

At the request of the principal, all knowledge obtained during the project will be treated as confidential and will not be disclosed to third parties without prior consent of the cooperating partner (principal). To this end, a cooperation agreement is signed by both the TU Delft and cooperating partner. This agreement also contains a nondisclosure agreement for the designer and TU Delft members of the supervising steering team. Detailed guidelines on how to act are given by the program management.

For the IDP, the EngD trainee will be mostly seconded to industry or to a research section at TU Delft. When working at the industrial partner, relocation may need to be considered.

14.3.3.9.2 Composition of steering group

The steering group of an IDP consists of a mentor from the industrial party (principal), a professor/mentor from the university with expert knowledge on the subject, and an experienced design mentor from the EngD program. The EngD program management will compose the steering team in close cooperation with the principal. The steering group members, together with an independent scientific member who is not involved in the IDP, form the IDP assessment committee.

The EngD trainee presents and defends the EngD design thesis to this committee, and the committee evaluates and grades the design and the design process. The steering group and the assessment committee are formally approved by the Board for Doctorates of TU Delft.

14.3.3.9.3 Milestones and implementation

During the IDP, at least four milestones can be identified: the kickoff, basis of design (BOD), the intermediate, and the final reviews (BOD, intermediate and final reports are to be submitted). They consist of meetings with both the industry and university supervisors and include presentations by the EngD trainee. The kickoff meeting is scheduled within the first two to four weeks after the start of the project. The BOD review is scheduled two to three months after the start of the project, the intermediate review is after 8 months, and the final review (defense) is at the end of the project. All members of the steering group attend these meetings. The independent assessor is only present during the IDP defense (final review).

Generally, the principal of the project will provide a project description, terms of reference, or proposal at the very beginning of the project. Often, the description is relatively brief (concise) and not clear on all details. One of the first activities of the designer will be to lay down the detailed project description in close cooperation with the principal. The project description addresses the objective of the project, deliverables, and a (time) plan how to obtain the deliverables (project approach). The project content will be formally discussed during the kickoff meeting in week two to four of the project.

The main purpose of the Kickoff meeting is to make clear the objectives and to agree upon them. The definition stage of a project leaves much room for misunderstandings between parties. Therefore, the description must be as explicit as possible. To avoid any misunderstandings, it is helpful to provide a description of the different project activities and the duration thereof. In this way, the parties involved will get a clear idea about the subject and how time is spent, and whether parties' expectations are satisfied. Focus points during the Kickoff meeting are:

- Project objective
- Scope of the project, delineation
- Approach
- Project planning, timetable
- First results

Several other meetings with the supervisors and/or experienced design engineers are organized throughout the year the design project takes. The meeting frequency depends on the need and opportunity. It is also the EngD trainee's responsibility to organize these meetings. At the minimum, it will be a monthly meeting with the project manager/design mentor from the EngD program (bi-weekly contact is advised).

Nontechnical skills needed for the execution of the project are to a large extent obtained during the first year, with dedicated courses on the social-economical aspects, and personal and professional skills as part of the core program. The designer is responsible to manage the project and to communicate adequately with all stakeholders involved. The designer has to organize and plan progress meetings. Moreover, the designer will also chair the meetings, and ensure distribution of the agenda and the minutes of meeting.

14.3.3.9.4 Reporting and oral presentations

The purpose of reporting is to document and transfer the obtained knowledge from the designer to the project principal. To give evidence of skills, the designer is requested to prepare a BoD report and an Intermediate report. In this way, the reporting effort is started at an early stage and is adequately spread over the project span.

Alongside the written report, the designer will also report orally during the steering group review meetings. At the end of the project, the IDP will be presented and defended for the assessment committee. This could take place at the university or at the principal. In the latter case also, company coworkers and management can attend the Final presentation (not the defense itself). The second presentation is held at TU Delft in the form of an open colloquium. As a rule, the colloquium is open to the public. Care must be taken to have the contents cleared by the principal beforehand so as not to disclose any confidential information.

14.3.3.9.5 Assessment

After the completion of individual design project, the work will be reviewed with respect to theory, implementation, and communication in accordance with the TU Delft EngD assessment criteria. For all three aspects, a mark will be given. The appraisal of the work will be carried out by the assessment committee (steering committee plus independent assessor) and is done under the responsibility of the professor/mentor.

14.3.4 Advanced courses (PhD, EngD, and participants from industry)

The TU Delft's EngD programs, BioTechDelft [11], the Safety and Security Institute [12], and Pro2Tech [3], alongside the Dutch Institute for Catalysis Research (NIOK) [13] and the J.M. Burgerscentrum (JMBC) research school for fluid mechanics [14] organizes two to five days advanced courses/workshops for PhD candidates, EngD trainees, postdocs, and for participants from the industry.

The Chemical Product Centric Sustainable Process Design course (by R. Gani) held at TU Delft and organized by the EngD program is described below in detail. The other courses are listed below.

14.3.4.1 Chemical product-centric sustainable process design (PhD/EngD course)

The objective of this PhD/EngD course is to give the participants a view of chemical product design and the important process design issues related to their development (product-centric process design). The course will highlight how to define the needs of a chemical product, how to identify the candidate chemicals and/or mixtures of chemicals, and how to quickly evaluate the important process design issues so that decisions related to product development can be made in the early stages of product development. The objective is also to highlight the currently available methods and tools that can be applied to solve various types of problems associated with product and process design in a systematic and integrated manner. Different case studies will be used as application examples. The ICAS software developed by Computer-Aided Process Engineering Center (CAPEC) at Technical University of Denmark (DTU) will be used. Also, two new software tools, *ProCAFD* (a tool for synthesis-design of chemical processes) and *ProCAPD* (a tool for synthesis-design of chemical products), made available through the PSE for SPEED Company Ltd. [15], will be used in the course.

In chemical product design and development, one first tries to find a product that exhibits certain desirable or targeted behavior and then tries to find a process that can manufacture it with the specified qualities. The candidate may be a single chemical, a mixture, or a formulation. For the latter product type, additives are usually added to an identified active ingredient (molecule or mixture) to significantly enhance its desirable (target) properties. Examples of chemical products, such as *functional chemicals* (solvents, refrigerants, lubricants, etc.), agrochemicals (pesticides, insecticides, etc.), pharmaceuticals and drugs, cosmetics and personal care products, home and office products, etc., can be found everywhere. The workshop will only cover chemical-based products and issues related to their design, development (includes also process design) and analysis (product performance). Even though it is possible to identify many chemicals and/or their formulations as potential chemical products, only a small percentage actually become commercial products. Finding a suitable process that can sustainably (i.e., safely, environmentally, reliably, efficiently, and economically) manufacture the identified chemical with the desired product qualities as well as evaluating the product performance during application and analyzing market trends play important roles in product design and development. From a process point of view, there are products where the reliability of the quality of the manufactured chemical may be the deciding factor (for example, drugs and agrochemicals), while there are others where the cost of manufacturing the product is at least as important as the reliability of the product quality (such as solvents, refrigerants, and lubricants). This means that product-centered sustainable process design is important because identifying a feasible chemical product is not enough. To make it sustainable, the process needs to be efficient, reliable, economically feasible, and environmentally acceptable. Also, while in the case of functional chemicals, the identified molecule or mixture is the final product, in the case of chemical-based consumer products (drugs, cosmetics, and personal care products), they are intermediate products from which the final products are obtained through additional processing. Therefore, the performance of the manufactured product, when applied, needs to be tested and validated. For some functional chemical products (such as solvents and refrigerants), this may be straight forward, but for some consumer products (such as drugs and food-products), it may not be so straight forward.

14.3.4.2 Other advanced courses
- TU Delft BioTechDelft [11]:
  - Downstream processing
  - Environmental biotechnology
  - Bioprocess design
  - Microbial physiology and fermentation technology
- Pro2Tech [3, 16]
  - Process technology for non-process technologists
- NIOK (Dutch institute for research in catalysis) [12]:

- Catalysis, an integrated approach
- Advanced catalysis engineering
- Environmental catalysis
- NIOK advanced topics in catalysis
- JM Burgerscentrum (JMBC) [13]:
  - Computational fluid dynamics 1
  - Particle-based modeling techniques

14.3.5 (Process) systems engineering (PSE, SE) education development

Recently, the (process) systems engineering (PSE) community has initiated efforts to redesign PSE courses, incorporating emerging topics such as carbon-neutral PSE, sustainable bio-based PSE, energy efficiency PSE, and artificial intelligence (AI) in PSE. These four areas offer opportunities for applying and developing PSE/SE design approaches, process control, and numerical methods [17].

14.4 Glossary

3TU.SAI	Federation of three Dutch technical universities – Stan Ackermans Institute
4TU.SAI	Federation of four Dutch technical universities – Stan Ackermans Institute
BSc	Bachelor of science
BOD	Basis of design
CAPEC	Computer-Aided Process Engineering Centre
CCTO	Certification Committee for (post-MSc) Technological Designer Programmes
CO_2	Carbon dioxide
DCF	Discounted cash flow
DDM	Delft design map
DPTI	Delft Process Technology Institute
DTU	Denmark's Technical University
ECTS	European Credit Transfer System
EngD	Engineering doctorate
GDP	Group design project
ICAS	Software tool developed at CAPEC/DTU
IDP	Individual Design Project
ISD	Inherent safer design
JMBC	J.M. Burgerscentrum, Research School for Fluid Mechanics
KIVI	Koninlijk Instituut Van Ingenieurs (Royal Dutch Society of Engineers)
LOPA	Layer of protection analysis
LST	Life science and technology (BSc and MSc programs)
MSc	Master of science
MST	Molecular science and technology (BSc program)
NIOK	Dutch Institute for Catalysis Research
OSPT	Onderzoeksschool Procestechnologie (Dutch for: Research School Process Technology)
PDEng	Professional doctorate in engineering

PhD	Doctor of philosophy
PIN-NL	Process Intensification Network – Netherlands
PPP	Public private partnerships
PSE	Process systems engineering
PSE/SE	Process systems engineering/systems engineering
Pro2Tech	Delft Process and Product Technology Institute
ProCAFD	Software tool for computer aided process synthesis-design
ProCAPD	Software tool for computer aided product synthesis-design
PSE-NL	Process Systems Engineering – Netherlands (network)
RUG	Rijksuniversiteit Groningen
SE	Systems engineering
TU	Technical University
TU/e	Technical University Eindhoven
UT	University Twente

References

[1] Molecular Science and Technology BSc curriculum, TU Delft, 2023. (Accessed November 30, 2023, at https://www.tudelft.nl/onderwijs/opleidingen/bachelors/mst/bsc-molecular-science-and-technology/onderwijsprogramma).

[2] Life Science and Technology BSc curriculum, TU Delft, 2023. (Accessed November 30, 2023, at https://www.tudelft.nl/onderwijs/opleidingen/bachelors/lst/bsc-life-science-and-technology/onderwijsprogramma).

[3] TU Delft Product & Process Technology Institute (Pro2Tech), TU Delft, 2023. (Accessed November 30, 2023, at https://www.tudelft.nl/pro2tech).

[4] QS world universities ranking by subject – Chemical engineering. QS Top universities, 2020, 2021, 2022, 2023). (Accessed November 30, 2023, at https://www.topuniversities.com/university-subject-rankings/chemical-engineering/2020; https://www.topuniversities.com/university-subject-rankings/engineering-chemical/2021; https://www.topuniversities.com/university-subject-rankings/chemical-engineering/2022; https://www.topuniversities.com/university-subject-rankings/chemical-engineering).

[5] Swinkels PLJ, Post-MSc technological design (PDEng) traineeships by Dutch universities of technology catalyze industrial innovation, QScience Proceedings (Engineering Leaders Conference 2014) 2015:21 (Accessed October 8, 2017, at http://dx.doi.org/10.5339/qproc.2015.elc2014.21.)

[6] Swinkels P, Grievink J, Hamersma P. Technical and human factors in chemical process design courses. In Tu S-D, Wang Z-D (Eds.). Total Engineering Education, Shanghai: East China University of Science and Technology Press, 2006, 59–65.

[7] Traineeships Engineering Design – EngD programme Process and equipment design, TU Delft, 2023. (Accessed February 4 2024, at https://www.tudelft.nl/en/faculty-of-applied-sciences/education/master-programmes/post-msc-programmes/product-process-and-equipment-design/objectives).

[8] Process Systems Engineering in The Netherlands, 2023. (Accessed November 30, 2023, at https://pse-nl.com).

[9] Process Intensification Network (PIN-NL), 2023. (Accessed November 30, 2023, at https://www.rvo.nl/onderwerpen/kennisnetwerken-industrie-en-chemie/kennisnetwerk-procesintensificatie-pin-nl).

[10] Dorst CH. Design problems and design paradoxes, Design Issues, 22(3), 2006, 4–17.

[11]   TU Delft BioTechDelft, postgraduate education, 2023. (Accessed 30 November 2023, at https://biotechdelft.com/).

[12]   TU Delft Safety & Security Institute, PhD, postdoc (and EngD) course 'Safe by Design', 2023. (Accessed 30 November 2023, at https://www.tudelft.nl/tu-delft-safety-security-institute/events/phd-course-safe-by-design).

[13]   NIOK Education, 2023. (Accessed 30 November 2023, at https://niok.nl/education/).

[14]   TU Delft JM Burgerscentrum, PhD programme, 2023. (Accessed 30 November 2023, at https://www.tudelft.nl/jmburgerscentrum/courses).

[15]   PSE for SPEED – Sustainable Product-Process Engineering, Evaluation & Design. (Accessed November 30, 2023, at http://www.pseforspeed.com.)

[16]   Pro2Tech course "Process technology for non-process technologists". (Accessed December 1, 2023, at https://www.tudelft.nl/evenementen/2021/tu-delft-process-technology-institute/process-technology-an-introduction).

[17]   Lewin DR, Zondervan E, Franke M, Kiss AA, Krämer S, Pérez-Fortes M, Schweidtmann AM, Slegers PM, Somoza-Tornos A, Swinkels PLJ, Wentink B. An educational workshop for effective PSE course development, Computer Aided Chemical Engineering, 52, 2023, 3495–3500. Accessed December 1, 2023, at https://doi.org/10.1016/B978-0-443-15274-0.50558-8.

A3 Appendix to Chapter 3

Group design with Belbin team roles

Project groups are mostly designed based on disciplines required only. However, if groups have to cooperate closely for the desired results, then they should get along with each other. In addition to discipline design, the Belbin team roles can be used to design the group or team [1].

Belbin role descriptions
The **Plant** role is to be highly creative and good at solving problems in unconventional ways. He may be the outsider planted in the team.
The **Monitor** provides a logical eye to make impartial judgements about options.
The **Coordinator** focusses on the team's objectives, and delegates work appropriately.
The **Resources investigator** provide knowledge from outside the team.
The Implementer plans a practical, workable strategy and carries it out as efficiently as possible.
The Finisher make sure to remove errors at the end of a task, subjecting it to the highest standards of quality control.
The Team worker helps the team to gel. He may also fill in gaps left by others.
The Shaper provides the necessary drive to keep the team moving to reach the goal.
The Specialist provides in-depth knowledge of key area. His weakness is to focus narrowly on their problem element and prioritize it for the team's progress. In the authors view, it is, in general, better to consult the specialist rather than make him a team member.

The preferred role of an individual can be determined himself using the Belbin self-perception method. This method can be found hereafter.
Arittzeta [2] provides a review of a large number of experimental studies on teams designed with Belbin team roles. His conclusion is that Belbin's team role theory has been validated. Some roles, however, show inter-correlation. With the inter-correlation information, the following major style grouping can be made:

Completer style: Completer-Finisher, Implementer, Monitor Evaluator
Bridge style: Team worker, Monitor Evaluator, Coordinator
Innovative style: Resource Investigator, Plant, Shaper

So, small teams should have one of each of these three major styles.
The author also had a powerful experience with groups designed using the Belbin self-perception team roles. The most complimentary team indeed produced the best

https://doi.org/10.1515/9783110782127-015

results, while the team with members of only one role (specialists) performed worse, by far.

Making inventory for Belbin team roles by self-perception

Teams with complementary Belbin roles can be easily composed as follows. Each potential team member fills in a questionnaire and concludes their preferred role(s). Then, a team or several teams can be composed with complementary roles. The questionnaire can be obtained from:

http://www.belbin.com/about/belbin-team-roles/

Creativity methods

The literature on creativity methods for innovation is vast. We limit ourselves here to methods for which the authors have experience in industrial product-process practice. A good introduction to further information is by Nijstad [3] and Hermann [4].

Brainwriting 6-3-5

Creating ideas within the product-process industries is probably slightly different from other industries, in that, process engineers, in majority, are introverts, who feel uncomfortable to directly express ideas. Brainwriting is sometimes a better method than brainstorming, in particular for introvert people. On average, brainwriting generates twice as many ideas as brainstorming [3]. For process engineers, this factor could be even higher.

The brainwriting 6-3-5 method is very simple to execute and needs no trained facilitator.

Step 1: Plenary, a clear problem to solve is defined and is stated to all participants.

Step 2: Groups of preferably 6 people are formed.
Each member gets a writing format with at least 3 columns and a number of rows at least equal to the number of members.

Step 3: In round 1, each group member writes down silently, three ideas in the first row on his writing format.

Step 4: After 5 minutes, the writing formats are rotated to the neighboring members and round 2 is executed in which each member enriches the three ideas of the first row by putting additional ideas in the second row.

Steps 5: This 5-minute round cycle is repeated until the members are exhausted of ideas. Of course, additional clean sheet writing formats can be provided to start new sessions and rounds.

Step 6: All filled-in sheets are gathered by the facilitator and documented outside the brainwriting session.

Potential problem analysis for startup

Potential problem analysis for the startup phase is a very useful for generating a robust startup plan. Here is a template that the author has prepared and used for making robust startup plans. It is derived from a problem analysis method [5] and turned into a potential problem analysis method.

Potential problem analysis template

Question	Reference product/process	New design	Different new design	Risk
What				
Where				
When				
Size				

References

[1] Belbin. 2017. Resourced from: www.belbin.com/rte.asp.
[2] Aritzeta A, et al. Belbin's team role model: Development, validity and applications for team building. Journal of Management Studies, 44(1), 2007, 96–118.
[3] Nijstad BA, Paulus PB. Group creativity. Group creativity: Innovation through collaboration. 2003, Sep 4, 326–39.
[4] Herrmann D, Felfe J. Effects of leadership style, creativity technique and personal initiative on employee creativity. British Journal of Management, 25, 2014, 209–227
[5] Kepner CH, Tregoe BE. The New Rational Manager, New |Jersey USA: Kepner-Tregoe Inc. 1981.

A4 Appendix belonging to Chapter 4

History of (chemical) product and process design, leading to Delft Design Map

(Chemical) products and their manufacturing processes have been around for a long time, long before formal product and process design methodologies were developed and taught at the universities of technologies since the end 1960s. Industry managed to do this based on heuristic knowledge of and operational experience with these products and processes. Experienced designers transferred their knowledge and experience through heuristic rules to novice designers. Design improvements were mostly incremental.

Design courses at universities in the 1970s–1980s focused heavily on the (very powerful) process "unit operations" design, with attention for shortcut methods for process integration and performance analysis, mainly based on cost estimation and techno-economic measures. The really good textbooks *Plant Design and Economics for Chemical Engineers* [1] and *Strategy of Process Engineering* [2] were used in many curricula in this period. The conceptual process design course taught at TU Delft by Prof. A. Montfoort (TU Delft 1970–1990) was based on this approach.

Process design methodology development for education started to take off in the 1980s. In this period, the steady-state process flowsheet simulators (like ASPEN™, ASPEN-PLUS™ (AspenTech) and PRO-II™ (Simulation Sciences Inc.) emerged from research work at universities (like Massachusetts Institute of Technology (MIT), and were launched as commercial tools for analyzing mass and energy balances of complete processes. Through these computational tools – including key thermodynamic models – a "systems engineering approach," built-up from unit operations and including process analysis and optimization became possible. The development of the process integration methodologies (thermal pinch analysis) at the University of Manchester Institute of Technology (UMIST) by Linnhoff in 1968 [3], with applications in Imperial Chemical Industries (ICI) in the 1980s, also gave the process systems engineering thinking a clear boost. Here, the systems engineering approach was used to arrive at considerable energy conservation.

After this, the attention for process design methods took off (see Fig. A.4.1). In the period 1985–1995, dynamic flow sheet simulators were also developed for modeling process dynamics, startup, and control: SPEEDUP and gPROMS® were both developed at the Imperial College London (ICL), whereas Jacobian was developed at MIT. The use of process flowsheet simulators allowed process systems engineering research and education to focus again on the fundamental process-synthesis problem: what unit operations and process structures were best suited for the required process task? The widely used textbook *Conceptual Design of Chemical Processes* by Douglas issued in 1988 [4] successfully attempted to address these questions. In this book, he

https://doi.org/10.1515/9783110782127-016

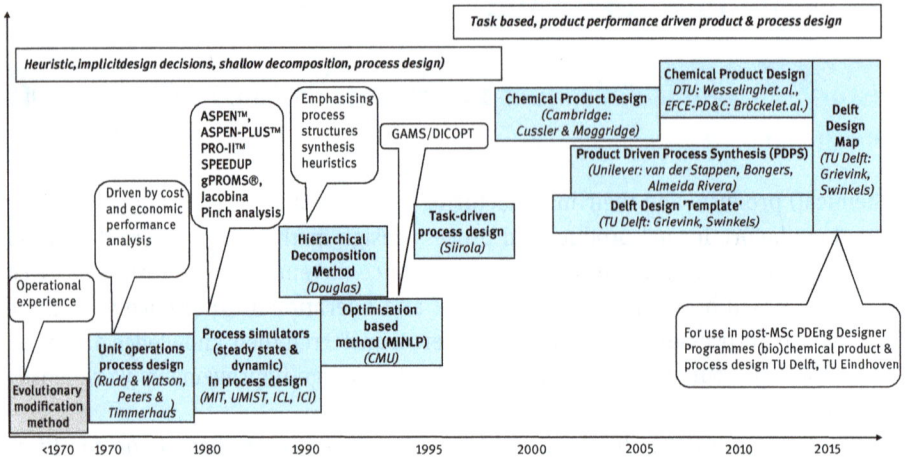

Fig. A.4.1: (Chemical) product and process design methods: a historical perspective.

successfully translated his vast wealth of industrial design experience into a hierarchical design decomposition scheme. Process structure options were generated/synthesized using plausible heuristic rules in combination with the unit operations concept. In 1999, Seider and Lewin published the first edition of their student textbook on process design principles [5], taking a similar approach but with more details on the process structure and equipment design and costing. In the new millennium, new editions of this textbook were upgraded over the years.

In the late eighties, computer-based optimization methods were also developed to solve process design problems. Based on predefined superstructures encompassing all process flowsheet structure options, mixed-integer non linear (MINLP) programming techniques were developed and applied to find the optimal flow sheet, given the optimization function. The GAMS/DICOPT MINLP algorithm was developed by Kocis and Grossman at the Carnegie Mellon University [6].

In the mid-nineties, a novel task-driven process design approach was published by Siirola [7, 8]. For the first time, unit operations were no longer the building blocks for process design, but rather individual tasks or functions, such as adsorption, reaction, heat and mass transfer, and diffusion. These tasks could be combined in single columns, generating novel design solutions, such as reactive distillation, in situ crystallizer fermentation, etc. Inside Eastman Chemical Company, this task-driven design approach had led to the design and implementation of large-scale reactive distillation columns in which many tasks were combined, the well-known methyl acetate production route [9]. This multifunctional design approach is now part of the process intensification design tool box.

For product design, let alone concurrent product-process design, still no generic design methods were developed and published in this period. For homogeneous products, such as food products and chemical products, designs were carried out inside companies, based on experience and tacit knowledge. In the nineties, the product design approach was at the phase where process design was in the late 1960s. Each industry has many heuristic rules, mostly limited to the existing product portfolio and based on in-house experimental validation. Industrial researchers are reaching out to academia, where attempts were made to develop and understand application and manufacturing of new molecules, structures, and materials.

With an increasing focus in the chemical industry and related academia on product innovation, tools were becoming available– once the desired product composition and structure is known – product-driven and task-based process design methods are available to tackle these design challenges. However, the product design phase – how to arrive from a desired function/performance of the product to its composition and structure – is another matter. This is a field where chemical engineers have historically not been educated and involved in, and is largely the domain of the chemistry and material sciences scientists.

In the first years of the millennium, the first (chemical) product design courses were initiated at several universities. Also, the first student textbooks were published at Cambridge by Cussler & Moggridge [10] at RU Groningen and DTU Copenhagen by Wesselingh et al. in 2007 [11]. In 2010, also the third edition of the textbook by Seider et al. [12] focused on product design aspects for the first time.

The approach adopted by Cussler et al. and Wesselingh et al. starts with describing and explaining how chemical products work in performing their task, and how changes can be made to improve them. Similar to the unit operation approach, this product-related knowledge was firstly presented and taught in a rather compartmentalized way, with little connection between the different classes of products and applications. Wesselingh (with his transport phenomena background) tried to place more emphasis on the importance of an engineering (quantification) approach, with transport phenomena as a key part of product performance. He attempted to capture product application properties in dimensionless numbers, and advocated the use of time-constant estimation of the ruling phenomena. This is a powerful tool process designers use in the "scale-up by scale-down" approach for processes, but is equally suited for the characterization of product-in-use processes.

Frens and Appel (both TU Delft) have also stressed in their product design courses, the importance of physical chemistry and complex rheology in the formulation and functioning of products in product design. Process systems engineering (PSE) scientists/engineers, with a focus on applying chemical thermodynamics in process synthesis & engineering (Gani at Technical University of Denmark (DTU)'s Computer-Aided Process-Product Engineering Center (CAPEC) and K. Ng at Hong Kong University of Science and Technology (HKUST) started to extend their work to product design, development, and simulation in the first decade of the new millennium.

In 2003, the European Federation of Chemical Engineering (EFCE) Section on Product Design & Engineering (PD&E) was founded in Granada, Spain. This happened at the first EFCE-sponsored PD&E Conference. In 2007, members of this section published two volumes of a book on *Product Design and Engineering: Best Practices*, edited by Bröckel et al. [13, 14]

Starting in 2001, the cooperation between Grievink (ex Shell) and Swinkels (ex Unilever) at TU Delft led to the combination of their complementary process and product design knowledge and experience. The oil, gas, and chemicals company Shell has a long history of designing and implementing large-scale continuous processes for homogeneous products. Unilever, the fast-moving consumer goods company, has a long history of designing and developing novel structured and multiphase products for its world markets. In this period, Grievink and Swinkels developed, what at that time was called hierarchically decomposed and task-oriented "Delft Design Template for Conceptual Process Design." In 2014, a simplified version of the Delft Design Template for Conceptual Process Design was published in 2014 [15].

In cooperation with TU Delft, Unilever developed and tuned the Delft Design Template approach to fast-moving consumer products (mainly foods) and proposed the "Product-Driven Process Synthesis (PDPS)" methodology, described by Meeuse in 2000 [16] and in 2017, by Almeida-Rivera [17].

Key developers van der Stappen, Bongers, and Almeida-Rivera at Unilever and Boom et al. (Wageningen University & Research (WUR)) also adopted this approach in their research into the design of new food products and their manufacturing.

A very similar approach in adopting a schematically conceptual model for chemical product design has been taken and published by Bernardo at the University of Coimbra in 2017– [18], in which the same design levels/ functions are proposed (quality function, property function, and process function) as in the Delft method, described in chapter 7. Bernardo also proposes a mathematical approach to solve the concurrent/ simultaneous product-process design optimization, as described in chapters 8 and 9.

The design methods, Delft Design Template (DDT) (now restructured and called Delft Design Map, DDM) and Unilever's Product Driven Process Synthesis (PDPS)), have been taught for many years at the post-MSc PDEng programmes at TU Delft (Process & Equipment Design, BioProcess Engineering, Chemical Product Design, and TU Eindhoven (Process and Product Design) and applied in the PDEng design projects executed for and in industry. The advantage is that PDEng graduates get a clear systematic approach to product-oriented process design, which is not dependent on product and process circumstances, and can be applied in the product-process industry sectors – chemical, biochemical, food, and others.

Delft Design Map for product and process design (DDM)

The Delft Design Map (DDM) is based on task-based design activity spaces, shaped in matrix form. The horizontal axis corresponds to innovation stages and the vertical axis corresponds to design steps, from problem definition to solution to reporting.

The empty blocks of the matrix are then filled during the design effort in a preplanned order until the matrix is filled. The design problem at hand (new product, new process & new supply chain, new reactor technology, and heat integration) determines where the emphasis of the design work should lie, but also emphasizes that designers should be aware that implications of apparent small design changes (in reactor technology or catalyst type, for instance) may have profound effects (positive or negative) on the entire system. The design team can also use the DDM to map the route of the search for suitable design solutions, encompassing formulating the design problem, planning the design tasks, following through these activities, including reporting decisions, and results. The DDM then becomes a planning, design management and communication tool. Therefore, the authors (after consulting Grievink) renamed the "Delft Template for Conceptual Process Design" into the "Delft Design Map for Product & Process Design."

This map will be even more helpful for students and industry designers than its predecessor. They get an overview of all design actions to be pursued to fill the matrix. This gives them confidence that they can make a design because the whole problem is cut up into small design action steps. The Delft Design Map is described in detail in chapter 4.

References

[1]   Peters MS, Timmerhaus KD. Plant Design and Economics for Chemical Engineers, 2nd edn, New York, NY, USA: McGraw-Hill, 1968.

[2]   Rudd DF, Watson CC. Strategy of Process Engineering, New York, NY, USA: John Wiley & Sons Inc, 1968.

[3]   Linnhoff B, Flower FR. Synthesis of heat exchanger networks: I. Systematic generation of energy optimal networks, AIChE Journal, 24(4), 1978, 633–642.

[4]   Douglas JM. Conceptual Design of Chemical Processes, New York: McGraw-Hill, 1988.

[5]   Seider WD, Lewin DR. Process Design Principles: Synthesis, Analysis and Evaluation, New York, NY, USA: John Wiley & Sons Inc, 1999.

[6]   Kocis GR, Grossman IE. Computational Experience with DICOPT: Solving MINLP problems in process systems engineering, Computers and Chemical Engineering, 13(3), 1989, 307–315.

[7]   Siirola JJ, An industrial perspective on process synthesis, 4th international conference on foundations of computer-aided process design (FOCAPD), AIChE Symp Series USA 1995, 91 (304), 222–233.

[8]   Siirola JJ. Industrial applications of chemical process synthesis. In Anderson JL, Bischoff KB (eds.). Advances in Chemical Engineering Volume 23: Process Synthesis, Academic Press, 1996, 1–92.

[9]   Agreda VH, Partin LR, Heise WH. High-purity methyl acetate via reactive distillation, Chemical Engineering Progress, 86(2), 1990, 40–46.

[10] EL, Moggridge GD. Chemical Product Design, Cambridge, UK: Cambridge University Press, 2001.

[11] Wesselingh JA, Kiil S, Vigild ME. Design & Development of Biological, Chemical, Food and Pharmaceutical Products, Hoboken: John Wiley & Sons Ltd, 2007.

[12] Seider WD, Seader JD, Lewin DR, Widagdo S. Process and Process Design Principles: Synthesis, Analysis and Evaluation, 3rd edn, Hoboken, NJ, USA: John Wiley & Sons Inc, 2010.

[13] Bröckel U, Meier W, Wagner G. Product Design and Engineering – Best Practices, Volume 1: Basics and Technologies, Weinheim, Germany: WILEY-VCH Verlag GmbH & Co. KGaA, 2007.

[14] Bröckel U, Meier W, Wagner G. Product Design and Engineering – Best Practices, Volume 2: Raw materials, Additives and Applications, Weinheim, Germany: WILEY-VCH Verlag GmbH & Co. KGaA, 2007.

[15] Grievink J, Swinkels PLJ, Van Ommen JR. Basics of Process Design. In De Jong W, Van Ommen JR (eds.). Biomass as a Sustainable Energy Source for the Future: Fundamentals of Conversion Processes, 1st edn, Hoboken NJ, USA: Wiley & Sons Inc, 2014, 184–229.

[16] Meeuse M, Grievink J, Verheijen PJT, Vander Stappen MLM. Conceptual design of process for structured products, AIChE Symposium Series USA, 96, 2000, 324–328.

[17] Almeida-Rivera C, Bongers P, Zondervan E. A structured approach for product-driven process synthesis in foods manufacture. In Martin M, Eden MR, Chemmangattuvalappil NG (eds.). Computer Aided Chemical Engineering Vol.39 Tools for Chemical Product Design – From Consumer Products to Biomedicine, Amsterdam, Netherlands: Elsevier BV, 2017, 417–41.

[18] Bernardo FP. Integrated process and product design optimization. In Martin M, Eden MR, Chemmangattuvalappil NG (eds.). Computer Aided Chemical Engineering Vol.39 Tools for Chemical Product Design – From Consumer Products to Biomedicine, Amsterdam, Netherlands: Elsevier BV, 2017, 347–72.

A9 Appendix belonging to Chapter 9: Cases of process modeling for simulation and optimization

This appendix presents 13 application cases. An upfront remark about a restrictive use of some terms is in order. The term *behavior* of a process or product relates to patterns of change in its chemical and physical quantities. The term *performance* is intentionally reserved for factors that express some kind of value and appreciation by the human society, including customers of a product and the operators and owners of a process. Such factors are related to safety, health, economy, and ecology. The literature references appearing in this appendix are listed at the end of this appendix, independently of those in Chapter 9. Also the commonly used mathematical symbols are listed at the end. However, many symbols are repeatedly in use in the various cases and sometimes with a different meaning. Therefore, to avoid ambiguity care has been taken for each case to have some additional description of its symbols as well.

A9.1 Linear models applications for *concept* stage

Before starting the application cases it is expedient to summarize five different mathematical forms of linear models, all of which occur in process modeling practice. A model is linear when each equation in the model is linear in all variables. Determin-

https://doi.org/10.1515/9783110782127-017

istic variables and models will only be covered, excluding stochastic and logical models. The five common linear deterministic model forms are:

1) Linear algebraic equations:

$$\mathbf{A} \cdot \underline{x} = \underline{b}$$

Symbols
A: $m*n$ matrix;
$\underline{x}$: n-vector; $\underline{b}$: m-vector.

2) Linear ordinary differential equations:

$$d\underline{x}(t) / dt = \mathbf{A} \cdot \underline{x}(t) + \mathbf{B} \cdot \underline{u}(t)$$

t: scalar coordinate
$\underline{x}$: n-vector function;
$\underline{u}$: m-vector function;
A: $n*n$ matrix;
B: $n*m$ matrix.

3) Linear partial differential equations:

$$\partial \underline{x}(z,t) / \partial t + v \cdot \partial \underline{x}(z,t) / \partial z = \mathbf{A} \cdot \underline{x}(z,t)$$
$$+ \mathbf{B} \cdot \underline{u}(z,t)$$

t, z: scalar coordinates;
v: scalar
$\underline{x}$: n-vector function;
$\underline{u}$: m-vector function;
A: $n*n$ matrix;
B: $n*m$ matrix.

4) Linear difference equations:

$$\underline{x}_{k+1} = \mathbf{A} \cdot \underline{x}_k + \mathbf{B} \cdot \underline{u}_k$$

k: integer index, $k \in \{0,1,2,\ldots\}$
$\underline{x}$: n-vector function;
$\underline{u}$: m-vector function;
A: $n*n$ matrix; B: $n*m$ matrix.

5) Linear integer equations with binary variables:

$$\mathbf{Z} \cdot \underline{Y} \leq \underline{\omega}$$

Z: $m*n$ matrix;
$\underline{Y}$: n-vector; $\underline{\omega}$: m-vector

The matrices **A** and **B**, vectors $\underline{x}$, $\underline{u}$, as well as the scalars t and z are all made up of real numbers. Matrix **Z** and vector $\underline{\omega}$ take integer values. The entries in vector $\underline{Y}$ take binary values {0,1}. Matrices A, B, and Z are constant, having fixed entries.

The application cases with linear models in this appendix deal with algebraic equations and real variables. An exception is the modeling of the structure of a process block diagram/flow sheet involving a linear model with binary variables.

Case A9.1: Exploring the stoichiometric space for an innovative chemical conversion

Reaction stoichiometry offers freedom in the design of a process concept. A simple example will be given on a search over a stoichiometric space to find a suitable reaction that serves a particular conversion goal best. In this case the goal is the chemical conversion of carbon dioxide as a greenhouse gas into carbon monoxide as a chemical reactant in syngas. The reuse of carbon in chemical products helps in closing the carbon cycle.

In analogy to steam reforming [$CH_4 + H_2O \Rightarrow CO + 3\,H_2$] one can apply dry reforming with CO_2 as is done in the steel industry [$CH_4 + CO_2 \Rightarrow 2\,CO + 2\,H_2$]. Unfortunately, the conversion to CO by these reforming reactions is reduced by the water-gas shift reaction [$CO + H_2O \Rightarrow CO_2 + H_2$], shifting CO partially back to CO_2. All reactions take place in the gas phase at high temperature. A disadvantage of the dry reforming reaction is the low molar ratio of $CO_2 : CH_4 = 1{:}1$. It would be nice to achieve a higher ratio of CO_2 to methane as a way of recycling more CO_2 molecules per methane molecule. This issue can be explored by means of modeling the possible reaction stoichiometry based on a given set of components. In this modeling exercise, the set {CH_4, CO_2, CO, H_2, H_2O} are the reacting species, comprising {C, H, O} as elements.

Synthesis of model

The molar reaction stoichiometry is expressed in a generalized way.

Generalized reaction: $\qquad$ a CH_4 + b $CO_2 \Rightarrow$ x CO + y H_2 + z H_2O

There are three likely reaction products at the right-hand side, excluding O_2 as a potential fourth candidate because the latter is not formed due to unfavorable thermodynamics.

The unknown stoichiometric coefficients {a, b; x, y, z} are constrained by three element balances:

C balance:	$a + b - x = 0$
H balance:	$4a - 2y - 2z = 0$
O balance:	$2b - x - z = 0$

The difference between the stoichiometric coefficients of CO_2 and CH_4 is introduced as a convenient, additional variable:

$$\delta = b - a \geq 0$$

These coefficients should be at least unity:

$$a \geq 1 \quad \text{and} \quad b \geq 1$$

The other coefficients must be nonnegative:

$$x \geq 0 \quad y \geq 0 \quad z \geq 0$$

When adopting the conventional form of writing chemical reaction equations, the stoichiometric coefficients take integer values only. From a mathematical point of view this integer condition is not strictly needed and one can take a = 1 as a pivot in the above reaction equation.

The objective is to find the reaction equation that maximizes the CO_2 consumption relative to methane consumption:

$$Max.\ F = b/a$$

Analysis of model

This model has six variables $\{a, b, x, y, z, \delta\}$ with four algebraic equality relations, resulting in two degrees of freedom. If a problem has a much larger size it can better numerically solved and optimized by applying linear programming. Due to the small size of this problem an analytical approach with graphical solutions will be followed here. The stoichiometric coefficients of the reaction products $\{x, y, z\}$ can be expressed as linear functions of a and b.

$$x = a + b \geq 0 \quad \Rightarrow \quad a \geq 0 \ \text{ and } \ b \geq 0$$

$$y = 3a - b \geq 0 \quad \Rightarrow \quad 3a \geq b$$

$$z = -a + b \geq 0 \quad \Rightarrow \quad b \geq a$$

Now the problem has been reduced to a two-dimensional search space $\{a, b\}$ for optimization.

Solutions of the model

Fig. A9.1 depicts the $\{a, b\}$ solution space with contour lines of the objective function for $(\delta = 0, 1, 2, ..)$ and two inequality constraints $\{3a \geq b; b \geq a\}$. Three solutions are shown in this space:

Dry reforming	$(a = 1; \quad b = 1; \quad \delta = 0)$;
Semi-dry reforming	$(a = 2; \quad b = 3; \quad \delta = 1)$;
Super-dry reforming	$(a = 1; \quad b = 3; \quad \delta = 2)$.

It is noted that the feasibility condition $3a \geq b$ imposes an upper bound on the objective function:

$$F = b/a \leq 3a/a \quad \Rightarrow \quad F \leq 3$$

One could wonder how many feasible solutions exist for $\delta = 1, 2, 3, \ldots$ It is convenient to take $\{a, \delta\}$ as the two independent variables. The remaining stoichiometric coefficients become:

$$b = a + \delta; \quad x = 2a + \delta; \quad y = 2a - \delta; \quad z = \delta$$

Critical feasibility condition is:

$$3a \geq b \quad \Rightarrow \quad 2a \geq \delta$$

Lower and upper bounds on F:

$$1 < F < 3$$

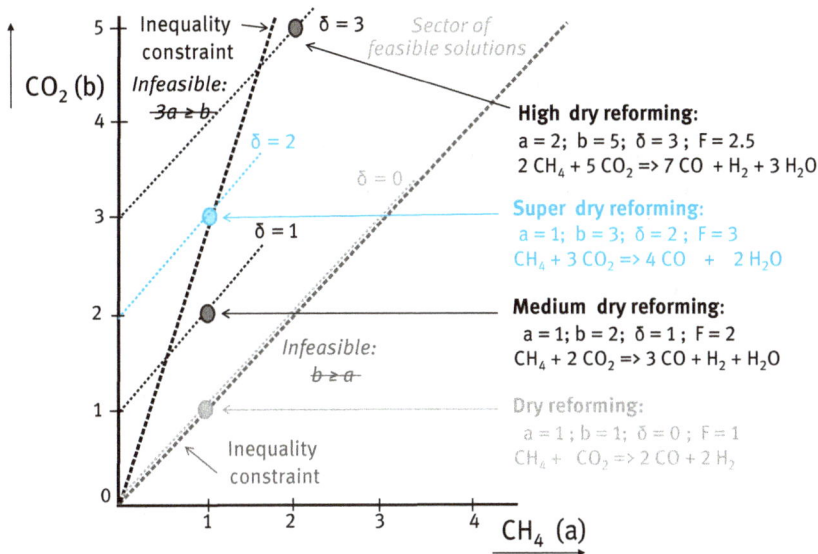

Feasible domain with three stoichiometric possibilities for CO_2 consumption. The highest consumption (***super dry reforming***) arises for a=1 and b=3.

Fig. A9.1: Searching for optimized solutions in the stoichiometric reaction space.

What are the results of a systematic exploration of the {a, δ} space for a = {1, 2, 3, .., 7} with 2a ≥ δ and δ = {0, 1, 2, 3, 4}? In fact, one finds a wide range of reaction equations. These reactions can be ordered in decreasing ratio of CO_2 consumption (=b/a). Tab. A9.1 shows the more striking reactions.

Tab. A9.1: Possible ranges of stoichiometry of dry reforming reactions with CO_2 excess.

a	b (= a + δ)	x (= 2a + δ)	y (= 2a – δ)	z (= δ)	a CH_4 + b CO_2 ⇒ x CO + y H_2 + z H_2O	b / a
1	3	4	0	2	1 CH_4 + 3 CO_2 ⇒ 4 CO + 2 H_2O	3.00
2	5	7	1	3	2 CH_4 + 5 CO_2 ⇒ 7 CO + 1 H_2 + 3 H_2O	2.50
1	2	3	1	1	1 CH_4 + 2 CO_2 ⇒ 3 CO + 1 H_2 + 1 H_2O	2.00
5	9	14	6	4	5 CH_4 + 9 CO_2 ⇒ 14 CO + 6 H_2 + 4 H_2O	1.80
4	7	11	5	3	4 CH_4 + 7 CO_2 ⇒ 11 CO + 5 H_2 + 3 H_2O	1.75
3	5	8	4	2	3 CH_4 + 5 CO_2 ⇒ 8 CO + 4 H_2 + 2 H_2O	1.67
5	8	13	7	3	5 CH_4 + 8 CO_2 ⇒ 13 CO + 7 H_2 + 3 H_2O	1.60
7	11	18	10	4	7 CH_4 + 11 CO_2 ⇒ 18 CO + 10 H_2 + 4 H_2O	1.56
2	3	5	3	1	2 CH_4 + 3 CO_2 ⇒ 5 CO + 3 H_2 + 1 H_2O	1.50
1	1	2	2	0	1 CH_4 + 1 CO_2 ⇒ 2 CO + 2 H_2	1.00

The upper one is super-dry reforming (b/a = 3); the bottom one (b/a = 1) is normal dry reforming. Theoretically, one can construct numerous reaction equations with a decreasing b/a ratio between 1.5 and 1.0. Being at the lower end of the performance spectrum, these reactions are of less interest.

In view of the many reactions in Tab. A9.1 one starts to wonder how many of these are actually independent ones from a stoichiometric point of view – a reaction within a given set of reactions has an independent stoichiometry if the latter cannot be construed as a linear combination of the stoichiometry of some other reactions within this set. In this dry-reforming case the outcome is that there are at most two independent reactions. A mathematical proof will be given below. To illustrate this outcome let us take the dry reforming and super-dry reforming reactions as the independent ones. It appears that all other reactions can be written as different linear combinations of the two independent ones. For instance, the medium-dry reforming reaction ($1\ CH_4 + 2\ CO_2 \Rightarrow 3\ CO + 1\ H_2 + 1\ H_2O$) is the sum of the super-dry reforming ($1\ CH_4 + 3\ CO_2 \Rightarrow 4\ CO + 2\ H_2O$) and dry-reforming reaction ($1\ CH_4 + 1\ CO_2 \Rightarrow 2\ CO + 2\ H_2$), divided by two. In fact, all reactions other than the super-dry and dry reforming reactions are linear combinations of these two.

It is interesting to note that the water-gas shift reaction [$CO + H_2O \Rightarrow CO_2 + H_2$] is also included in the set of dependent reactions. If one subtracts the super-dry reforming stoichiometry from the regular dry reforming stoichiometry and divides by two one gets the water-gas shift reaction.

Can it be mathematically proven there are at most two independent reactions?

Independence of reaction stoichiometry has its mathematical basis in linear algebra and vector spaces. The stoichiometry of a reaction can be treated as a single vector in a real-valued vector space ($\Re^N$) where N is the number of chemical species between which a specified set of reactions occurs. N is the dimension of the "species space". A vector is a dependent one if it is a linear combination of some other vectors in this space. In ($\Re^N$) one can choose always a set of N independent vectors which span the entire space. Any other vector in this space is a linear combination of these N independent ones. To determine the number of reactions with an independent stoichiometry (= "independent reactions") one must also take into account the conservation of elements in the reactions. The coefficients in the element (C, H, O) balance equations also make up vectors of dimension N in the species space. Let E be the *element-species* matrix for the dry reforming case:

Elements		CH_4	CO_2	CO	H_2	H_2O	⇐ Species
	C [	1	1	1	0	0	]
MatrixE =	H [	4	0	0	2	2	]
	O [	0	2	1	0	1	]

Each row in this matrix corresponds to a vector in the five-dimensional "species" space. Such a vector is called an "element" vector. It is easily verified that above matrix E has three linearly independent vectors in the "species space"; i.e., none of the rows can be written as a linear combination of the two other rows. Due to the principle of conservation of elements in reactions, each of these "element" vectors must be orthogonal to the reaction "stoichiometric" vectors. Be reminded that a row in stoichiometric matrix S is the stoichiometry of a reaction, while a column in S shows the occurrence of a particular chemical species in the reactions. So, due to the conservation principle the inner product of an "element" vector and a "stoichiometric" vector is always zero. In matrix notation:

$$E * S^T = 0 \quad \text{with} \quad N_{species} \geq N_{indep.elements} + N_{indep.reactions}$$

This inequality relation between the number of species and the numbers of independent elements and reactions is a useful one in chemical reaction network analysis. In the case of the dry reforming reactions, there are three element vectors in a five dimensional "species" space. Only two other independent reaction vectors are left to fully span the "species" space. These two vectors can be arbitrarily taken from any set of reactions with independent stoichiometry. Here, the dry reforming and the super-dry reforming reactions are selected as the independent ones. This result completes the proof of having only two reactions with independent stoichiometry. For the sake of clarity it is mentioned that stoichiometric dependence between reactions implies also dependence between their reaction equilibrium equations through the law of mass action.

Evaluation of model and solution

The derived stoichiometric model enables exploring for potentially interesting and useful reactions in a given species space. The super-dry reforming reaction is the highest performing reaction for CO_2 to CO conversion, corresponding to the (b/a = 3) entry in Tab. A9.2. An experimental study of this super-dry reforming reaction and its underlying, innovative catalytic scheme was first reported in [1] by Buelens LC et al. Our stoichiometric analysis also revealed the high-dry (b/a = 2.5) and medium-dry reforming reactions (b/a = 2) as interesting side-catches. In principle these reactions may occur simultaneously with super-dry reforming. The stoichiometric scheme also implicitly covers the water-gas shift reaction, so one might worry indeed about a negative effect of this reaction in the super-dry reforming. Buelens LC et Al. [1] have developed a novel experimental conversion scheme. The shift reaction is avoided by means of clever sequencing of catalytic reactions. In the first reaction the methane is fully oxidized to H_2O and CO_2, using an oxygen-carrying solid-state medium:

$$CH_4 + 4[O] \Rightarrow CO_2 + 2H_2O$$

All CO_2 generated by the reaction and in the feed is chemically adsorbed in another solid phase, while the gas-phase reaction water is flushed out of the reaction system. In the next step all adsorbed CO_2 is reduced to CO:

$$4\,CO_2 \;\Rightarrow\; 4\,CO + 4\,[O]$$

The oxygen is adsorbed by a solid medium for subsequent use in the next oxidation cycle of CH_4, as in a chemical looping fashion. Jointly these reactions satisfy the stoichiometry of super-dry reforming.

As a final step it is useful to reflect on uncertainties and limitations playing a role in formulating and solving a model. In this case, the model structure and the outcome of the analysis are determined by the chosen set of chemical species and reactions, which co-determine the set of independent reactions. In this case, one could possibly argue to add solid carbon (C_{solid}) along with the Boudouard reaction ($2\,CO \Leftrightarrow CO_2 + C_{solid}$) if the CO_2 is released as a gas at lower temperatures.

Case A9.2: A conversion–separation–recycle structure using lumped units

A linear algebraic model of a common reactor-separator-recycle processing structure operating at steady-state is derived and solved. This case may serve as an example on how to tackle such cases in general.

Synthesis of process block diagram and model

The arrangement of the main units in a process and their connecting mass streams is shown by means of process block diagram (Fig. A9.2). Each main unit is represented as a block in the diagram. The process block diagram shows:

- A *"conversion"* unit, which turns by two parallel reactions the reactants A and B into the main product P (by reaction 1) and a side product Q (by reaction 2).

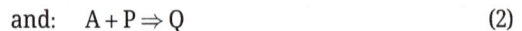

$$\text{Reactions:} \quad A + B \Rightarrow P \tag{1}$$

$$\text{and:} \quad A + P \Rightarrow Q \tag{2}$$

- A *"separation"* unit recovers unconverted reactants from the products for recycling.
- A *"split"* unit purges an inert trace component (C) in the feeds from the process so as to avoid its accumulation through the recycle.
- A *"mixing"* unit combines the recycle stream with fresh feed into a feed to the conversion unit.
- The connecting *"streams"* to, in between, and from the process units.

This process counts five chemical species {A, B, C, P, Q} occurring in varying amounts in seven streams {$S^{(1)}$, $S^{(2)}$, . . ., $S^{(7)}$}. The labeled streams are indicated in the block diagram.

Synthesis of the model

At this stage of design and modeling each process unit is defined by its steady-state processing functionality (e.g., conversion) rather than as a piece of equipment (e.g., reactor). The units in a block diagram will be modeled in an input-output fashion. The transformations of the incoming streams (or mass) into the outgoing streams will be captured in a lumped way by means of some parameters, without expanding yet on detailed mechanistic workings inside. For example, a chemical conversion unit can be characterized by the degree of conversion of a feed component as well as by selectivity factors. The function of a separation unit can be simply represented by separation factors. Each factor indicates how much of a component in the feed to the separator is going to a particular exit stream of this unit.

The variables and equations for all streams and units will be specified. To avoid modeling errors it is important to maintain modeling clarity by keeping the notations uniform. It is customary to work with a uniform representation of all streams and strive for similarity in the model representation of the process units. Furthermore a deliberate attempt is made to keep the model equations linear in the variables to facilitate ease of computation and interpretation of solutions.

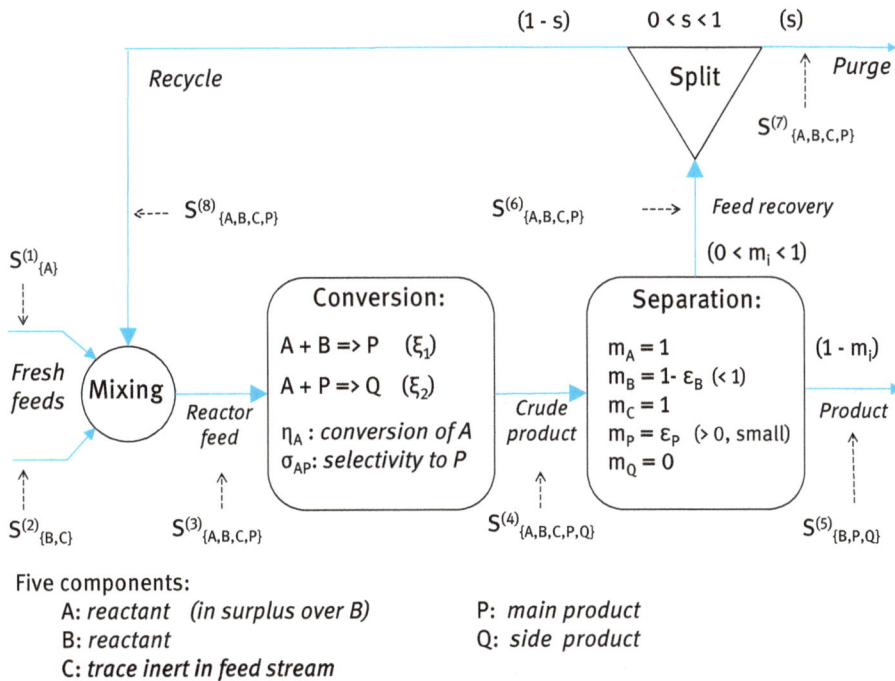

Five components:

- A: *reactant* (in surplus over B)
- B: *reactant*
- C: *trace inert in feed stream*
- P: *main product*
- Q: *side product*

Figure A9.2: Block diagram of a conversion–separation–recycle system.

Streams

Each stream is characterized by a vector of molar flows (*f*) of the five species. When a component is absent in a particular stream its molar flow is set to zero.

$f^{(j)}_i$:   molar flow of species i [lower index] in stream j [upper index] j; $j = 1, .., 8$

Note that this process model assumes fixed stream connections of the units, i.e., the process unit models contain the flows of these pre-assigned stream connections. There is no separate flow sheet structure model in this case. Such a structure model would allow for adjustable connections between units by being able to switch streams connections on or off.

Conversion unit

One wants to favor the conversion to main product P over the formation of the side product Q. The reaction stoichiometry shows that A is consumed by both reactions; therefore it should not be in short supply relative to B, which is involved in the first reaction only. The molar balances are given in a matrix-vector notation by:

$$\underline{f}^{(3)} - \underline{f}^{(4)} + \mathbf{S}^T. \, \underline{\xi} = 0$$

S is the stoichiometric matrix: $[-1-10+10]$ and extents of reactions $\underline{\xi}$: $[\xi_1]$

$$[-100-1+1] \quad [\xi_1]$$

The conversion of a component and the selectivity to products can be related to the extents (ξ) of reactions in which it participates. Choosing limiting reactant A as a reference, its degree of conversion of A (η_A) and the selectivity of reactant A to product P (σ_{AP}) are related to the two extents of reaction:

Conversion: $f^{(3)}_A \cdot \eta_A - (\xi_1 + \xi_2) = 0$
Selectivity: $(\xi_1 - \xi_2) - \sigma_{AP} \cdot (\xi_1 + \xi_2) = 0$

At this stage of design one can use the degree of conversion and the selectivity as specified model parameters. The required numerical parameter values (or ranges for these values) can be derived from the literature or experimental conversion data. Having specified the parameter values, the two equations remain linear in the variables ($f^{(3)}_A$, ξ_1, ξ_2).

Separation unit

The amounts of chemical species entering the unit are distributed (by diffusional type of separation) over two exit streams, according to their thermodynamic preferences for either of the exit streams. These distributions are characterized by means of a single distribution parameter per species: $0 \le m_i \le 1$. This parameter m_i indicates the fraction of the incoming molar flow going to the recycle stream $S^{(5)}$, while the remaining fraction $(1 - m_i)$ goes to the product stream $S^{(4)}$. In this process the components A and C will go fully to the recycle stream, along with the bulk of B and a small fraction of P. A bit of B, the bulk of P, and all of Q go to the product stream.

$$f^{(6)}{}_i - m_i.f^{(4)}{}_i = 0 \text{ and } f^{(5)}{}_i - (1-m_i).f^{(4)}{}_i = 0 \quad i = \{A, B, C, P, Q\}$$

The numerical order of magnitude (or the numerical range) of the distribution coefficients must be derived from literature or experimental data.

Splitting and mixing

The entrance stream is split in two streams of identical composition. All components are distributed by the same fraction (s) over the two exit streams. In this case:

$$f^{(7)}{}_i - s.f^{(6)}{}_i = 0 \quad \text{and} \quad f^{(8)}{}_i - (1-s).f^{(6)}{}_i = 0 \quad i = \{A, B, C, P, Q\}$$

The fraction s is a (specified) design decision parameter.

In a mixer two or more incoming streams are added together to the exit stream. In this case:

$$f^{(8)}{}_i + f^{(1)}{}_i + f^{(2)}{}_i - f^{(3)}{}_i = 0 \quad i = \{ A, B, C, P, Q \}$$

Analysis and computation of model solution

Having achieved a compact linear form of the model, one should apply a degree of freedom analysis. How many unknowns occur in this model and how many equations have been specified? The difference between the two is the degree of freedom in the model. Regarding the unknowns, there are five species flows in eight streams plus two extents of reactions, resulting in 42 variables. These variables can be ordered into a single vector of unknowns $\underline{x}$ of dimension 42 in a vector space of real variables. The number of equations adds up to: $5 + 2$ (conversion unit) + 2^*5 (separator) + 2^*5 (splitter) + 5 (mixing) = 32 equations. These equations are linearly independent, containing no redundancies. Thus, the degree of freedom of this model is $42 - 32 = 10$. In order to make the model compact for a convenient solution procedure the coefficients of the variables in these linear equations are put into a matrix A_M of dimension 32 (rows) * 42 (columns). The right-hand sides of the equations and can be put into the right-hand side vector $\underline{b}_M$ of dimension 32. In this case all right-hand sides are zero and the vector $\underline{b}_M$ is the null vector. The resulting matrix-vector form is non-square [A_M (32 * 42); $\underline{x}$ (42); $\underline{b}_M$ (32)], falling short by 10 degrees of freedom to finding a solution to:

$$A_M \cdot \underline{x} = \underline{b}_M$$

To fill the gap one needs an external scenario of how this process is supposed to run with target feed and/or product flows. Such an external scenario must specify the remaining 10 degrees of freedom.

Specification of an external scenario

The first step is to specify the feed streams. These streams come from elsewhere and usually one has no direct control over its composition. Having two fresh feed streams for the two main reactants, A and B, one may readily assume that feed stream (1) is rich in A and stream (2) is rich in B. The molar compositions of these steams will be given by known molar ratios $\{r^{(1)}_{i|A}, r^{(2)}_{i|B}\}$. Here $r^{(1)}_{i|A}$ is the given molar ratio of component i relative to main component A in stream (1), while $r^{(2)}_{i|B}$ is the given molar ratio of component i relative to main component B in stream (2). Knowing these ratios one can express the species feed rates relative to the feed rate of the main component:

$$f^{(1)}_i - r^{(1)}_{i|A}.f^{(1)}_A = 0 \qquad i = \{B, C, P, Q\}$$

$$f^{(2)}_i - r^{(2)}_{i|B}.f^{(2)}_B = 0 \qquad i = \{A, C, P, Q\}$$

These specifications of the two feed streams take away eight degrees of freedom, leaving two degrees of freedom to be determined.

Now there can be several alternative scenarios. For instance:

Specify the fresh feed streams:

$$f^{(1)}_A = b^{(1)}_A = \text{given flow spec of component A in stream (1)};$$

$$f^{(2)}_B = b^{(2)}_B = \text{given flow spec of component B in stream (2)}.$$

Set targets for the output flows of the main and side products:

$$f^{(5)}_P = b^{(5)}_P = \text{target rate of product P in stream (5)};$$

$$f^{(5)}_Q = b^{(5)}_Q = \text{target rate of side product Q in stream (5)};$$

Set a target rate for the main product with a ratio for the side product rate:

$$f^{(5)}_P = b^{(5)}_P = \text{target rate of product P in stream (5)};$$

$$f^{(5)}_Q - r^{(5)}_{Q|P}.f^{(5)}_P = 0 = r^{(5)}_{Q|P}.\text{is molar ratio of Q over P in stream (5)}$$

These scenario specifications can be cast into a matrix vector form as well:

$$A_{Scenario} \cdot \underline{x} = \underline{b}_{Scenario}$$

Dimensions: $A_{Scenario}$: 10(rows)*42 and $\underline{b}_{Scenario}$: 10(rows)

The molar ratios $r^{(*)}_{i|j}$ become elements in the $A_{Scenario}$ matrix while the flow specifications $b^{(*)}_*$ are put in the right-hand side $\underline{b}_{Scenario}$ vector.

Comments on model solution procedure

The 32 equations for the process units are combined with 10 equations for the scenarios. Together they form a square problem $A \cdot \underline{x} = \underline{b}$. The matrix A should be regular; if it is singular the most likely cause is a modeling error, such as having dependent mass balances in the set of equations.

A square linear problem $A \cdot \underline{x} = \underline{b}$ can be quickly solved by any numerical linear equation solver. It may be tempting to solve this model analytically, computing the variables in a sequential manner. For instance, one can start by eliminating the extents of reactions from the variable and equation sets and work only with the species flows as unknowns. Then, one can solve for the flows of species A in the process as well as for the flows of the other species. However, such model reductions and transformations are not effective in terms of the engineering work efforts. With the exception of the model developer, for all other users of the model working with the analytical solution to the transformed model it will be hard to comprehend the solution equations, unless the derivations from the linear full model are reproduced again as an expensive repeat. Maintaining the transparency of the original process model is more important to effective engineering work than sacrificing it for small potential gains in computational speed. In engineering practice the application of analytical model transformations to enhance speed of computation are often counterproductive in the long run.

Evaluation of the solution and the model

The model parameters (degree of conversion, selectivity, separation factors, and split ratio) as well as the external scenario specifications can be assigned numerical values. These values will be called the reference values. One can solve the linear $A \cdot \underline{x} = \underline{b}$ model and find the corresponding molar flows of the species $(\underline{x})$ in the process. However, some model parameters may not be very accurate yet at this stage of design; these may vary over a range. Similarly, the external specifications (e.g. production rates) may show variability. So, it makes sense to analyses the model solutions for a wider range than only the reference values of the model parameters. How to select the parameter perturbations within their specified ranges of uncertainties? There are at least three options:

- To apply local sensitivity analysis by means of small parameter perturbations around the reference values
- To apply a nonlocal sensitivity analysis by tracking a trajectory of parameters in the feasible parameter space
- Nonlocal analysis by statistical sampling in the parameter space, governed by the probability density functions of the parameters

Apart from the local sensitivity analysis the two other methods are too involved to be discussed here. A suitable reference for further reading is found in [2] and in a reactor design application [3]. One can easily perform a numerical local sensitivity analysis.

Given a parameter reference case, one can introduce a sequence of small perturbations in model parameters, $\{\underline{p}^{(q)}, q = 0, 1, 2, \ldots, Q\}$, where $q = 0$ is the unperturbed reference. The model expands into multiple instances of matrix A and vector $\underline{b}$:

$$\{A(\underline{p}^{(q)}), \underline{b}(\underline{p}^{(q)})\}, \qquad q = 0, 1, 2, \ldots, Q.$$

One can generate a sequence of dimension-free sensitivity matrices:

$$\Omega^{(q)} = [(\underline{x}^{(q)} - \underline{x}^{(0)})/\underline{x}^{(0)}]/[(\underline{p}^{(q)} - \underline{p}^{(0)})/\underline{p}^{(0)}] \qquad q = 1, 2, \ldots, Q.$$

In view of the linear differences between $A^{(q)}$ and $A^{(0)}$ and $\underline{b}^{(q)}$ and $\underline{b}^{(0)}$ one would expect to have avail a computational short cut to easily compute $\underline{x}^{(q)}$ from $\underline{x}^{(0)}$. This is not the case. For every perturbation (q) the full set $A^{(q)} \cdot \underline{x}^{(q)} = \underline{b}^{(q)}$ must be solved. But it is not a burden given the efficiency of numerical solvers for linear equation sets. For further analysis one may vertically stack all sensitivity matrices in a main sensitivity matrix:

$$\Omega^{(\text{main})} = \left[\Omega^{(1)} / \Omega^{(2)} / \ldots / \Omega^{(q)} / \ldots \Omega^{(Q)} \right]$$

Application of Singular Value Decomposition to $\Omega^{(\text{main})}$ yields the dominant singular value(s); the corresponding right singular vectors indicate which parameters or parameter combinations are the more influential ones.

Maintaining model linearity when adding an enthalpy balance

The enthalpy flow of a stream ($F^{(j)}{}_H$) is computationally the product of a flow and a temperature. It spoils the desired linear structure of the model equations. Yet, it is important for process design to get some quantitative understanding of the main enthalpy flows and the main enthalpy sources and sinks. By means of some reasonable assumptions about temperature levels one can induce a linear structure for the enthalpy balance equations. This is done in the following way.

An enthalpy flow is obtained by taking the sum over all species in the stream of the product of the specific enthalpy, $h_i(T)$ and the flow rate of each component $f^{(j)}{}_i$:

$$F^{(j)}{}_H = \sum_{i=1, \ldots, N} h_i(T) \cdot f^{(j)}{}_i$$

The specific enthalpy $h_i(T)$ accounts only for the sensible heat effects to get to the temperature T from the thermodynamic reference condition; it excludes the enthalpy of formation of a component from its elements at reference conditions. The product of two process variables, temperature (T) and flow rate ($f^{(j)}{}_i$), destroys linearity. However, if one would be able to fix the temperatures of the streams and process units – and so the specific enthalpy – one can retain the linear form of the enthalpy flow variables. It is also supposed that the linear mass balance model has been solved and the local flows $\{f^{(j)}{}_i\}$ are all available for the computation of the enthalpy flows. The enthalpy balance over the conversion unit to which a heat flow is added (Q_{exch}) for cooling or heating would then read:

$$F^{(3)}{}_H - \Delta H_{R,1} \cdot \xi_1 - \Delta H_{R,2} \cdot \xi_2 - F^{(4)}{}_H + Q_{\text{exch.}} = 0$$

$$F^{(3)}{}_H - \sum_{i=1, \ldots, N} h_i(T^{(3)}) \cdot f^{(3)}{}_i = 0$$

$$F^{(4)}{}_H - \sum_{i=1, \ldots, N} h_i(T^{(4)}) \cdot f^{(4)}{}_i = 0$$

When accounting for the reaction enthalpies {$\Delta H_{R,1}$, $\Delta H_{R,2}$} as being given, these three equations match with three additional unknowns {$F^{(3)}{}_H$, $F^{(4)}{}_H$, Q_{exch}}.

This approach of using preset temperatures is justified when the effects of the enthalpy sources and sinks (such as a reaction enthalpy $\Delta H_R \cdot \xi$ or a phase change enthalpy) are much larger than the change in a stream enthalpy $\Delta F^{(i)}{}_H$ due to a variation in temperature.

Obviously, this linear approach by means of use of guesstimated temperatures will break down for an adiabatic operation of a process unit, because of the additional specification $Q_{\text{exch}} = 0$. A new variable needs to be introduced: it is the temperature of the unit outlet stream $T^{(4)}$ within this enthalpy balance. By adjusting this temperature the right amount of enthalpy is carried away to ensure adiabatic operation.

Case A9.3: Alternative connectivity in a block diagram by use of a structure model

The units in the preceding block diagram in Fig. A9.2 are directly connected by a stream. The behavior models of the process units include the connecting streams. There is no distinction between the behavioral models for the units and a structural model for describing how units are connected by streams. The behavioral and structural model are lumped together. This lumping is disadvantageous if a process block diagram need to be rearranged with rearranged connections between process unit. A unit may have gotten different streams. Then, the model equations of a unit involved in a reshuffle needs to be rewritten because ingoing and outgoing streams are relabeled. Is it logical that a change in connectivity of units is reflected in having to change the unit models? No, because unit models are the more complex ones in the overall process model and such model rewriting is error-prone and may involve quite some effort. It would be better to have instead some changes in the stream connections only. This preferred way of modeling can be realized at a slight expense of introducing more linear connectivity equations in the model. The principle of creating flexible connections is explained by means of Figure A9.3.

The top part of this diagram shows a direct connection between units. For instance, stream $S^{(4)}$ connects directly the conversion and separation units. The lower part assigns to each unit's generic slots to pass inlet and outlet streams. These slots

Direct stream connections

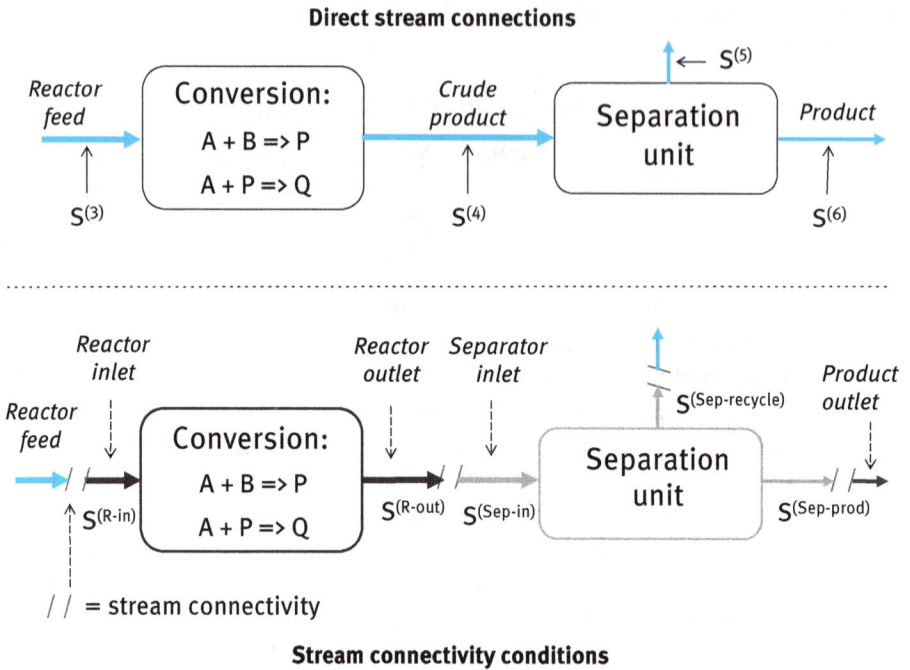

Stream connectivity conditions

Fig. A9.3: Stream connectivity conditions between process units.

can be labeled as pertaining to a particular unit. For instance, the generic entrance slot for a feed stream to the conversion unit is now labeled as $S^{(R\text{-in})}$, while its generic outlet slot is $S^{(R\text{-out})}$; similarly, the generic inlet slot of the separator is $S^{(Sep\text{-in})}$.

The model equations for the conversion unit are written in terms of streams pertaining to it slots:

Species balances: $\underline{f}^{(R\text{-in})} - \underline{f}^{(R\text{-out})} + \underline{\underline{S}}^T.\underline{\xi} = \underline{0}$

Conversion: $f^{(R\text{-in})}{}_{A}.\eta_A - (\xi_1 + \xi_2) = 0$

Selectivity: $\xi_1 - \sigma_{AP}.(\xi_1 + \xi_2) \quad = 0$

Similarly, the separation equations in the separation unit can be written for all species (i) as:

Recycle stream $\qquad f^{(Sep\text{-recycle})}{}_i - m_i \cdot f^{(Sep\text{-in})}{}_i \quad = 0$

Product stream $\qquad f^{(Sep\text{-prod})}{}_i - (1 - m_i) \cdot f^{(Sep\text{-in})}{}_i = 0$

The unit models have a generic form now. It is independent of their connectivity to other units. To connect units, the outlet slot of one unit is connected to the corresponding inlet slot of another unit. Making such a connection must be explicitly done by means of stream connectivity equations. Here is the example of connecting the stream through the outlet slot of the conversion unit with the stream though the inlet slot of the separation unit.

$$\text{Connectivity:} \quad \underline{f}^{(R\text{-out})} - \underline{f}^{(Sep\text{-in})} = \underline{0}$$

The set of all connectivity equations forms a **structure model** of the block diagram.

Is there a price to pay for this convenient generalization by separating behavioral and structure models? The number of variables related to streams that connect two process units will double and for every additional stream variable there is an additional connectivity equation. Fortunately, the price in terms of additional computing effort is very low. Modern numerical equation solvers are very efficient in dealing with such simple linear connectivity equations and variables. They are internally quickly eliminated by a solver, reducing the problem to its core structure of multiterm model equations. Overall one can say that keeping a process model transparent to its developer and users is much more important for efficient work processes than small amounts of additional computing costs.

Creating models of alternative block diagram structures by use of binary variables

The benefit of having explicit connectivity equations can also be exploited when inserting alternative process units. These alternatives exercise the same function but with different efficiencies and costs. Let assume that the current separation block (now labeled as Sep-A) will be put in competition with an alternative separation unit, labeled Sep-B. It is possible now to flexibly connect the outlet stream of the conversion unit to the inlet stream of either of both separation units. This flexible connection is realized with help of binary variables $Y_{sep,A}$ and $Y_{sep,B}$; a binary variable takes a value as 0 or 1 only.

$$\text{Connectivity:} \quad \underline{f}^{(R\text{-out})} - Y_{sep,A} \cdot \underline{f}^{(Sep\text{-A-in})} - Y_{sep,B} \cdot \underline{f}^{(Sep\text{-B-in})} = \underline{0}$$

Since only one of two types of separation units can be present, this exclusivity condition is expressed as a constraint on the binary variables:

$$\text{Exclusivity of units:} \quad Y_{sep,A} + Y_{sep,B} = 1$$

For $Y_{sep,A} = 1$ the conversion unit is connected with separation unit type A. For $Y_{sep,B} = 1$ a connection is made with separation unit type B. If the binary variables would be treated as variables along with continuous variables in the design computations the model would become nonlinear due to the product form of a binary with a continuous variable in terms, such as: $Y_{sep,A} \cdot \underline{f}^{(Sep\text{-A-in})}$. Therefore, the explicit use of binary design variables will be postponed till the treatment of mixed integer, nonlinear programming approaches. The latter are applied to formulating and solving process synthesis and design problems in Section 9.4] and in Section A9.3.

Case A9.4: Optimization of process performance by linear programming (LP)

For linear process design optimization one has to do five things:
a) Open up some degrees of freedom in the design model for optimization.
b) Select a linear performance metric.
c) Introduce linear inequality constraints on process variables to demarcate the physical domain in which a process design remains physically feasible.
d) Select a reference production scenario and compute its performance.
e) Compute the optimum solution and assess its relevance over the reference solution.

This will be applied to the conversion-separator-recycle (C-S-R) process shown in block diagram Figure A9.2. The species balances give rise to a linear model with some residual degrees of freedom.

Re (a): The feed rates of species A and B in streams (1) and (2) are set free for optimization. The internal process parameters {degree of conversion, selectivity, separation fraction, and split factors} are kept fixed in order to maintain model linearity. The process performance can be optimized by varying an external production scenario (i.e., feed rates).

Re (b): The selected performance metric (E_{profit}) is an economic one, the profit rate. This rate is obtained by deducting from the proceeds of the products:
– the costs of the feeds per unit time; and
– a cost term for cleaning of the purge stream

The resulting profit rate expression is:

$$E_{profit} = f^{(5)}_P * p_P + f^{(5)}_Q * p_Q - f^{(1)}_A * c_A - f^{(2)}_B * c_B - f^{(7)}_B * c_{purge}$$

The economic parameters {p_P, p_Q, c_A, c_B, c_{purge}} are respectively, the unit sales prices of products P and Q, the unit costs of feed A and B, and the purge cost of component B. Since a cost is associated with removing the excess of reactant B via the purge, one may expect that an optimization will push for reduced use of reactant B.

Re (c): Two inequalities are imposed:
– a market constraint on the amount of Q that can be sold: $F_{Q,max} - f^{(5)}_Q \geq 0$
– The exit flow of B from the reactor cannot go below zero: $f^{(4)}_P \geq 0$

Re (d): The base case (reference) scenario is:
– $f^{(1)}_A = 0.9 * f^{(2)}_B$
– $f^{(5)}_P = 0.1$ [kmol/s]

Re (e): Comparing the results of the base case and the optimized case.

Table A9.2 lists numerical values of the design, economic, and feed parameters in the C-S-R process model.

Table A9.2: Design, economic and feed stream data for C-S-R process.

Design parameters (symbol)	Value (–)	Economic parameters (symbol)	Value ($/kmol)
Conversion of feed A (η_A)	0.98	Sales price of product P (p_P)	75.0
Selectivity of A to P (σ_{AP})	0.80	Sales price of product Q (p_Q)	50.0
Separation fraction A (m_A)	1	Cost of feed A (c_A)	15.0
Separation fraction B (m_B)	0.95	Cost of feed B (c_B)	30.0
Separation fraction C (m_C)	1	Cost of purging B (c_{purge})	5.0
Separation fraction P (m_P)	0.05	**Production parameters**	**(kmol/s)**
Separation fraction Q (m_Q)	0.0	***Nominal*** production rate P ($f^{(5)}{}_P$)	0.10
Split factor (s)	0.05	Maximum sale of Q ($F_{Q,max}$)	0.02
Feeding ratios in stream (1)	**Ratio (–)**	**Feeding ratios in stream (2)**	**Ratio (–)**
A ($r^{(1)}{}_{A\mid A}$)	1.0	A ($r^{(2)}{}_{A\mid B}$)	0.0
B ($r^{(1)}{}_{B\mid A}$)	0	B ($r^{(2)}{}_{B\mid B}$)	1.0
C ($r^{(1)}{}_{C\mid A}$)	0	C ($r^{(2)}{}_{C\mid B}$)	0.02
P ($r^{(1)}{}_{P\mid A}$)	0	P ($r^{(2)}{}_{P\mid B}$)	0.0
Q ($r^{(1)}{}_{Q\mid A}$)	0	Q ($r^{(2)}{}_{Q\mid B}$)	0.0

Given these input data the "mass" balances and economic performance function can be computed for the nominal and the optimized case. The "mass" balances are expressed by means of the **molar** flows of the five species in the process streams. The "mass" balance results are presented in Table A9.3 and the key performance data in Table A9.4 covering both the nominal and optimized case.

The results in Table A9.3 show a substantial increase in the output rate of product P, while side product Q is pushed to its maximum level for attainable sale. It is also remarkable that reactant B is reacted nearly to extinction in the optimum case, so reducing the cost of purging excess B. Also the recycle flows (stream 8) have become much smaller in the optimized case by better balancing the feed intakes, relative to the base case. In the optimized case there is a slight surplus of intake of A over B in order to push the production of side product Q to its sales limit.

The results in Table A9.4 show that the economic performance has been substantially improved by optimization. For instance, the profit per unit of product P increases from 19,98 $/kmol P to 28,61 $/kmol. It is also seen that the optimum is indeed bounded by two active inequality constraints. The production of Q reaches its maximum (sales) level, while the component B is completely reacted away and has zero flow in the effluent.

Figure A9.4 shows the shift from the nominal case to the optimum point in the plane of the two (independent) feed flows, A and B. In the nominal case, the production

Table A9.3: Molar flows in C-S-R process at nominal and optimized cases.

Species	Stream 1	Stream 2	Stream 3	Stream 4	Stream 5	Stream 6	Stream 7	Stream 8
Nominal case								
A (kmol/s)	**0.12546**	0.00000	0.12789	0.00256	0.00000	0.00256	0.00013	0.00243
B (kmol/s)	0.00000	**0.13940**	0.38562	0.27282	0.01364	0.25918	0.01296	0.24622
C (kmol/s)	0.00000	0.00279	0.05576	0.05576	0.00000	0.05576	0.00279	0.05297
P (kmol/s)	0.00000	0.00000	0.00500	0.10526	**0.10000**	0.00526	0.00026	0.00500
Q (kmol/s)	0.00000	0.00000	0.00000	0.01253	*0.01253*	0.00000	0.00000	0.00000
Optimum case								
A (kmol/s)	**0.20020**	0.00000	0.20408	0.00408	0.00000	0.00408	0.00020	0.00388
B (kmol/s)	0.00000	**0.18001**	0.18007	*0.00007*	0.00000	0.00007	*0.00000*	0.00007
C (kmol/s)	0.00000	0.00360	0.07200	0.07200	0.00000	0.07200	0.00360	0.06840
P (kmol/s)	0.00000	0.00000	0.00798	0.16798	**0.15958**	0.00840	0.00042	0.00798
Q (kmol/s)	0.00000	0.00000	0.00000	0.02000	*0.02000*	0.00000	0.00000	0.00000

Table A9.4: Comparison of nominal and optimum cases in C-S-R process.

	Profit ($/s)	$f^{(5)}_P$ (kmol/s)	$f^{(5)}_Q \leq 0.02$ (kmol/s)	$f^{(5)}_B \geq 0$ (kmol/s)	Feed A (kmol/s)	Feed B (kmol/s)	$\xi_{reaction\ 1}$ (kmol/s)	$\xi_{reaction\ 2}$ (kmol/s)
Nominal	1.99810	0.10000	0.01253	0.01364	0.12546	0.13940	0.11280	0.01253
Optimal	4.56521	0.15958	0.02000	0.00000	0.20020	0.18001	0.18000	0.02000

of the maximum sales volume of Q is not reached yet. The optimum shifts along the profit gradient towards the intersection of two constraints, where profit takes its achievable maximum value.

Analysis of sensitivities of process performance for parameter changes
Suppose one would like to further improve the performance of the process. The question is where to have the most effective of further technological or commercial developments. To address this question a parametric sensitivity analysis will be made. The following three parameters have been chosen for further analysis for their critical role in process performance:

Technological
Selectivity factor of reactions from A to P (σ_{AP}) because it will enhance production of P.

Separation factor of component P (m_P) because it will also enhance output of P.

Commercial

An increase of the maximum sales volume of Q.

Starting from the optimized situation the values of these three model parameters are sequentially changed by 2% relative to their nominal values. With each parameter change the process design is re-optimized for maximum profit. The effect of the parameter change on the profit is expressed as a scaled sensitivity of the profit with respect to that parameter (p):

$$S_{par} = P_{ref} * \{\Delta\,Profit/\Delta\,p\}/Profit_{ref}$$

Selectivity factor of reactions from A to P (σ_{AP}): 0.11109
Separation factor P (m_P):   -0.00009
Larger sales volume of side product Q: 0.01948

The conclusion from these sensitivity data is that further performance enhancement, from among these three options, should focus on improving the selectivity of the reactions. However, this priority may be reversed when the cost of further technological development well exceeds the benefits of exercising the commercial option.

In a concluding reflection on this optimization case one could argue that it is hypothetical example case with nice results to show. Indeed, it is a hypothetical case with

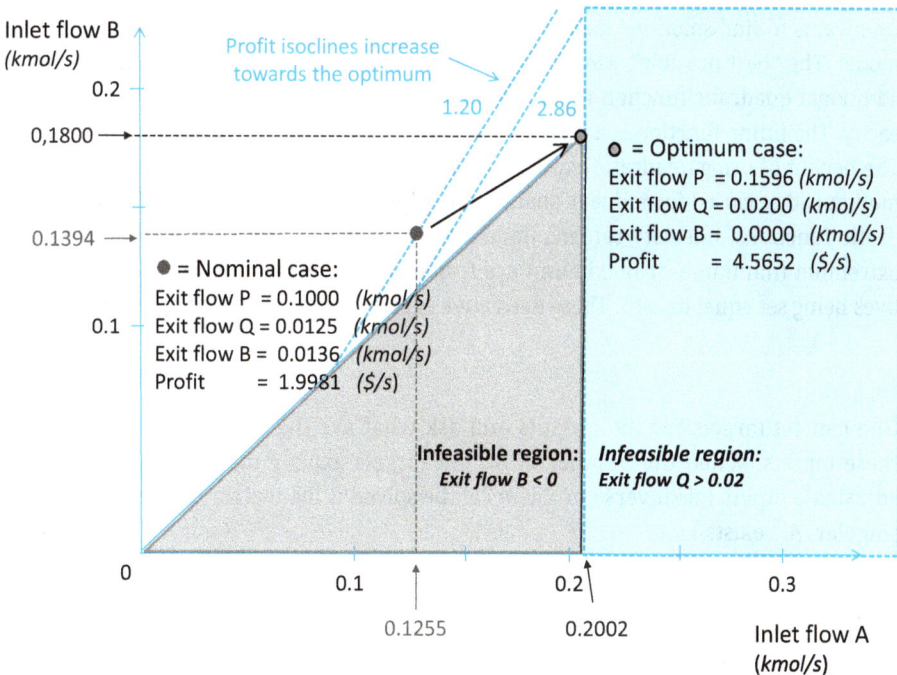

Figure A9.4: Graph of outcome of optimization of reactor-separator-recycle process.

illustrative results. Yet, the current way of presenting and analyzing the results between a base case and an optimized case is representative of practical applications.

Case A9.5: Inverse use of linear models for process targeting and estimation

Consider a linear algebraic process model, $A \cdot \underline{u} = \underline{y}$, where $\underline{u}$ is the N-dimensional vector of process inputs and $\underline{y}$ the M-dimensional vector of process outputs. Matrix A $(M*N)$ contains the internal process parameters. The matrix is non-square, making direct algebraic inversion impossible. Now there are three different ways to proceed in using such a model:

1) Input-forward: given A and u, compute $\underline{y}$. This case is straightforward and will be skipped
2) Output-targeting: given A and targets for y, compute $\underline{u}$.
3) Estimation: given a sequence of experimental observations of inputs $\{\underline{u}_1, \underline{u}_2, ..,\underline{u}_K\}$ and of outputs $\{\underline{y}_1, \underline{y}_2, .., \underline{y}_K\}$ find the unknown process parameters which are the elements of matrix A.

Both the targeting and the estimation are inverse problems. These are hard(er) to solve if the numbers of inputs and outputs are different. Then one has to deal with a non-square matrix A ($M > N$: more inputs than outputs or $M < N$, more outputs than inputs). One wants to find solutions match in some "best possible" way with the linearity of the model. The "best possible" way has to be defined: here it will be done by means of an additional quadratic function that aims for minimization of deviations from model linearity. The fitting function is a quadratic function over the model residuals: $\underline{\varepsilon} = A \cdot \underline{u} - \underline{y}$. The benefit of using quadratic expressions as fitting functions in both targeting and estimation is that closed-form linear analytical expressions result as answers. By using quadratic functions, linearity returns immediately when the necessity conditions for an extremum (minimum or maximum) are formulated in terms of the first-order derivatives being set equal to zero. These derivative expressions are linear in all variables.

Targeting

One can set targets for the outputs and ask what are the inputs needed to realize these targets. When the number of output targets exactly matches the number of adjustable inputs this inverse problem can be solved if the matrix A is regular (= non-singular; A^{-1} exists):

$$\underline{u} = A^{-1} . \underline{y}_t \qquad \text{with } t = \text{target}$$

What to do when there are more targets M than adjustable inputs N, (M > N)

Obviously, it will be impossible to exactly match the targets through the few inputs available. Then a strategy could be to minimize the differences between the targets and what the inputs can achieve on the outputs. If one takes a quadratic criterion with a (diagonal) weighting matrix W for scaling the different outputs one obtains the following minimization problem. It has a closed analytical expression for the best matching $\underline{u}^*$:

Minimize over $\{\underline{u}\}$ the matching function

$$\Omega = t^{1/2} \, (\mathbf{A}\underline{u} - \underline{y}_t)^T . W_y^{-1} . (\mathbf{A}\underline{u} - \underline{y}_t)$$

Result:

$$\underline{u}^* = \left[\mathbf{A}^T . W_y^{-1} . \mathbf{A}\right]^{-1} . \left(W_y^{-1} . \mathbf{A}^T\right) . \underline{y}_t$$

The condition for $(A^T.W.A)^{-1}$ to exist is that matrix A must have full-row rank, meaning no linear dependencies between the outputs. The residual errors $(\underline{e}^*)$ of the matching are:

$$\underline{e}^* = \mathbf{A} \cdot \underline{u}^* - \underline{y}_t$$

What to do when there are more adjustable inputs N than targets M, (M < N)

In such a situation a number of inputs will remain undetermined by the target outputs. That is, unless one follows a policy of minimizing the changes in the input vector $\underline{u}$, taking zero as the reference value. The result is again a mathematical optimization problem. One may scale the various inputs by means of the diagonal matrix W_u.

Minimize the Lagrangian function over $\{\underline{u}\}$

$$L = 1/2 \, (\underline{u})^T \cdot W_u^{-1} \cdot (\underline{u}) - \underline{\lambda}^T \cdot (\mathbf{A} \cdot \underline{u} - \underline{y}_t)$$

subject to:

$$(\mathbf{A} \cdot \underline{u} - \underline{y}_t) = \underline{0}$$

The outcome is in a closed form for $\underline{u}^*$:

$$\underline{u}^* = W_u \cdot \mathbf{A}^T \cdot \left[\mathbf{A} \cdot W_u \cdot \mathbf{A}^T\right]^{-1} \cdot \underline{y}_t$$

The Lagrange multipliers are:

$$\underline{\lambda} = \left[A \cdot W_u \cdot A^T\right]^{-1} \cdot \underline{y}_t \quad (\underline{\lambda} \text{ has dim. M}).$$

A Lagrangian multiplier indicates the sensitivity of the objective function L with respect to a change in the corresponding output target $y_{t,m}$: $\lambda_m = d\,L\,/\,d\,y_{t,m}$

The interpretation of the solution is that the optimized inputs take minimized sizes while jointly meeting the output targets.

Estimation

Let a sequence of K experimental observations of inputs $\{\underline{u}_1, \underline{u}_2, \ldots, \underline{u}_K\}$ and outputs $\{\underline{y}_1, \underline{y}_2, \ldots, \underline{y}_K\}$ be given. Experiment k yields the combined measurements of inputs and outputs $\{\underline{u}_{k,e}, \underline{y}_{k,e}\}$. The goal is to obtain statistical estimates of the process parameters. These parameters are in this case the elements of matrix A, satisfying the linear model. The usual approach to obtain parameter estimates is a weighted least squares fitting to the experimental data. The first step is always to look into the scaling of the output variables. These variables may take very different magnitudes (e.g., the temperature of a stream is 300 K while the molar fraction of a trace component in a stream can be $< 10^{-4}$. To ensure more robust solutions it is prudent to scale all output variables to the same order of magnitude of unity. Let the scaling factors (s) be ordered in a diagonal matrix D_y. Similarly one is advised to scale the inputs by means of another diagonal scaling matrix D_u. Then the scaled outputs, outputs and parameter matrix become:

$$\underline{y}' = D_y^{-1} \cdot \underline{y}; \underline{u}' = D_u^{-1} \cdot \underline{u} \quad A' = D_y^{-1} \cdot A \cdot D_u^{+1}$$

The goal is to find the parameters in the linear equation that match the experimental observations best according to some criterion.

How to select a suitable criterion?

The selection of a proper criterion relates to the nature of the probabilistic distribution of the measurement errors. Here, a *simplified approach* is taken by assuming that the error distribution of the residuals of the model equation using experimental input and output data is a multivariate, normal distribution with mean zero and uncorrelated spreads (i.e., the covariances of the residuals are zero). That is, the error in the residual of one equation is probabilistically independent of the error in another equation. Due to the scaling of the variables one can also assume that the spreads of all residuals are the same. A quadratic fitting criterion can be used, corresponding to the quadratic argument of a normal error distribution. If there are reasons to treat the measurement error distributions of outputs and inputs separately, one can use a more refined total least squares approach (see below for a formulation). For now a weighted least squares approach is applied to the residuals of the equations, working with scaled quantities. The objective function is defined as the sum over all experiments of the squared residuals of the equations, using the (scaled) experimental data for the inputs and outputs. One can assign also a weight per experiment $\{w_k, k = 1, \ldots K\}$ if a different degree of trust is put in the quality of the data per experiment. If one would put $w_k = 1/K$, the total sum (Ω) is scaled as the averaged squared residual over all experiments.

$$\text{Error sum:} \quad \Omega = \frac{1}{2} \sum_{k=1}^{K} \{\underline{y}'_{k,e} - A' \cdot \underline{u}'_{k,e}\}^T \cdot w_k \cdot \{\underline{y}'_{k,e} - A' \cdot \underline{u}'_{k,e}\}$$

Minimization of this sum function with respect to the elements of matrix A results in an analytical closed expression for matrix A'. This least squares solution can be written in a compact form if the measured observations are ordered in data matrices:

$$\mathbf{Y}'_K = [\underline{y}'_{1,e}, \underline{y}'_{2,e}, \ldots, \underline{y}'_{K,e}] \quad \{M * K\} \qquad \text{data matrix}$$

$$\mathbf{U}'_K = [\underline{u}'_{1,e}, \underline{u}'_{2,e}, \ldots, \underline{u}'_{K,e}] \quad \{N * K\} \qquad \text{data matrix}$$

$$\mathbf{W}_K = \text{diagonal}\,[\,w_k, k = 1, \ldots, K\,]\,\{K * K\} \quad \text{weights matrix}$$

Solution: $\mathbf{A}'_{\text{estim}} = [\mathbf{Y}'_K \cdot \mathbf{W}_K \cdot \mathbf{U}'_K{}^{\mathrm{T}}] \cdot [\mathbf{U}'_K \cdot \mathbf{W}_K \cdot \mathbf{U}_K{}^{\mathrm{T}}]^{-1}$

Matrix dimensions: $(M * N) \quad (M * N) \quad (N * N)$

This solution can be computed as long as the *input information* matrix $[\mathbf{U}'_K, \mathbf{W}_K, \mathbf{U}'_K{}^{\mathrm{T}}]$ is regular; meaning that matrix $\mathbf{U}'_K$ has must have rank N. This requirement translates into the following practical conditions:

- The measurements of the components of $\underline{u}$ should be linearly independent, or in other words, no measurement of a component should be a linear combination of measurements of some other components.
- There should be at least N experiments ($K \geq N$).
- No experiment should be a perfect linear combination of other experiments.

Having stated these requirements the problem of matrix parameter estimation based on squaring the residuals of the equations is completed.

Refined estimation by total least squares

With the conventional approach to estimating the model parameters it is common to lump all measurement errors of input and output variables occurring in an equation into one remaining residual error per model equation. This is conceptually a bit rough, though computationally convenient. For those liking a more refined statistical approach the following improvements are made:

The measurements errors of inputs and outputs are accounted for separately. Normal distributions with zero mean and covariances: Σ_u and Σ_y are assumed.

A distinction is made between the *experimental* values of variables $\{\underline{y}'_e; \underline{u}'_e\}$ and their *theoretical* counterparts, $\{\underline{y}'_m; \underline{u}'_m\}$. The latter exactly satisfy the linear model equations:

$$\underline{y}'_m - A' \cdot \underline{u}'_m = \underline{0}.$$

The relation between the experimental and model variables is given by the presence of experimental and modeling errors, lumped in the gross error vectors $\underline{\varepsilon}_y$ and $\underline{\varepsilon}_u$:

$$\underline{y}'_e = \underline{y}'_m + \underline{\varepsilon}_y$$

$$\underline{u}'_e = \underline{u}'_m + \underline{\varepsilon}_y$$

Given these relations one can formulate the total least squares estimation problem involving a minimization of the sum over all experiments of the squared errors between experimental and model variables, subject to the model equations:

Minimize (over A')

$$\Omega_{TLS} = \frac{1}{2} \cdot \sum_{k=1}^{K} \{ \varepsilon_{y,k}^T \cdot \Sigma_y^{-1} \cdot \varepsilon_{y,k} + \varepsilon_{u,k}^T \cdot \Sigma_u^{-1} \cdot \varepsilon \varepsilon_{u,k} \}$$

Subject to:

$$\underline{y}'_{e,k} - \underline{y}'_{m,k} - \varepsilon_{y,k} = \underline{0} \qquad k = 1, \ldots, K$$

$$\underline{u}'_{e,k} - \underline{u}'_{m,k} - \underline{\varepsilon}_{u,k} = \underline{0} \qquad k = 1, \ldots, K$$

$$\underline{y}'_{m,k} - A' \cdot \underline{u}'_{m,k} = \underline{0} \qquad k = 1, \ldots, K$$

This problem objective function can be reformulated as a Lagrangian function with Lagrangian multipliers dealing with each of the associated algebraic equations.

$$L = \Omega_{TLS} - \sum_{k=1}^{K} [\lambda_{y,k}^T \cdot (\underline{y}'_{e,k} - \underline{y}'_{m,k} - \underline{\varepsilon}_{y,k}) - \lambda_{u,k}^T \cdot (\underline{u}'_{e,k} - \underline{u}'_{m,k} - \underline{\varepsilon}_{u,k}) - \lambda_{u,k}^T \cdot (\underline{y}'_{m,k} - A' \cdot \underline{u}'_{m,k})]$$

The solution to this (large scale) optimization problem is quicker determined by numerical means. A closed form, analytical solution exists but it contains many matrix and vector operations with matrix inversions. This will require numerical support anyway. It is also possible to find estimates for the covariance matrices of the unknown quantities, though the expressions become very involved.

In any case it is recommended to consult a book on engineering statistics, e.g., [4] or an expert statistician on the nature of suitable error distributions, the corresponding choice of a suitable fitting function, and aspects of interpretation of the estimation results to avoid conceptual pitfalls.

Case A9.6: Notes on the use of other types of linear models

(a) Computing dynamic responses of a process (unit) by means of dynamic state-space models

Inputs to process: $\underline{u}(t)$; Internal states of process: $\underline{x}(t)$; process outputs: $\underline{y}(t)$
Model equations:

$$d\underline{x}(t)/dt = \mathbf{A} \cdot \underline{x}(t) + \mathbf{B} \cdot \underline{u}(t) \text{ with } \underline{x}(t_0) = \underline{x}_0 \text{ and } t > t_0$$

$$\underline{y}(t) = C \cdot \underline{x}(t)$$

The analytical solution can be expressed in a closed form:

$$\underline{x}(t) = \exp\{\mathbf{A} \cdot (t - t_0)\} \cdot \underline{x}_0 + \int_{t_0}^{t} \exp\{\mathbf{A} \cdot (t - \tau)\} \cdot \mathbf{B} \cdot u(\tau) \cdot d\tau$$

In the *concept* stage such dynamic modeling will be useful when changes in the feed supply to the process are expected to occur and its response is of interest.

(b) Distributed process/product dynamics

Inputs to process: $\underline{u}(z,t)$; Internal states of process: $\underline{x}(z,t)$
Model equations:

$$\partial\underline{x}(z,t)/\partial t + v \cdot \partial\underline{x}(z,t)/\partial z = \mathbf{A} \cdot \underline{x}(z,t) + \mathbf{B} \cdot \underline{u}(z,t) \text{ for } t > t_0 \text{ and } z > z_0$$

Initial condition:

$$\underline{x}(z,t_0) = \underline{x}_0(z) \text{ for } t = t_0 \text{ and } z > z_0$$

Boundary condition:

$$\underline{x}(z_0,t) = \underline{x}_{IN}(t) \text{ for } t > t_0 \text{ and } z = z_0$$

This is a Cauchy type of problem and an analytical solution in a closed form exists. While working with the analytical solution will give more insight, direct numerical equation solving will be quicker. In the *concept* stage this form of dynamic modeling is useful for describing the change in the mass (or size) distribution of a particulate product in a process unit (e.g., crystals, polymer). In that case vector $\underline{x}$ becomes a scalar function $x(z,t)$. Quantity $x(z,t)$ is the number of particulates in per unit volume of the process unit, having mass z at time t. The quantity v is a fixed mass growth rate. Similarly, matrix $\mathbf{A}$ reduces to a scalar and usually it takes a negative value $(A \Rightarrow -a)$, representing the mass decay rate of particulates. The final term, $\mathbf{B} \cdot \underline{u}(z,t)$, also becomes a scalar one, representing the distributed mass birth rate of particulates with mass z at time t. Commonly, only particulates with small mass stand a chance to be born; e.g., $u(z,t) = u(z_0,t) \cdot e^{\{-\lambda \cdot z/z0\}}$.

$u(z_0,t)$ is the birth rate of particulates with minimum mass (z_0) while λ (> 0) is a decline factor for the mass birth rate of bigger mass. The boundary condition $x(z_0,t) =$ $x_{IN}(t)$ indicates how many particulates of minimum mass z_0 are introduced from an external source over time t. If there is a nonzero birth rate term, $u(z,t)$ one usually takes $x_{IN}(t) = 0$. And the other way around.

This model enables computing the evolution of the mass size distribution of the particulates under different scenarios, such as different initial distributions, the impact of changes in the physical parameters (V, A, B, $u(z_0,t)$, λ) on the mass distribution. It is of practical importance to know if the mass distribution becomes a steady one over long enough time. Or that the distribution will show sustained oscillations.

(c) Discrete time model for the dynamics of an FMCG warehouse inventory

Inputs to process at time k:

$$\underline{u}_k; \quad \text{internal states at time } k: \quad \underline{x}_k; \quad \text{Outputs:} \underline{y}_k$$

Model equations:

$$\underline{x}_{k+1} = A \cdot \underline{x}_k + B \cdot \underline{u}_k - C \cdot \underline{y}_k \quad k = 1, 2, 3, \ldots, K \Rightarrow \infty$$

$$A: N{*}N \text{ matrix;} \quad B: N{*}J \text{ matrix;} \quad C: N{*}M \text{ matrix}$$

Initial condition:

$$\underline{x}_k = \underline{0} \quad \text{for } k = 1$$

This model can be applied in *concept* stage to determine the dynamics of the inventory of a warehouse for Fast-Moving Consumer Goods (FMCG). Knowing the dynamics is important for the design of the storage capacity of the warehouse. A closed analytical expression for the solution x_{k+1} can be derived but it is more cumbersome to use than doing a direct numerical simulation.

Suppose N types of products are kept in stock in the warehouse. For each type of product the stored mass may be different. The mass (x_i) of each type of product n, {$n = 1, ..,$ N} that is stored at instant k is accounted for in the mass hold-up vector $\underline{x}_k = [x_{1,k}, x_{2,k} ..,$ $x_{n,k} .., x_{N,k}]^T$. The N products are supplied to the warehouse in J different types of packages, at every instant k. All packages contain the same N products; however the mass fraction of a particular product may vary over the input packages. Each type of input package {u_j} has a fixed composition of the N products. This composition is stored in the columns of delivery matrix B (= N rows $*J$ columns). An entry u_j {$j = 1, .., J$} in the input vector $\underline{u}$, represents the mass of a package of type j. The vector $\underline{u}_k = [u_{1,k} \; u_{2,k} .., u_{j,k} .., u_{J,k}]^T$ shows the masses of all supplied packages at instant k.

In the warehouse the products are rearranged in new packages of different compositions for sales purposes. The products are sold in multiple packages: there are M different sales packages to customers. The composition of sales package m is specified by the data in column m of matrix C (=N rows $*$ M columns). There are M types of packages leaving at any instant k. The mass of sales package m at instant k is $y_{m,k}$. The column vector $\underline{y}_k = [y_{1,k} \; y_{2,k} .., y_{m,k} .., y_{M,k}]^T$ is made up of the masses of all sold packages at instant k.

Product storage and handling in the warehouse leads to loss of product quality. So, a small fraction of the products deteriorates and has to be discarded. The decay fraction can vary per type of product but it remains constant over time. This product decay is expressed as a fractional loss in matrix A:

$$A = I_N - D.$$

I_N is the identity matrix of dimension N. The diagonal matrix D $(0 < D < I_N)$ of dimension N contains the decay fractions of the product mass stored.

It is a *design question* how big the warehouse should be to store in a steady state, its hold-up $\langle \underline{x} \rangle$, given time-averaged masses of the input and output packages, $\langle \underline{u} \rangle$ and $\langle \underline{y} \rangle$. The answer can be given in a closed form:

$$\langle \underline{x} \rangle = (I_N - D) \cdot \langle \underline{x} \rangle + B \cdot \langle \underline{u} \rangle - C \cdot \langle \underline{y} \rangle$$

The steady state hold-up is:

$$\langle \underline{x} \rangle = D^{-1} \cdot \{ B \cdot \langle \underline{u} \rangle - C \cdot \langle \underline{y} \rangle \}$$

Analysis of this solution:

The first question is what happens when a product n does not decay: $D_{nn} = 0$. An infinitely large stored mass of product n can only be avoided when the total input of product n taken over all supply packages is exactly balanced by its total output over all the sales packages:

$$B|_n \cdot \langle \underline{u} \rangle - C|_n \cdot \langle \underline{y} \rangle = 0$$

For D_{nn} small the total supply of product n must slightly exceed the total sales of this product in order to keep a modest storage $\langle \underline{x}_n \rangle$:

$$B|_n \cdot \langle \underline{u} \rangle = C|_n \cdot \langle \underline{y} \rangle + D_{nn} \cdot \langle \underline{x}_n \rangle$$

The design question about the size of the warehouse can be addressed now the total averaged mass hold-up $\langle M_{\text{warehouse}} \rangle$ can be derived from the typical supply $\langle \underline{u} \rangle$ and demand $\langle \underline{y} \rangle$ pattern:

$$\langle M_{\text{warehouse}} \rangle = \sum_{n=1}^{N} \langle x_n \rangle$$

The warehouse hold-up should be designed larger than the average mass content because of the dynamic fluctuations in supply and demand. The dynamic model $\underline{x}_{k+1} = A \cdot \underline{x}_k + B \cdot \underline{u}_k - C \cdot y_k$ can assist in performing scenario analysis for expected varying supply and demand patterns around the prior estimated steady states values $\langle \underline{u} \rangle$, $\langle \underline{y} \rangle$ and starting from the steady-state hold-up $\langle \underline{x} \rangle$ for $k = 0$.

One can do further analysis of this design problem with demand and supply treated as stochastic processes; this is beyond the scope of this section.

(d) Linear integer equations with binary variables

Model equations:

$$Z \cdot \underline{Y} \leq \underline{\omega}$$

Symbols:

Z: $m*n$ matrix of integers; $\underline{Y}$: n-vector of binaries; $\underline{\omega}$: m-vector of integers.

This type of algebraic equation relates binary variables to each other. It plays an essential role in defining structure models that numerically represent superstructure flow-sheets, harboring many alternative specific flow-sheets. A small example is given as an illustration of the use of these equations. Suppose that in flow-sheet synthesis one can apply either reactor type A or reactor type B. It is known that reactor B performs better than reactor A (by achieving higher yields) but B is also substantially more expensive than A. Yet reactor type A may outperform reactor type B by placing two reactors A in cascade. The superstructure for the conversion section is one reactor B in parallel to two reactors A in cascade. One of the following structures has to be extracted in the computer-assisted process synthesis:
(1) One reactor B, no reactor(s) A
(2) One reactor A, no reactor B
(3) Two reactors A in cascade, no reactor B
(4) Exclusion: Reactor A2 cannot be placed in cascade after reactor B

Binary variables are assigned for the presence of a reactor ($Y = 1$: present) in the flow-sheet:

Y_{A1} belongs to reactor A1; Y_{A2} to reactor A2; Y_B to reactor B.

The following conditions enforce the above selection options:

$$Y_{A1} + Y_B = 1 \quad \text{(either reactor A or B)}$$

$$Y_{A1} + Y_{A2} \leq 2 \quad \text{(maximum of two reactors A in cascade)}$$

$$-Y_{A1} + Y_{A2} \leq 0 \quad \text{(reactor A2 only possible when A1 present)}$$

The optional structures mentioned above are all covered as solutions to this structure model:
(1) One reactor B, no reactor(s) A: $Y_B = 1$, $Y_{A1} = 0$, $Y_{A2} = 0$
(2) One reactor A, no reactor B: $Y_B = 0$, $Y_{A1} = 1$, $Y_{A2} = 0$
(3) Two reactors A in cascade, no reactor B: $Y_B = 0$, $Y_{A1} = 1$, $Y_{A2} = 1$
(4) Reactor A2 not in cascade after reactor B: $Y_B = 1$, $Y_{A1} = 0$, $Y_{A2} = 1$ is not a solution.

It requires a process design optimization effort to find out which of the three options is the more attractive one in terms of overall process performance.

A9.2 Nonlinear models and process simulations for *development* stage

Case A9.7: Note on ill-conditioned constitutive equations in design models

This case is a simple illustration of a locally ill-conditioned model by the use of ill-conceived reaction kinetic equation. It violates the computational requirement of having bounded derivatives in the model equations everywhere in the domain of model application. The model involves a continuous flow, well-mixed tank reactor (CSTR) with a simple isothermal reaction $A \Rightarrow B$. The reaction rate law is a commonly occurring type of kinetic expression. The powers of the species concentrations in the rate expression are allowed to deviate from the stoichiometric coefficients (Law of mass Action, Guldberg Waage) and become empirical ones. Often, such a power is allowed to take any real value (negative, zero, positive) when fitting the rate parameters to experimental kinetic data. Such type of rate law is applied here in the design equation for a CSTR.

Variables: x exit concentration of reactant A $x \geq 0$
            x_f inlet concentration of reactant A $x_f > 0$
            r reaction rate $r \geq 0$
            τ residence time of reacting medium $\tau > 0$
            k reaction rate constant $k > 0$
            a exponent of concentration in rate law $0 \leq a < 1$

It is assumed that this exponent a takes a value less than one, but remains nonnegative. The inlet concentration (x_f), the residence time (τ), the rate constant (k) and exponent (a) are all known quantities. The outlet concentration (x) is the unknown to be determined as function of the residence time.

Equations:

$$\text{Reactor design}: \quad m \equiv x_f - x - \tau \cdot r(x) = 0$$

$$\text{Rate law}: \quad r(x) = k \cdot x^a$$

$$\text{Derivative of reactor equation}: \quad \partial m / \partial x = -1 - \tau . k . \left[a . x^{(a-1)} \right]$$

$$\text{For } x \Rightarrow 0: \quad x^{(a-1)} \Rightarrow \infty \quad \text{(numerical singularity)}$$

Indeed, if one of the powers of a species concentration in a rate law is below one, the derivative of the rate law with respect to that species concentration becomes unbounded if this species concentration goes to zero. The smaller the concentration x the smaller the reaction rate r gets, but it slows down disproportionally much faster. In fact, the rate slows down accelerated to minus infinity when the concentration approaches zero. The rate itself stops at zero concentration. The derivative of the reaction rate with respect to the concentration goes to infinity as the concentration goes to zero. A numerical solution of the reactor design equation $m(x) = 0$ will fail when x gets close to zero. Beyond a mathematical singularity it is also a physical singularity. This kind of modeling problem appears at the lower boundary of the physically feasible domain; e.g., the boundaries of zero concentrations. *This case illustrates that constitutive laws must be formulated with due regard to correct physical behavior in boundary situations. Otherwise, it will bring process engineering computations to unnecessary trouble.*

This particular singularity problem could have been easily avoided by a physically more realistic modeling of the reaction rate equation:

$$\text{Refined rate law: } r(x) = k \cdot x / (b + x)^{(1-a)} \quad \text{with } a < 1,\ k > 0,\ b > 0 \text{ (small)}$$

Parameter a is the empirical exponent while parameter b is an empirical threshold concentration (small). When the concentration x gets below this threshold value b and would go to zero, both the reaction rate and its derivative start to level down to zero.

$$\text{Derivative: } \mathrm{d}r(x)/\mathrm{d}x = \left\{ k \cdot [b + x] - k \cdot x \cdot (1 - a) \cdot (b + x)^{-a} \right\} / \left\{ (b + x)^{(2-a)} \right\}$$

It is noted that for $b > 0$ the derivative remains finite in lim. $x \Rightarrow 0$:

$$\mathrm{d}r(x)/\mathrm{d}x = k \cdot b^{(-1+a)}$$

For the upper limit $a \Rightarrow 1$ one gets the normal exponential decay rate:

$$\mathrm{d}r(x)/\mathrm{d}x = k.$$

Case A9.8: Consistency of design specifications with thermodynamics

It is rather easy to overlook the (implicit) limits imposed by thermodynamics when setting design targets for process units. If targets are set beyond the limit of thermodynamic equilibrium the computations with the process model and the thermodynamically infeasible design specifications will always fail due to this underlying inconsistency. The failure can manifest by lack of convergence in the computations or yielding unphysical design parameters for the unit. It seems like the numerical solver is failing. Actually, an infeasible problem is impossible to solve and any solver will fail!

A very simple design example illustrates this point. In a continuous flow, well-mixed tank reactor a single, reversible first-order reaction $A \Rightarrow B$ is carried out at isothermal conditions. The reactor design question is what residence time is needed in order to meet a target conversion of A. The required residence time will determine the volume of the reactor. The reactor design model is given by the following process design equations and process data:

$$[M]: \quad C_{A,f} - C_A - \tau \cdot \{\, k_f \cdot C_A - k_b \cdot C_B \,\} = 0$$

$$C_{B,f} - C_B - \tau \cdot \{\, k_f \cdot C_A - k_b \cdot C_B \,\} = 0$$

Feed concentrations are known:

$$C_{A,f} = 1 \text{ and } C_{B,f} = 0$$

Known reaction rate parameters:

$$k_f = 1 \text{ and } k_b = 1/3$$

The outlet concentrations $\{C_A, C_B\}$ and the residence time τ are the unknowns. Apart from the process model equations [M] another equation is needed for computing three unknowns. A design target must be added. The design target is set at 90% conversion of reactant A, leaving 10% unconverted:

$$[S_{perf}]: \quad C_A - 0.1\, C_{A,f} = 0$$

The process design equations [M] are reworked to render the residence time explicit:

$$\tau = (C_{A,f} - C_A) / \{\, k_f \cdot C_A - k_b \cdot (C_{f,A} - C_A) \,\}$$

Computing the residence time with the given data yields a *negative* value:

$$\tau = -0.5$$

This answer is a bizarre outcome; what went wrong?

Analysis of the design problem

The reaction is reversible, i.e., conversion cannot be pushed beyond reaction equilibrium. The equilibrium condition is:

$$k_f / k_b = K = C_{B,eq} / C_{A,eq} \text{ with } C_{A,eq} + C_{B,eq} = 1$$

The equilibrium concentration of A:

$$C_{A,eq} = C_{A,f}.1/(1+K) = 1/(1+3) = 0.25$$

When approaching reaction equilibrium the denominator $\{k_f \cdot C_A - k_b \cdot (C_{f,A} - C_A)\}$ goes to zero. As a result the residence time goes to infinity and it turns negative beyond the equilibrium condition. This infinity situation may give rise to a declaration of a singularity in computations with a numerical tool.

The initial target of 90% conversion of A results in an exit concentration for A of 0.1. This is smaller than the equilibrium concentration of 0.25. A more realistic target for C_A at 110% of the equilibrium concentration gives a positive residence time:

$$C_A = 0.275 > 0.25 \quad \Rightarrow \quad \tau = 21.75$$

Setting a more lax conversion target (160%):

$$C_A = 0.400 > 0.25 \quad \Rightarrow \quad \tau = 3.0$$

Moving away from the equilibrium boundary reduces the residence time and vessel volume, though at a loss of having a lower conversion. This analysis illustrates an inherent trade-off between size of reaction vessel and conversion, which is in fact quite general. The resulting heuristic is:

Physical design targets should be located within a thermodynamically feasible domain.

The distance between a target and the thermodynamic equilibrium boundary is negotiable in design.

Case A9.9: Analysis of operability aspects in process design

The operational success of a process plant depends to large extent on the quality of its design. Therefore, operational characteristics of process should be taken into consideration in process design. This can be done by both experimental testing and by model-based simulation of the behavior of a process design under various operational scenarios. The experimental testing of vulnerable equipment parts in a process often focuses on the effects of active operation over a long time horizon; e.g., the wear and tear of equipment over time under a range of processing conditions as well as loss of functionality of active agents in a process (e.g., catalysts, membranes, solvents) by deactivation, fouling, or decompositions. Such long term effects play out on a time scale that is much longer than the total pass-through time of feed material through the entire process. These effects are often hard to observe, model, and simulate. It involves phenomena having low amplitudes and slow rates, occurring below the threshold level of detail in a process (unit) model. Yet, the effects of such phenomena consistently push in the same direction. By accumulation they cause deterioration of process performance over a long time horizon (months, year).

Model based operability analysis is suitable to study the effect of the dominant phenomena on the behavior and performance of a process over smaller time scales, e.g., the time scale of daily production and below. Operability analysis is needed to assess the effects of deterministic and random changes outside and inside a process. The changes in the surroundings of the process (e.g., feed quality, production rates, product quality requirements, and environmental conditions) can be a switch from one steady level to another steady level as well as dynamic fluctuations. Inside a process (model) there will be some uncertainties in its physical parameters. The outcomes of an operability analysis may give reason to change a design as in making it easier to operate and control a process plant. Conditions for proper operability are often included as conditions to be satisfied (inequality constraints) in an optimization of process economics. Without respecting such conditions it will be hard to realize in reality the computed economic performance by coping with an unwilling process plant. Safety analysis of a design always ranks high in any operability analysis of the intended production practice.

Operability analysis

Operability analysis of a process has become an extensive subfield of process engineering, linking process design, control and operations. It is too extensive to cover it in any degree of detail in this section. The aim is to create awareness of the main aspects of operability while giving some references to pertaining literature for further study. Marlin [5] reviews eight topics of operability from the perspective of use in process design teaching:

Safety

Process units can fail and people operating a process make mistakes from time to time, resulting in incidents or even accidents. Layers of protection (LOPA) can be designed to prevent failures or to mitigate the effects of failures.

The **operating window** of process units and of the entire process:

The operating window is the joint ranges of processing conditions (T, P, x, throughput) for which the process (unit equipment) is supposed to function normally, according to specifications. Are there any peculiar features of the operating window? Does it offer enough room for flexible operation?

Flexibility and controllability of the process

A process is flexible when there is enough room to adjust some manipulated process variables within their operating window to counteract the effects of external changes or disturbances to keep the production goals on target. Controllability implies that enough manipulated inputs (by means of actuators such as valves for flows) are in place in the process to effectively keep the controlled output variables close to their target values; e.g., set points for product quality, physical conditions (P, T), and the rate of production. Often a distinction is made between steady state (looking at steady-state changes) and dynamic controllability (the nature of time responses by outputs to changes in inputs). A process that is both flexible and controllable is resilient to external changes and disturbances. Care must be taken that process integration and intensification do not result in a loss of controllability. Integration often takes away operational degrees of freedom in the process adversely affecting controllability, while intensification will reduce time constants and increase speed of response.

Reliability

Process unit equipment will malfunction from time to time and the production rate of a process and/or the product quality will be adversely affected. One can express the reliability of a process unit by means of a probability of failure. By putting identical process units in parallel one has backups in case of failure and the reliability of the entire process increases. An increase in reliability of a process with more product output per year improves income from product sales. But it requires higher investment costs due to the extra units, leading to a trade-off between enhanced production from higher reliability and additional investment costs.

Transitions

During process operations there will be transitions from one steady level of operation to another level. This can involve startup and shut down, regeneration or cleaning and sterilization, switches between two successive steady-state production modes, and load following. The process design should address such transitions, making sure these can be accomplished over a practical time interval and at a reasonable cost.

Dynamic behavior

This involves the time responses of process units from inputs to outputs, aspects of stability, and runaway. These can all be affected by process design variables such as hold-up volumes and surface areas for heat and mass transfers. It involves also an identification of multiple steady states if these would occur in the process (units) and making a decision at which state the process should be run.

Efficiency

This includes effective use of resources in a process, including its utilities. Are there ways to maintain product quality (avoiding production losses) and steering on profit (avoiding loss of economic opportunity)?

Monitoring and diagnosis

This involves the continuous monitoring of processing and equipment conditions by means of sensors. The purpose is to offer information to the process operators to assist them in running the process, both short-term for rapid analysis and actions and long-term for identifying slow changes in the process such a fouling, plugging, and deactivation.

In which respects can process models and simulations contribute to operability analysis?

Let us suppose that a process model covers steady-state and dynamic situations within the specified operating window. Then this model is applicable to all of above operability cases, except reliability and safety. Reliability is an exception as it requires probabilistic models for failure rates. Models for reliability engineering are covered in, among others, [6] and [7]. Safety is another exception because it often involves process events going beyond the normal operating window. The conventional process models are rarely tuned to such exceptional conditions beyond the normal operating window. Separate dedicated models are required to study the physical behavior of a process unit under abnormal conditions, such as flow reversal in pipelines, a thermal runaway event in a reactor or flows through pressure relieve valves.

A9.3 Nonlinear models and process optimization for *feasibility* stage

Case A9.10: Constraint analysis to identify options for technological innovations

For reasons of keeping optimization complexity at bay only a case of nonlinear programming (NLP) is presented here. If a process design model would contain both continuous and discrete decision variables, it is assumed the discrete variables are kept fixed during the optimization over the continuous variables. Prior to presenting a small chemical engineering example on constraint analysis to identify options for process improvements, some necessary optimization theory is presented to set the scene for the application case.

Necessary conditions for an extremum in optimization

The generic format of the optimization reads as:

$$
\begin{array}{ll}
\text{Minimize} & F(\underline{x}) \\
\text{over} & \{\underline{x}\} \\
\text{Subject to} & \underline{g}(\underline{x}) = \underline{0} \\
\text{and} & \underline{h}(\underline{x}) \geq \underline{0}
\end{array}
$$

All functions are algebraic ones, supposed to be continuous, and at least differentiable once. When analyzing and solving optimization problems the Lagrange function plays a key role; see [8, 9].

Lagrange function:

$$
L(\underline{x}, u, v) = F(\underline{x}) + \underline{g}(\underline{x})^{\mathrm{T}} \cdot \underline{u} + \underline{h}(\underline{x})^{\mathrm{T}} \cdot \underline{v}
$$

$\underline{u}$ is a vector of Lagrange multipliers, while $\underline{v}$ is a vector of Kuhn-Tucker multipliers. In an NLP optimization one searches for the optimum point(s) in the solution space $\{\underline{x}\}$ where:

(a) the gradient of the Lagrange function vanishes:

$$\nabla_x L^* = \nabla_x F(\underline{x}^*) + \nabla_x \underline{g}(\underline{x}^*)^T \cdot \underline{u}^* + \nabla_x \underline{h}(\underline{x}^*)^T \cdot \underline{v}^* = \underline{0},$$

(b) the equality and inequality equations are satisfied:

$$\underline{g}(\underline{x}^*) = \underline{0} \quad \text{and} \quad \underline{h}(\underline{x}^*) \geq \underline{0},$$

(c) the *complementary slack* conditions apply:

$$\underline{h}(\underline{x}^*)^T \cdot \underline{v}^* = \underline{0} \quad \text{and} \quad \underline{v}^* \geq \underline{0},$$

(d) there are no linear dependencies between the equality and the active inequality constraints.

These four conditions are jointly called the *Karush-Kuhn-Tucker necessity* conditions [9]. These conditions need to be satisfied in solution point $\{\underline{x}^*, \underline{u}^*, \underline{v}^*\}$. Vector $\underline{u}$ contains the Lagrange multipliers associated with the equalities $\underline{g}(\underline{x}) = \underline{0}$. Vector $\underline{v}$ contains the Kuhn-Tucker multipliers associated with the inequality constraints $\underline{h}(\underline{x}^*) \geq \underline{0}$. A Kuhn-Tucker multiplier can be interpreted as the derivative of the objective function F with respect to active inequality constraint h_i:

$$v_i^* = \partial L / \partial h_i \text{ for } \underline{x} = \underline{x}^*$$

The *complementary slack conditions* state that (a) if an inequality constraint is not active (> 0), its corresponding Kuhn-Tucker multiplier v_i is zero, i.e.,

$$h_i(\underline{x}^*) > 0 \quad \text{and} \quad v_i = 0.$$

And (b) the other way around: active inequality constraints have positive Kuhn-Tucker multipliers:

$$h_i(\underline{x}^*) = 0 \quad \text{and} \quad v_i^* > 0.$$

In addition to these necessity conditions for an extremum there exist sufficiency conditions for the existence of a (local) minimum. For detailed, fundamental explanations, see [9].

Constraint analysis

Suppose one has performed an optimization and a minimum is obtained:

$$\{\underline{x}^*, \underline{u}^*, \underline{v}^*\} \quad \text{and} \quad \{F^*, \underline{g}(\underline{x}^*) = \underline{0},\ \underline{h}(\underline{x}^*) \geq \underline{0}\}$$

A subset of the inequality constraints will be active in the optimum and co-determine the position of the minimum. The associated Kuhn-Tucker multipliers indicate how much the objective function F^* would change if a pertaining inequality constraint would shift a bit: $h(x) \geq 0 \Rightarrow h(x) - \varepsilon \geq 0$. Why is this relevant to know the effects of inequality constraints on the objective function? Realistic process models will reflect current limitations of the technologies applied in the process. Such limitations are often expressed as inequality constraints in the optimization problem. These inequalities delineate the domain in which a technology can trustfully be applied. The technologies that are constraining an optimum by means of active inequality constraints can be identified by means of their associated Kuhn-Tucker multipliers. When analyzing the results from a successfully completed optimization, one can put the Kuhn-tucker multipliers in order of decreasing magnitude. In that way one obtains the order by which active inequality constraints are restraining the objective function. That is, the active inequality constraint with the largest Kuhn-Tucker multiplier has the biggest influence on the objective function of all active inequality constraints. If one wants to enhance future performance of a process, one may decide to push away some active inequalities and particularly those with a high impact on the performance function. One may want to improve those technologies that have one or more constraining inequalities with a high impact on the performance. One may widen the domain of application and so relax tight inequality constraints and increase the performance. The message is that an analysis of the active inequality constraints and their Kuhn-Tucker multipliers in an optimization outcome can help find incentives and clues for further technology development and process innovations.

Interpretation of the results at a computed optimum of a simple reactor design case

Chemical engineering textbooks on optimization present mainly small-scale, mathematically motivated examples dealing also with multiplier values. However, no further use is made of such multipliers for a posterior analysis, probably because these mathematical examples have no application context. Therefore, a small-scale chemical engineering example is presented here to illustrate the use of constraint analysis. The example involves the optimization of a continuous flow, well-mixed reactor with an isothermal reaction $A \Rightarrow B$. The density of the reaction mixture is assumed constant. The reactor has two free design variables, both with an upper bound: reaction volume and temperature. So, there are two inequality constraints (upper bounds). The rate of production of B is fixed. The objective of the process optimization is to maximize the profit rate. The profit rate accounts for the proceeds from product sales, minus the cost of feed A, minus the cost of waste treatment of unconverted A, minus a capital charge term related to the volume-dependent capital investment of the reactor. The capital investment is assumed to scale linearly with reactor volume. The reactor optimization problem will be cast in the standard NLP format assuming the objective function will be minimized. This can be achieved by changing the sign of the profit rate.

Optimization problem (NLP-1):
Objective function (F(x)):

$$\text{Minimize} \qquad F = -P \ (= \text{maximize profit rate P})$$

Equality constraints (g(x) = 0):

Profit rate expression:
$$P = p_B \cdot f_{B,\,out} - c_A \cdot f_{A,\,feed} - c_{waste} \cdot f_{A,\,out} - cc.Inv_{ref} \cdot (V_{reac}/V_{ref})/(\theta * 3600.0)$$

Reactor design equations:
$$f_{A,\,feed} - f_{A,\,out} - V \cdot r = 0$$
$$f_{B,\,feed} - f_{B,\,out} + V \cdot r = 0$$
$$r - k \cdot x_{A,\,out} = 0$$
$$k - k_{T=300} \cdot \exp\{-\gamma \cdot [\,(T_{ref}/T) - 1\,]\} = 0$$
$$f_{A,\,feed} - \{\Phi \cdot x_{A,\,feed}\} = 0$$
$$f_{A,\,out} - \{\Phi \cdot x_{A,\,out}\} = 0$$
$$f_{B,\,feed} - \{\Phi \cdot x_{B,\,feed}\} = 0$$
$$f_{B,\,out} - \{\Phi \cdot x_{B,\,out}\} = 0$$
$$f_{B,\,out} - B_{production} = 0$$

Inequality constraints (h(x) ≥ 0):
Reactor temperature: $310.0 - T \geq 0$
Reactor volume: $500.0 - V \geq 0$

Variables (x):

$f_{A,feed}$	Flow rate of A to reactor	[kg s^{-1}]
$f_{A,out}$	Flow rate of A out of reactor	[kg s^{-1}]
$f_{B,feed}$	Flow rate of B to reactor	[kg s^{-1}]
$f_{B,out}$	Flow rate of B out of reactor	[kg s^{-1}]
k	First-order reaction rate constant	[s^{-1}]
r	Reaction rate	[kg m^{-3} s^{-1}]
P	Profit rate	[€ s^{-1}]:

T	Reactor temperature	[K]
$X_{A,out}$	Concentration of A in effluent	[kg A m^{-3}]
$X_{B,out}$	Concentration of B in effluent	[kg B m^{-3}]
V	reactor design volume	[m^3]
Φ	volume flow (in and out)	[m^3 s^{-1}]

Parameters with numerical values:

c_B	Price of product B	1.200	[€/kg B]
c_A	Cost of fresh feed A	1.000	[€/kg A]
c_{waste}	Cost of waste processing	0.500	[€/kg A]
cc	Capital charge factor	0.250	[€/(€ year)]
B_{prod}	Production rate of B	2.0	[kg B s^{-1}]
Inv$_{ref}$	Reference capital investment reactor	6.000 10^{+6}	[€]
k_{Tref}	Reaction kinetic constant at T_{ref}	2.000 10^{-3}	[s^{-1}]
T_{ref}	Reference temperature for kinetics	300.0	[K]
V_{ref}	reference reactor volume	260.0	[m^3]
$X_{A,feed}$	Concentration of A in feed	10.0	[kg A m^{-3}]
$X_{B,feed}$	Concentration of B in feed	0.0	[kg B m^{-3}]
γ	kinetic parameter { $= E_{act} / (R_{gas}.T_{ref})$ }	50.0	[–]
Θ	hours of production per year	8,000.0	[h/y]

This optimization problem has 12 variables and 10 equalities, including one economic function and nine reactor design equations. The difference between number of variables and number of equalities leaves two degrees of freedom. These two degrees of freedom are reactor volume and temperature; both have an upper bound.

Results of optimization

Having two degrees of freedom one can take values for T and V and easily compute all reactor variables as well as the profit rate P. The outcome of an optimization in the {T, V} plane is that the profit rate is constrained by the upper bounds of the reactor temperature and volume. The numerical results are:

P^*	Profit rate	0.175	[€ s^{-1}]
$f_{A,feed}{}^*$	Flow rate of A to reactor	2.083	[kg s^{-1}]
$f_{A,out}{}^*$	Flow rate of A out of reactor	0.083	[kg s^{-1}]
$f_{B,feed}{}^*$	Flow rate of B to reactor	0.0	[kg s^{-1}]
$f_{B,out}{}^*$	Flow rate of B out of reactor	2.000	[kg s^{-1}]
k^*	First-order reaction rate constant	1.003 10^{-2}	[s-1]
r^*	Reaction rate	4.012 10^{-2}	[kg m^{-3} s^{-1}]
T^*	Reactor temperature	310.0	[K]
$X_A{}^*$	Concentration of A in effluent	0.4	[kg A m^{-3}]

$x_B{}^*$	Concentration of B in effluent	9.6	[kg B m^{-3}]
V^*	reactor design volume	500.0	[m^3]
Φ^*	volume flow (in and out)	0.208	[m^3 s^{-1}]

The profit rate is higher in the infeasible area; e.g., $T = 310$ [K]; $V = 562$ [m^3]; $P = 0.177$ [€ s^{-1}]

The corresponding Kuhn-Tucker multipliers in this constrained optimum are:

$v_T{}^*$	Temperature multiplier in optimum	0.0175	[€ K^{-1}]
$v_V{}^*$	Volume multiplier in optimum	0.0001	[€ m^{-1}]

This result suggests one should aim for an increase of the upper bound of the temperature.

The constraint analysis for this small-scale optimization example can be carried over to large-scale optimization problems that are solved by means of numerical mathematical programming algorithms. Upon extracting the Kuhn-Tucker multipliers of the active inequality constraints from the computed solution and putting these in decreasing order of magnitude, one can identify the corresponding dominating inequality constraints. Relaxation of the bounds of the more dominating inequalities will improve performance. These dominating inequalities often express underlying process technological restrictions. Such restrictions may be removed by focused improvements in technology development, enhancing process performance.

Case A9.11: An example of a Pareto trade-off between two performance metrics

A simple chemical engineering example is given of a Pareto set of design solutions for two performance criteria. Commonly, two criteria in the realms of economy and ecology are often used. The latter may cover also efficiency of transfer of a critical chemical element from reactants to products.

Example
The chemical reactor problem of case A9–10 is used as a vehicle to demonstrate the effects of optimizing two criteria simultaneously. The two criteria are the profit rate (P) and the mass efficiency (ME) of the process. The latter indicates which fraction of the intake mass is usefully transferred to the product. The reactor temperature and volume are the two design variables. The results are shown in Figure A9.5. There is a region to the left in the diagram (where the profit rate rises to its peak) where both the mass efficiency and the profit rate get better. To the right of the peak starts the Pareto set where an improvement in the mass efficiency is met by a decrease of the profit rate.

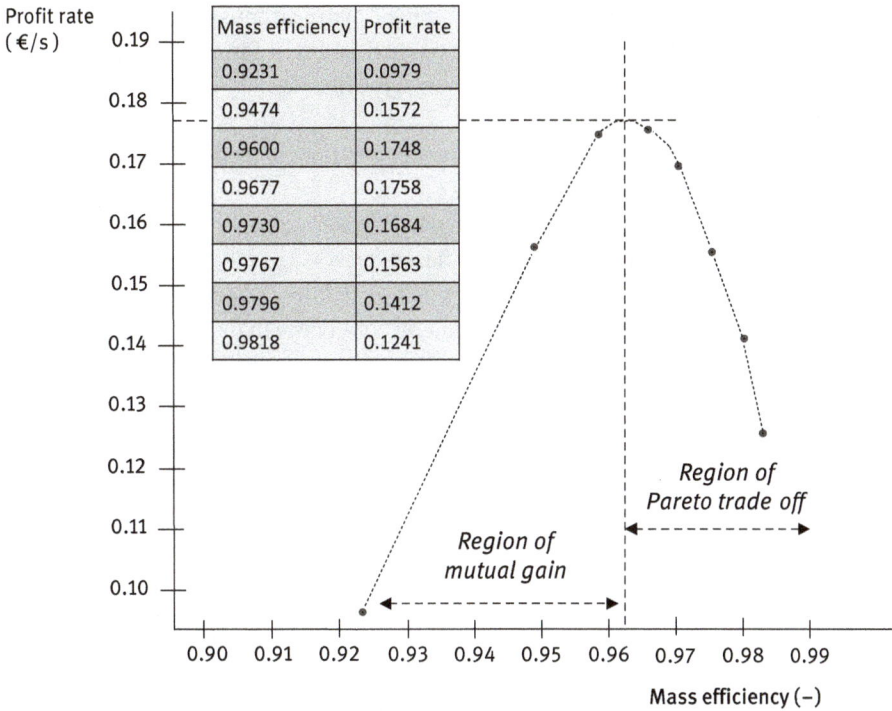

Mass efficiency	Profit rate
0.9231	0.0979
0.9474	0.1572
0.9600	0.1748
0.9677	0.1758
0.9730	0.1684
0.9767	0.1563
0.9796	0.1412
0.9818	0.1241

Figure A9.5: Pareto trade-off between profit and product mass efficiency of process.

Case A9.12: Optimization of designs of industrial processes: GTL example

Due to production scale and complexity GTL processes are prime candidates for design optimization.

The main conversion steps in the GTL process are:

(1) Syngas manufacturing, using steam (by reforming) or pure oxygen (by catalytic partial oxidation) to turn methane into syngas (CO: n H_2; $2 < n < 4$).

(2) Fischer-Tropsch synthesis of wax (paraffinic hydrocarbons with some olefins and alcohols)

Reaction stoichiometry: n CO + $(2n + 1)$ $H_2 \Rightarrow C_nH_{2n+2} + n$ H_2O n: $1 \Rightarrow \sim 300$.

Low temperature (200–240 °C) conversion with cobalt catalyst is used for NG conversion.

(3) Hydro-cracking of the wax into a slate of transportation liquids.

The Fischer-Tropsch (FT) reactors cannot fully convert the syngas due to technological limitations of catalyst and reactor equipment. The question arises on what to do with the unconverted syngas, being enriched with light hydrocarbons ($\leq C_4$) formed by FT synthesis. This is jointly called off-gas. Four possibilities exist for off-gas routing:

- Short recycle over the FT reactors to control the hydrogen to carbon ratio in reactor feeds.
- Long recycle to the syngas manufacturing section for reuse of carbon in CO and CO_2.
- Fuel to the utility system for steam and power generation
- Stack (a fraction of the off-gas must be flared to get rid of inerts in the feeds).

The design complexity is mainly due to the interactions between the syngas manufacturing, the FT conversion due to the syngas recycle structure, as well as the heat exchanges between SGMU and FT sections, and the utility system. The design of the hydro-cracking is similar to existing refinery designs.

The list of process synthesis decision variables for SGMU and FT sections is a long one:

- Type of syngas manufacturing units (SGMU), with choices between Steam Methane Reforming (SMR), Catalytic Partial Oxidation (CPOx) and Autothermal reforming (ATR).
- Use of multiple SGMU technologies in parallel to control syngas composition.
- Oxygen-to-carbon ratio in feeds to SGMUs and their operating temperatures.
- Type of FT reactors (e.g., fixed bed, slurry, structured reactors) with cobalt catalyst.
- Amount of catalyst in a standardized FT reactor type.
- Number of sequential stages for FT reactors and number of parallel reactors per stage.
- Distribution of syngas from the SGMUs over the FT stages.
- Mixing ratio of fresh and recycle syngas to control H_2:CO ratio in the combined feed to an FT reactor stage (short recycle).
- FT reactor temperature and pressure, uniform per stage:
- Recycle fraction of off-gas to the SGMUs with distribution over these units (long recycle)
- Fraction of off-gas to the utility system for steam and power generation
- Transfer rates of high quality thermal energy (high T) to lower quality levels in utility system
  The energy quality levels have pre-assigned temperature and pressures.
- Production rate of liquid wax from FT section. If feed driven, intake of NG to SGMU section.
- Recycling of FT reaction water to SMR.

In addition, there will be multiple performance metrics:

- Profitability, often expressed as Net Present Value (NPV)

- Carbon efficiency, expressed as amount of carbon in product over amount of carbon in feeds
- Energy efficiency

In view of the complicated processing structure it is also an option to place a lower bound on the availability and reliability of the process to assure a high-enough annual product output.

An industrial case study has been published [10] on model-based, optimization-driven synthesis of a future GTL process, using profitability and carbon efficiency as performance measures. The process units are modeled in a parsimonious way focusing on their main features of behavior. Each unit model makes use of one or more constitutive equations (kinetics, phase equilibria). For each unit the domain of applicability of the constitutive equations has been outlined by means of inequality constraints. The resulting outcome of a GTL case study shows a design region in which both performance metrics can be improved simultaneously. However, after the peak of the profitability there is a Pareto curve with a trade-off between the profitability and carbon efficiency. In a separate study [11] the operational reliability of a GTL process was modeled and optimized. One can go even one step further by jointly optimizing the design of a process along with planning of production and process maintenance. This has been successfully done for a multipurpose process [12].

Case A9.13: Process performance optimization by model and experiments

Data-driven process model development is sometimes cheaper and quicker than developing a (semi-) mechanistic model. However, it is required that one can indeed measure all input and output variables of a product / process, relevant to the characterization of its behavior and performance. This case story will focus on experimental process optimization, using a response surface approach. The procedure that is followed applies equally well to the optimization of product performance.

The inputs to a process model must comprise the feed and key operating conditions of the process. The inputs may also include some quantitative geometric design decision variables as long as these can be implemented and adjusted in an experimental set-up; e.g., by adjusting volume or a contact surface between two different phases. The relevant measurable outputs are those that characterize the behavior and performance of a process and which can be measured. It is a key condition for data-driven black-box modeling that process inputs can be adjusted at will in a wide enough region to be able to collect information-rich output data. Making a design of experiments will be instrumental to assure that measured output data is spread over the entire output space, rather than located in a sub-space. Such a broad coverage is needed for good quality estimation of the unknown parameters in the black box model, which in turn will determine the descriptive accuracy of the model. Now, first a sequential, multistep procedure, represented in Figure A9.6, will be explained.

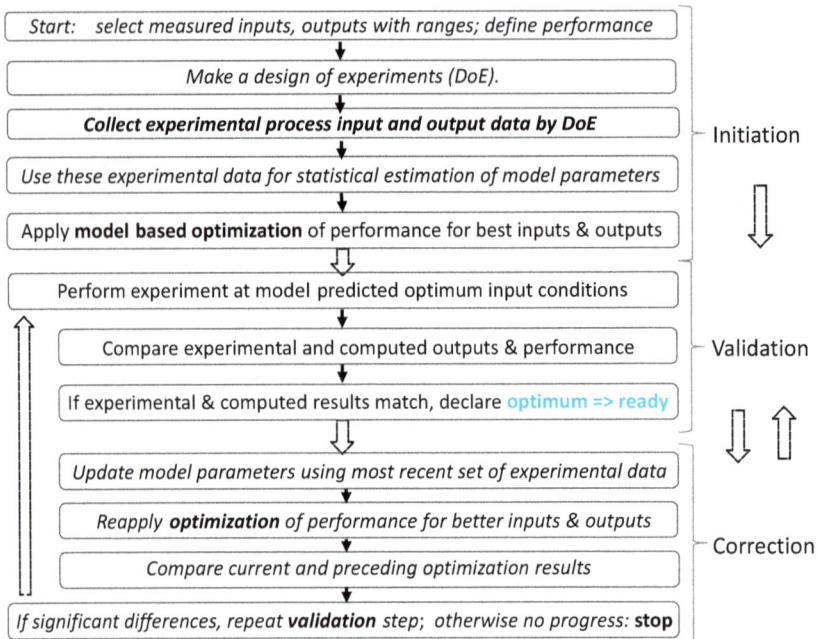

Fig. A9.6: A work procedure for an experimental approach to optimum process performance.

This work procedure combines modeling and experimentation in optimization of the process performance. After having explained the procedure, the mathematical features of a black box model will be outlined. The objective of this work procedure is to closely approximate the point of optimum performance of the process through a sequence of model computations and experimentation towards. It is a strong feature of this approach that the point of optimum performance point is experimentally confirmed. The procedure gradually moves the performance of the process to the optimum in an alternating play between:

(a) *prediction*: let the current model "predict" at which inputs the optimum outputs are obtained.

(b) *validation*: an experiment is performed at the model-predicted optimum conditions. If the measured and predicted performances are close enough the optimum is found and the search stops.

(c) *correction*: if there is too much of a discrepancy between model and experimental performances the model needs adjustment. The newly generated experimental data under (b) will be used to improve the statistical estimates of the model parameter.

The sequence {prediction, validation, correction} is needed because any model, and specifically a black box one, remains an approximation to the real process with its unknown features. The structure of the procedure is shown in Fig. A9.6.

The work procedure and its three phases are explained.

Initiation phase:

Step 1: *Define process boundaries with measurable inputs and outputs and performance*
The selected inputs and outputs must be relevant for performance.

Step 2: *Make a (statistical) Design of Experiments (DoE)*
The objective of the DoE is to create a diverse, rich-enough data set, which allows for statistical parameter estimation with good accuracy.

Step 3: *Perform experiments according to DoE plan and collect measured input and output data.*

Step 4: *Apply a statistical estimation procedure to derive the first set of parameter estimates.*
Check for poorly estimated parameters and collinearity of outputs.

Step 5: *Use the model with new parameter estimates to compute the optimum conditions*
This computation is in fact a model-based NLP optimization, requiring a model (available), bounds on the ranges of all inputs and outputs (given as inequality constraints), and a performance function in input and output variables.

Validation phase:

Step 6: *Perform an additional experiment with the computed (predicted) optimum input conditions.*

Step 7: *Compare the experimental and computational data {inputs, outputs, performance}*
This can be done, for instance, by looking at the difference per variable over all variables.

Step 8: *Check how close the computed and experimental optimum outputs and performance are*
If all outcomes are close indeed (some quantitative criterion may be needed), then one reasonably assumes that the computed and the experimental results indeed approximate the true but unknown optimum. The search is successfully finished. If not close enough yet a next corrective phase is entered.

Corrective phase:

The corrective phase acts on the difference between the model and the real process.

Step 9: *Update the model parameters*
Use the most recent experimental data to update the model parameters by means of a statistical estimation procedure updates, preferably in a Bayesian fashion: prior ⇒ posterior info.

Step 10: *Compute the next (new) optimum conditions, using the updated parameters.*

Step 11: *Check if the difference between the new and preceding optimum is significant*

If so, start the next validation phase. If not so, further experimentation is likely not going to improve the optimization results. It is advised to terminate the procedure and reanalyze the model and the experimental approach.

It is likely that this sequence of validation-corrections needs to be repeated a couple of times.

The outcome of this procedure is a set of input conditions for which the performance of the process is optimal and experimentally confirmed. As a side effect a process model is obtained, having parameter estimates that are accurate enough to predict the outputs and the performance metric. The strength of this successive approach of gradual refinement of parameter estimates is that model predictions and experimental data should become consistently close in the region of optimum performance. Thus, a key benefit of this multistep approach is that it refines the accuracy of the computed model outcomes particularly in a region that surrounds the optimum performance point of the process.

The process model in general will be a nonlinear multidimensional one, $\underline{Y}^{(k)} = \underline{f}\,(\underline{U}^{(k)}; \underline{p})$. Here, k is an index of an experimental situation. The structure of the equations is case-dependent and should reflect as much as possible the basic understanding of the how the process works. In spite of its black box nature, the model (equations) should at least respect conservation principles of mass, the chemical elements and energy. Regarding accuracy of measurements, the conventional assumption that inputs ($\underline{U}$) are set with 100% accuracy is often not realistic. There are always some fluctuations around the intended target settings. The assumption here is that multiple inputs ($\underline{U}$) can be set with small known adjustment errors. These errors are characterized by means of an error covariance matrix (V_U). Usually the errors in the adjustments of the inputs are smaller than the measurement errors of the outputs ($\underline{Y}$).

The experimental data and the associated model variables will be fitted by means of a quadratic norm (a weighted least squares approach), consistent with the common assumption of normally distributed measurement errors. The input variables in the model ($\underline{U}$) are to be jointly fitted to the measured inputs as well as to model. Similarly, the process outputs ($\underline{Y}$) in the model must be jointly fitted to measured data (experimental outputs) and obey the model equations. That is, the model variables must obey the model equations exactly, while the corresponding experimental data do not due to measurement errors and differences between model and the real process. While inputs and outputs are constrained by the model equations and their link to the experimental data, the remaining degree of freedoms in the fitting are the adjustable model parameters ($\underline{p}$).

This approach falls into the category of response surface methods [13]. There are two extensions to the above chosen approach:
– The usual assumption in statistical textbooks that inputs have no errors is unrealistic for most chemical processes and must be relaxed. The result is a practical generalization of the least-squares solution, commonly called total least squares or "error in variables approach."

– Application of a multistep sequential procedure to move to the optimum performance of the process in an alternating play between (a) doing an experiment at model-predicted optimum conditions and (b) using the new experimental data to correct the model parameter estimates. The purpose of the multistep approach is to refine the accuracy of the model equations particularly in a region that surrounds the optimum performance point of the process.

The final outcome is a process model having parameter estimates that are accurate enough to predict the optimum performance point. This point is also confirmed by experiments. The strength of this successive approach of gradual refinement of parameter estimates is that model outputs and experimental data would become consistent in the region of optimum performance.

List of symbols, subscripts, and abbreviations

Symbols

A	real (non-)square matrix (real values as entries)
b	real vector
B	real (non-)square matrix, often mapping process inputs to states
c	cost related parameters in economic model expressions
c	molar concentration
d	symbol for design parameters (continuous) in a process model
E	set of quantitative performance metrics for a process design
E	matrix of elemental composition of a set of chemical species
E	reaction activation energy
f	algebraic function
f	molar flow rate
F	objective function in an optimization problem
g	set of equality functions in an optimization problem
h	mole-based specific enthalpy of a component
h	set of inequality functions in an optimization problem
ΔH_R	molar reaction enthalpy
i	generic index; e.g. for chemical species
j	generic index; e.g., for chemical reactions
k	index for discrete time stepping
k	reaction rate constant
K	reaction equilibrium constant
L	Lagrangian function
m	phase distribution factor
N	dimension of a vector
p	generic symbol for a set of physical parameters in a process model
p_P	proceeds from product sales, in economic model equations
P	performance function such as a profit rate
Q	heat exchange rate

r	reaction rate
R_{gas}	universal gas constant
s	split factor, distributing an incoming stream into two outgoing streams
S	stoichiometric matrix
S	sensitivity factor
t	time coordinate in dynamic models
t	target value for a design quantity or performance function in process design
T	function expressing a metric (function) related to a target
T	temperature
u	generic symbol for inputs to a process model
v	Kuhn-Tucker multipliers in an optimization problem
V	volume
w	disturbance variables in a process model
W	matrix of weight factor for scaling of process output variables
x	generic symbol for physical states in a process model
x	molar fraction of a chemical species
y	set of measured outputs of a process (model)
Y	generic symbol for structural design parameters (discrete) in a process model
z	generic symbol for a set of adjustable inputs to a process model
z	spatial coordinate in distributed models
Z	matrix taking integer values only
$\{\alpha,\beta,\gamma\}$	indices for thermodynamic phases in models
γ	scaled reaction activation energy
ε	deviation between computed and experimental value of a process output variable.
η	degree of conversion of a feed component in a reactor
ξ	extent of reaction
θ	number of production hours per year
λ	Lagrangian multiplier
Ω	sum of squared deviations between model and experimental outputs
τ	residence time
Φ	volume flow
σ	selectivity factor of a reaction network, relative to a particular reaction product
$\mathfrak{R}^N$	mathematical space of dimension N for real variables
∇	Spatial derivative

Subscripts

average	averaged value of a quantity
design	design related
ecol	ecology related
ecom	economy related
oper	operations related
perf	performance related
ref	reference value of a quantity when making a comparison
tech	(process) technology related

Abbreviations

ATR	Autothermal reforming
CPOx	Catalytic partial oxidation (of natural gas)
C-S-R	Conversion-Separation-Recycle (process)
DAE	Differential algebraic equation set
GTL	Gas to Liquid
FT	Fischer-Tropsch
LP	Linear Programming
MILP	Mixed-integer linear programming
MINLP	Mixed-integer nonlinear programming
MTBE	Methyl *tertiary* butyl ether
NG	Natural Gas
NLP	Nonlinear Programming
SG	Syngas
SGMU	Syngas manufacturing units
SMR	Steam-Methane Reforming
XTL	X = {Biomass, Coal, Natural Gas} to Liquid fuels

References and further reading

[1] Buelens LC, Galvita VV, Poelman H, Detavernier C, Marin GB. Super-dry reforming of methane intensifies CO2 utilization via Le Chatelier's principle, Science, 354(6311), 2016, 449–452.
[2] Varma A, Morbidelli M, Wu H. Parametric Sensitivity in Chemical Systems, Cambridge: Cambridge University Press, 1999.
[3] Haaker MPR, Verheijen PJT. Local and global sensitivity analysis for a reactor design with parameter uncertainty, Chemical Engineering Research & Design, 82, 2004, 591–598.
[4] Ogunnaike BA. Random Phenomena. "Fundamentals and Engineering Applications of Probability and Statistics, Boca Raton: CRC Press, 2010.
[5] Marlin TE. Teaching "operability" in undergraduate chemical engineering design education, Computers & Chemical Engineering, 34, 2010, 1421–1431.
[6] Mannan S. 'Lees' Process Safety Essentials, Hazard Identification, Assessment and Control, Ch.5, Amsterdam: Elsevier, 2014.
[7] Ogunnaike BA. Random Phenomena. "Fundamentals and Engineering Applications of Probability and Statistics", Ch.21, Boca Raton: CRC Press, 2010.
[8] Biegler LT, Grossmann IE, Westerberg AW. Systematic Methods of Chemical Process Design, New York: McGraw Hill, 1997.
[9] Biegler LT. "Nonlinear Programming. Concepts, Algorithms and Applications to Chemical Processes", Ch. 4, Philadelphia: SIAM, 2010.
[10] Ellepola JE, Thijssen N, Grievink J, Baak G, Avhale A, Van Schijndel J. Development of a synthesis tool for Gas-To-Liquid complexes, Computers and Chemical Engineering, 42, 2012, 2–14.
[11] Blanco H, Grievink J, Thijssen N, Herder PM. Reliability integration to process synthesis applied to GTL processes, Computer-Aided-Process-Engineering, 33, 2014, 79–84.

[12] Goel HD, Grievink J, Weijnen MC. Integrated optimal reliable design, production, and maintenance planning for multipurpose process plants, Computers & Chemical Engineering, 27, 2013, 1543–1555.

[13] Ogunnaike BA. Random Phenomena. "Fundamentals and Engineering Applications of Probability and Statistics", Ch.19, Boca Raton: CRC Press, 2010.

[14] Franceschini G, Macchietto S. Model-based design of experiments for parameter precision: State of the art, Chemical Engineering Science, 63, 2008, 4846–4872.

[15] Klebanov N, Georgakis C. Dynamic response surface models: A data-driven approach for the analysis of time-varying process outputs, Industrial & Engineering Chemistry Research, 55, 2016, 4022–4034.

A13 Appendix to Chapter 13: communicating

A13.1 Activity report, example

TUDelft	Design project: 'name & code' confidential	
Activity Report title:		Reference
AR title: 'AR – PC–PF/Knowledge – PSw'		'PC–PF/Kn – PSw/20171015'

Version	Author	Date	Description
V1.0	Name	15/10/2017	Text

Summary

Text

Contents
- Objectives
- Approach, methods, tools
- Results & discussion
- Conclusions & recommendations
- References
- Appendices

Objectives

Text

Approach, methods, tools

Text

Results & discussion

Text

Conclusions & recommendations

Text

References

Appendices

https://doi.org/10.1515/9783110782127-018

A13.2 Stream specification (passing battery limit)

Table A13.1 shows the stream specification of the process stream passing the battery limit.

Tab. A13.1: Stream specification (passing battery limit) [2].

Stream name:	Propylene/propane				
Component	Units	Specification			Additional information
		Available	Used in the design	Notes	(also ref. note numbers)
Propylene	%wt	50–70	69.0–70.0	(1)	(1) Values taken in consultation with
Propane	%wt	50–30	29.0–30.0	(1)	principal.
Light ends	%wt	<1.0		(2)	(2) As "worst case" scenario, LEs and HEs
Heavy ends	%wt	<1.0		(2)	are taken as ethane and butane,
Sulfur	ppm wt	<20		(3)	respectively.
Metals	ppm wt	<4		(3)	(3) Contaminants not harmful for the
Ethane	%wt		<1.0	(2)	process compounds not included in mass
Butane	%wt		<1.0	(2)	balance.
Total			100.0		(4) Not yet known; will be provided by
					principal later.
		Process conditions and price			
Temp.	°C		25		Basis: Mixture of values for
Press.	Bara		8		C3 = /PG and C3o.
Phase	V/L/S		V		Delivery per pipeline/rail car/
Price	Eu/ton		tbd	(4)	Tank truck/1000 L container/drum: indicate!

A13.3 Concepts/criteria matrix

Table A13.2 shows an example of "concepts/criteria matrix," or Pugh matrix [1]. This type of matrix supports the design team's discussion on the design alternatives, stimulates creativity to further improve on the concepts and stimulates the decision-making process. It should not be used as a quantitative decision-making tool. Careful selection of the specifications (to avoid interrelations/overlap) and suppression of the tendency to quantify the subjective scores to a single score for each concept is needed. Inferior and superior concepts can be easily identified and the remaining concepts should be further developed and improved to get closer or surpass the superior concept(s) that fulfill and/or exceed the specifications.

Tab. A13.2: Concepts/criteria matrix.

Concepts/criteria matrix	Concept A	Concept B	Concept C
Specification 1	+	o	–
Specification 2	–	o	+
Specification 3	+	–	+
Specification 4	o	+	–
Specification 5	+	–	+

A13.4 Pure components properties table

Table A13.3 shows how to report pure component properties.

A13.5 Agenda and minutes of meeting (MOM) templates

Agenda template

Agenda
- Meeting information
  - Meeting title:
  - Project:
  - Date:
  - Starting time:
  - End time:
  - Location:
- Meeting preparation:
  - Please read:
  - Please bring:
- Invited attendees:
  - Names, affiliation:
- Agenda items
  - Attendance and introductions
  - Arrangement of agenda
  - Confirmation of MOM (previous meeting)
  - Matters arising from the MOM (action points)

- Meeting topics
  - Topic 1, owner, allocated time
  - Topic 2, owner, allocated time
  - Etc.
- Any other business
- Next meeting

Minutes of meeting (MOM) template

Minutes of meeting
- **Meeting information**
  - Meeting title:
  - Project:
  - Date:
  - Starting time:
  - End time:
  - Location:

- **Agenda items**
  - Attendance and introductions
    - Attendees: names, affiliations
    - Absentees: names, affiliations
  - Arrangement of agenda
  - Confirmation of MOM (previous meeting)
  - Matters arising from the MOM (action points)
  - Meeting topics
    - Topic 1, owner, decisions, action points and timing
    - Topic 2, owner, decisions, action points and timing
    - Etc.
  - Any other business
  - Next meeting
  - Reports, handouts presented, discussed during the meeting
  - List of decisions, action points and timing

Tab. A13.3: Pure components properties table [2].

PURE COMPONENT PROPERTIES

Component name		Formula	Mol. weight g/mol	Phase	Boiling point [1] °C	Melting point [1] °C	Flash point [1] °C	Liquid density [2] kg/m3	Vapour density [3] kg/m3	Auto-ignit temp. [1] °C	Flammable limits % by vol in air	Lower Explosion Limit (LEL) %	Upper Explosion Limit (UEL) %	LC$_{50}$ in air/ water mg/m3	MAC value mg/m^3	LD$_{50}$ oral [4] g	Chemical reactivity	Notes
Design	**Systematic**				*Technological data*							*Health & safety data*						
Propylene	Propylene	C$_3$H$_6$	42.08	>	-48.0	-185.2	-108	609.0	1.50	497.0	?	2.0	11.1	n.a.	n.a.	?	Extremely flamable,	(2),(5)
Propane	Propane	C$_3$H$_8$	44.09	>	-42.2	-187.6	-104	585.0	1.56	450.0	?	2.1	9.5	n.a.	1800	?	flamable, Explosive with air	(2),(6)

Notes:
[1] At 101.3 kPa
[2] Density at 25 °C, unless specified otherwise
[3] At 0 °C
[4] Oral ingestion in (g) for a male of 70 kg weight
[5] Density at -47 °C from H$_2$O at 4°C
[6] Density at -45 °C from H$_2$O at 4°C

Project ID number: GDP-PED-2017-1/1
Completion date: October 1st 2017

*Converting mg/m3 -->ppm & vice versa: mg/m3 to ppm calculator

A13.6 Frequently Occurring Opportunities for Improvement (FOOFI): list for design project presentations

This FOOFI list is composed in the form that frequently occurring errors/mistakes are indicated, based on [2].

General
– Page numbering missing
– Slides not readable (too small font size); minimum 16 pt
– University logo missing (use standard TU Delft or company format)
– Very few pictures, tables, or figures are used; this makes the presentation boring
– Use of long sentences instead of keywords (use a maximum of seven lines of seven words in each)
– Presenter reads information from screen (mainly because information is in "sentence form")
– Spelling mistakes (use spell checker)
– Presentation too long (aim at 1–1.5 slides/per minute)
– Too many colors used
– (Black) text invisible on colored background. Be very selective of colors; only use black text on white (or *very* light) backgrounds. White text works well on black/ dark gray or (dark) blue backgrounds.

Title page
– Project principal (owner/customer) not mentioned
– Logo of customer (in case of external project) missing
– Names of project team members missing
– Presentation date and location missing

Table of contents
– Missing table of contents
– Present presentation chapters at slide bottom, highlighting the current chapter

Introduction
– References to data, figures, and tables missing

Project objectives
– Presentation has too much detail, without highlighting the objectives.
– Conclusions & recommendations are not linked to objectives.

Basis of design (BOD)
- BOD section, including all aspects agreed with customer, is missing, e.g.:
  - Product performance specifications
  - Sales volume
  - Plant capacity (t/annum)
  - Number of operating hours/year (including shift pattern)
  - Location
  - Battery limit of system to be designed
  - Specifications (composition, delivery form & conditions, prices) at battery limit of:
    - products
    - byproducts
    - raw materials
    - available utilities
    - waste streams handling
  - Pure component properties table (including major SHE aspects)

Process block scheme (PBS)
- Missing
- No indication of conversion and split factors for separation blocks
- No indication of t/a, t/t values
- No indication of stream names

Process flow scheme (PFS)
- Standard symbols not used
- Not all raw materials enter at the left
- Not all products, byproducts, and waste streams leave at the right
  - See separate information on making process flow diagrams
- PFS not readable: split into smaller (enlarged) parts

Mass balances
- No overall mass balance given (e.g., in the form of an input/output diagram showing all input and output streams, utilities and verification of *total in = total out*).

Tables
- Too many decimals after the decimal point
- Comma is used instead of decimal point
- Wrong units are specified: too many significant numbers are presented (copied from computing software)
- Numbers are not adjusted "right" or at the decimal point (numbers should be easily viewed/compared)

– Units are not mentioned in the column title
– No SI units are used ("kcal", "j" instead of "J", "kJ/h" instead of "kW" or "MW", "bars", "bar" used instead of "bara" or "barg", etc.)

Units
– SI units not used ("kcal", "j" instead of "J", "kJ/h" instead of "kW" or "MW", "bars", "bar" instead of "bara" or "barg",' etc.)

Comparison of alternatives (conversion routes, separation methods, unit operations, etc.)
– Inappropriate comparison methods can be improved easily:
  – When comparing two different product and/or process design options, the comparison criteria should be mentioned in the first column, and the performance of the alternatives should be scored for each criterion.
  – The criteria should be formulated in such a way that a score of " +" or "++" means that the performance for that criterion is good or very good.
  – The criteria should be ordered in order of importance.
    – Use a 5-level scale: "– –", "–", "0", " +", "++". Alternatively, and much preferred, present actual performance data in the cells and use a 5-scale color code: "– –": red, "–": orange, "0": white, " +": light green, "++": dark green.
    – This ensures that all criteria will be evaluated for each alternative and missing information is immediately evident.

Equipment design
– Design specifications per product component or equipment unit are not clear.
– The design variables to reach the design specifications for the product or process component are not made clear.
– It is unclear how the design variables are chosen/fixed to achieve the design specifications of each product or process component and the whole product or process.

Safety, health, environment, economy, technology, social (SHEETS) evaluation
– General: not a complete SHEETS evaluation given (mainly focused on economics).
  – Focus on all SHEETS aspects is required by stakeholders and de-sign is evaluated accordingly.
– On economics:
  – Gross Income is not presented, or only at the very end; mention this upfront.
  – No distinction made between operating costs and investment costs.
  – No conclusions formulated on the economical data presented and the objectives.
  – No sensitivity analysis carried out: what are the financial risks for the project?

Conclusions and recommendations
- No reference is made to the project objectives: have they all been met?
- No presentation of the design's strengths and weaknesses: how can it be improved?
  - What would be possible next steps?
  - Difference between project objectives and achieved results should feed into the recommendations.

A13.7 Equipment summary and specification sheets

Tables A13.4-13.7 show examples of equipment summary sheets.

Tab. A13.4: Equipment summary sheet: reactors, columns and vessels – summary [2].

Equipment number.:	C-01	V-01
Name:	P/P splitter	C-01 reflux accumulator
–	Tray column	Horizontal
Pressure [bara]:	20.9 / 21.5	20.8
Temp. [°C]:	50.0 / 57.0	50.0
Volume [m3]:		105.1 *(1)*
Diameter [m]:	4.5	3.5
L or H [m]:	69.0	11.0
Internals		
– Tray type:	Sieve Trays	n.a.
– Tray number:	130	n.a.
– Fixed packing		
Type:	n.a.	n.a.
Shape:	n.a.	n.a.
– Catalyst		
Type:	n.a.	n.a.
Shape:	n.a.	n.a.
Number		
– Series:	1	1
– Parallel:	–	–
Materials of construction *(2)*:	Trays: SS314 Column: CS	CS
Other:		

Remarks:
(1) V-01: effective volume = 68.8 m^3 for residence time of 5 min.
(2) SS = Stainless steel; CS = Carbon steel; n.a.: not applicable

Designers:	P.P. Kolom H.P. Pomp	W. Wisselaar R.G. Klep	Project ID-Number: Date:	*CDP3201* January 1st 1998

Tab. A13.5: Equipment summary sheet: heat exchangers and furnaces – summary [2].

Equipment number.: Name:	E-01 C-01 Ovhd condenser	E-02 C-01 reboiler	E-03 propylene cooler	E-04 propane cooler
	Finned tubes Air cooled	Thermosyphon	Fixed tube sheet	Fixed tube sheet
Substance				
– Tubes:	$C_3 = /n - C_3$	$n - C_3/C_3 =$	$C_3 = /n - C_3$	$n - C_3/C_3 =$
– Shell:	n.a.	L.P. Steam	Cooling Water	Cooling Water
Duty [kW]:	91,450	91,450	344	378
Heat Exchange area [m^2]:	12,760	2,850	75	62
Number				
– Series:	–	–	1	1
– Parallel:	–	6	–	–
Pressure [bara]				
– Tubes:	20.9	21.7	26.0	26.0
– Shell:	Atm.	4.0	4.0	4.0
Temperature				
In / Out [°C]				
– Tubes:	50.0 / 50.0	57.0 / 57.0	50.0 / 30.0	57.0 / 30.0
– Shell:	25.0 / 41.0	190.0 / 133.0	20.0 / 40.0	20.0 / 40.0
Special Materials of Construction: *Other:*	Tubes: CS Shell: n.a. Plot Area: 40 × 40 m	Tubes: CS Shell: CS	Tubes: CS Shell: Al-Br	Tubes: CS Shell: Al-Br
Designers:	P.P. Kolom H.P. Pomp	W. Wisselaar R.G. Klep	Project ID-Number: Date:	*CDP3201* January 1st 1998

Tab. A13.6: Equipment summary sheet: pumps, blowers and compressors – summary [2].

Equipment number.: Name:	P-01 C-01 Reflux:	P-02 A/B C-01 Bottom:	P-03 A/B Propylene Transfer:
Type: *Number:*	Centrifugal 2	Centrifugal 2	Centrifugal 2
Medium transferred:	$C_3 = /n - C_3$	$n - C_3/C_3 =$	$C_3 = /n - C_3$
Capacity [kg/s] [m^3/s]:	0.239	0.011	0.015

Tab. A13.6 (continued)

Equipment number.: Name:	P-01 C-01 Reflux:	P-02 A/B C-01 Bottom:	P-03 A/B Propylene Transfer:
Density [kg/m³]: – Pressure [bara]	450	450	450
Suct. / Disch.:	20.8 / 23.0	21.7 / 26.0	20.8 / 26.0
Temperature In / Out [°C]:	50.0 / 50.0	57.0 / 57.0	50.0 / 30.0
Power [kW] – Theor.: – Actual:	53 75	5 8	8 13
Number – Theor.: – Actual:	2 (1)	2 (1)	2 (1)
Special Materials of Construction:	MS casing	MS casing	MS casing
Other:	Double mechanical seals	Double mechanical seals	Double mechanical seals

Remarks:
(1)   Tray numbering from top to bottom

Designers:	P.P. Kolom H.P. Pomp	W. Wisselaar R.G. Klep	Project ID-Number: Date:	CDP3201 January 1st 1998

Tab. A13.7: Equipment specification sheet: distillation column [2].

Equipment number: Name:		C-01 Propylene/Propane Splitter	
General Data			
Service:		– distillation / ~~extraction~~ / ~~absorption~~ /	
Column Type:		– ~~packed~~ / tray / ~~spray~~ /	
Tray Type:		– ~~cap~~ / sieve / ~~valve~~ /	
Tray Number (1)			
– Theoretical:		90	
– Actual:		130	
– Feed (actual):		85	
Tray Distance	0.500	Tray Material: SS314	(2)
(HETP) [m]:			
Column Diameter [m]:	4.500	Column Material: CS	(2)
Column Height [m]:	69.000		
Heating: ~~none~~ / ~~open steam~~ / reboiler /			(3)

Tab. A13.7 (continued)

Equipment number:				C-01				
Name:				**Propylene/Propane Splitter**				

Process Conditions

Stream Details		Feed		Top		Bottom		Reflux / Absorbent		~~Extractant~~ / ~~side stream~~
Temp.	[°C]	:	53	:	50	:	57	:	50	
Pressure	[bara]	:	21.5	:	20.8	:	21.7	:	20.9	
Density	[kg/m³]	:	450	:	450	:	450	:	450	
Mass Flow	[kg/s]	:	11.9	:	6.88	:	5.02	:	107.74	

Composition	mol%	wt%	mol%	wt%	mol%	wt%	mol%	wt%	mol%	wt%
Propylene	60.0	58.8	95.0	94.8	10.0	9.6	95.0	94.8		
Propane	40.0	41.2	5.0	5.2	90.0	90.4	5.0	5.2		

Column Internals (4)

Trays (5)	Packing: Not Applicable
Number of	Type:
~~caps~~ / sieve holes /: ...	Material:
Active Tray Area [m²]: ...	Volume [m³]:
Weir Length [mm]: ...	Length [m]:
[m]: Diameter of	Width [m]:
chute pipe / hole / [mm]: ...	Height [m]:

Remarks:
(1) Tray numbering from top to bottom.
(2) SS = Stainless Steel; CS = Carbon Steel.
(3) Reboiler is E-01; operates with LP steam.
(4) Sketch & measures of Column & Tray layout should have been provided.
(5) Tray layout valid for whole column.

Designers:	P.P. Kolom	W. Wisselaar	Project ID-Number:	CDP3201
	H.P. Pomp	R.G. Klep	Date:	January 1st 1998

A13.8 FOOFI (Frequently Occurring Opportunities For Improvement) list for design project reports

FOOFI list for Design Project Reports (Concept Stage, Feasibility Stage, Final Report)

This FOOFI list is composed in the form that frequently occurring errors/mistakes are indicated, and is based on [2].

General
- Page numbering missing
- Text not readable (too small font size); minimum 10 pt.
- University logo missing (use standard TU Delft or company format)
- Very few pictures, tables or figures are used: this makes the report very boring
- Use of very long sentences
- Spelling mistakes (use spell checker)
- Report main text too long (aim at 60 pages main text; remainder in appendices
- (Too) many colors used in figures
- (Black) text invisible on colored background: be very selective of colors; only use black text on white (or VERY light) backgrounds; white text works well on black/dark gray or (dark) blue backgrounds.

Front page
- Design project stage (concept stage, feasibility stage, final report) omitted from front page
- Project principal (owner/customer) not mentioned
- Logo of customer (in case of external project) missing
- Names of project team members missing
- Report issue date missing
- Insufficient or nondescriptive key words
- Date on report review meeting missing on front page

Page numbering, table of contents, appendix table of content
- No page numbering before Chapter 1. (it should be: i, ii, iii, etc.)
- No page numbering for appendices (should be Appendix – 1 –, etc.)
- No page numbering of appendix table of contents, (it should be: Appendix – i –, Appendix – ii –, etc.)
- No information on appendices in (main) table of contents: list the detailed contents and pages of the appendices
- No "Appendix Table of Contents" before Appendix 1

Chapters
- No symbol list and no list of abbreviations. These lists should be positioned before chapter references.
- Abbreviations are not explained the first time they appear.
- No chapter references present.

Tables and figures
- Tables and figures do not have titles (above table, below figure).
- Tables and figures cannot be understood without the main text. Insufficient information is provided in the title and/or figure to be able to read/understand the table/figure.

Presentation of numbers in tables
- Too many decimals are present after the decimal point.
- Comma is used instead of decimal point.
- Wrong units are specified: too many significant numbers are presented (copied from computing software).
- Numbers are not adjusted "right" or at the decimal point (numbers should be easily viewed/compared).
- Units are not mentioned in the column title.
- No SI-units are used ("kcal", "j" instead of "J", "kJ/h" instead of "kW" or "MW', "bars", "bar" instead of "bara" or "barg", etc.).

Units
- SI units not used ("kcal", "j" instead of "J", "kJ/h" instead of "kW" or "MW", "bars", "bar" instead of "bara" or "barg", etc.)

Pure component properties table
- Table omitted from report
- Some data categories missing

Block flow diagram (BFD)
- Not all streams enter at the *left* and leave at the *right hand* side of the block scheme.
- No descriptive title of each individual stream. The name should indicate whether this is a liquid or a solid. Indication of the phase and whether it is a solution or dispersion or solids is necessary, especially when dealing with multiple phases.
- Individual streams do not have numbers: < 1 >, etc. (For referral in process stream summary (PSS)).
- No use of bold lines for flows containing main raw materials/products.
- Lines do cross; they should not; use interrupted lines.
- No mass flow rate (t/a), no mass flow rate, relative to main product or feed (t/t), for each individual stream is present.
- No indication of phase, T, P of entering and exiting streams (for completeness and later use for battery limit).
- Missing data on total *in* (t/a, t/t), total *out* (t/a, t/t).
- Blocks represent equipment instead of tasks (heat, cool, separate, recover).

- No indication of pressure, temperature, or pH of the task blocks.
- The block scheme should have a 1:1 relationship to the stream summary table.

Batch cycle diagram
- Batch cycle diagram is not included (for batch process).

Process stream summary, overall component mass balance and stream heat balance
- The process stream summary is not according to the prescribed format (listing for each stream: stream number, stream name, components names, including molecular weights (kg/kmol), enthalpy (kW), phase (V/L/S), pressure (bara), temperature (K), mass flow rate (kg/s), and moles flow rate (kmol/s). Show overall mass balance and stream heat balance.
- Pressure, temperature, phase, and enthalpy have been forgotten.
- The stream summary table for the process block scheme or process flow scheme should have a 1:1 relationship to these schemes; they do not.
- No overall mass and energy balance is given (e.g., in the form of an input/output diagram) showing all input and output streams, utilities, and verification of *to- tal in = total out*.
- New formats are created/invented; easiest way is to download provided template files.

(Reaction) equations
- (Reaction) equations do not have numbers (number making it difficult to refer to). (e.g., per chapter: (Eq. 1.1), (Eq. 2.10).
- No phases (s), (l), (g), (aq.) indicated with species in reaction equations.
- For complex molecular structure formulas, no component names are given.
- For complex molecules, no molecular structure formula is given (stoichiometry cannot be verified).

Description of the design and design approach
- Principal not mentioned and what is his/her interest in this design project.
- In what form will the product be delivered to what type of customers? What is/will be the supply pattern of the feedstocks and delivery of the product?
- No information given about competitor activities and patent situation.
  - No mention of the main challenges you have/will need to overcome during the design: availability of data, uncertainties. How will you go about solving these issues?

Basic assumptions
– No clear definition of the battery limit (e.g., By drawing envelope in block scheme). Are storage and waste treatment included?
– No clear definition of streams entering and exiting the battery limit using the correct format (Tab. A13.1)
– No conclusions drawn on thermodynamic and kinetic literature data: region of validity, accuracy, demonstrated with experimental data.

Product and process designs
– Hierarchical decomposition method from Delft Design Map (DDM) is not used. You should first view the overall supply chain, stakeholder and consumer needs and main conversion routes, and define rather broad tasks for the process blocks.
– Design choices (at different design levels) not clear or not clearly justified.
– No clear justification for the choice of the product composition & structure, and process options. No use is made of comparing alternatives against predefined selection criteria.
– A lot of qualitative statements are made without referring to the sources. Try to be as quantitative as possible to accept/reject choices, effectively using your estimation skills.

Comparison of alternatives (conversion routes, separation methods, unit operations, etc.)
– Inappropriate comparison methods are applied:
  – When comparing two different product and/or process design options, the comparison criteria should be mentioned in the first column, and the performance of the alternatives should be scored for each criterion.
  – The criteria should be formulated in such a way that a score of " +" or "++" means that the performance for that criterion is good or very good.
  – The criteria should be organized in order of importance.
  – Use a five-level scale: "—", " –", "0", " +", "++". Alternatively, and much preferred, present actual performance data in the cells and use a five-scale color code: "—": red, " –": orange, "0": white, " +": light green, "++": dark green.
  – This ensures that all criteria will be evaluated for each alternative and missing information is immediately evident.

Equipment design
– Design specifications per product component or equipment units are not clear.
– The design variables to reach the design specifications for the product or process components are not made clear.

- It is unclear how the design variables are chosen/fixed to achieve the design speci-fications of each product or process component and the whole product or process.

Safety, health, environment, economy, technology, social (SHEETS) evaluation
- General: a complete SHEETS evaluation is not given (mainly focused on economics).
  - Focus on all SHEETS aspects required by stakeholders and the design evaluated accordingly.
- On economics:
  - Gross income is not presented, or only at the very end; mention this upfront.
  - No distinction made between operating costs and investment costs.
  - No conclusions formulated on the economic data presented and the objectives.
  - No sensitivity analysis carried out: what are the financial risks for the project?

Conclusions and recommendations
- No reference is made to the project objectives: have they all been met?
- No presentation of the design's strengths and weaknesses: how can it be improved?
  - What would be possible for the next steps?
  - Difference between project objectives and achieved results should feed into the recommendations.

References
- Findings/assumptions/data, etc., are mentioned without referring to the sources. This makes it difficult for the future reader to check the validity and is considered plagiarism.

References and further reading

[1] Pugh S. Creating Innovative Products Using Total Design, New York, NY, USA: Addison-Wesley-Longman, 1996.
[2] De Haan AB, Swinkels PLJ, de Koning PJ. Instruction Manual Conceptual Design Project CH3843 – From Idea to Design. Delft, Netherlands: Department of Chemical Engineering, Delft University of Technology, 2016.

Index

https://doi.org/10.1515/9783110782127-019

490 —— Index

www.ingramcontent.com/pod-product-compliance
Lightning Source LLC
Chambersburg PA
CBHW060958210326
41598CB00031B/4863